MINORITY INTERNAL MIGRATION IN EUROPE

International Population Studies

Series Editor: Professor Philip Rees,
School of Geography, University of Leeds, UK

This series provides an outlet for integrated and in-depth coverage of innovative research on population themes and techniques. International in scope, the books in the series will cover topics such as migration and mobility, advanced population projection techniques, microsimulation modeling, life course analysis, demographic estimation methods and relationship statistics.

The series will include research monographs, edited collections, advanced level textbooks and reference works on both methods and substantive topics. Key to the series is the presentation of knowledge founded on social science analysis of hard demographic facts based on censuses, surveys, vital and migration statistics.

Other titles in this series:

Demography at the Edge
Remote Human Populations in Developed Nations
Edite by Dean Carson, Rasmus Ole Rasmussen, Prescott Ensign,
Lee Huskey and Andrew Taylor
ISBN 978-0-7546-7962-2

Educational Opportunity
The Geography of Access to Higher Education
Alexander D. Singleton
ISBN 978-0-7546-7867-0

Geographies of Ageing
Social Processes and the Spatial Unevenness of Population Ageing
Amanda Davies and Amity James
ISBN 978-1-4094-1776-7

Population Projections
Methods, Practice and Applications
Tom Wilson and Phil Rees
ISBN 978-1-4094-1989-1 (hardback)
ISBN: 978-1-4094-1991-4 (paperback)

Minority Internal Migration in Europe

EDITED BY

NISSA FINNEY
University of Manchester, UK

and

GEMMA CATNEY
University of Liverpool, UK

Routledge
Taylor & Francis Group

LONDON AND NEW YORK

First published 2012 by Ashgate Publishing

2 Park Square, Milton Park, Abingdon, Oxon OX14 4RN
711 Third Avenue, New York, NY 10017, USA

Routledge is an imprint of the Taylor & Francis Group, an informa business

First issued in paperback 2016

British Library Cataloguing in Publication Data
Minority internal migration in Europe. -- (International
   population studies)
   1. Migration, Internal--Europe--Case studies.
   2. Minorities--Europe--Case studies.
   I. Series II. Finney, Nissa. III. Catney, Gemma.
   304.8'089'0094-dc23

ISBN 978-1-4094-3188-6 (hbk)
ISBN 978-1-138-25099-4 (pbk)

Library of Congress Cataloging-in-Publication Data
Finney, Nissa.
   Minority internal migration in Europe / by Nissa Finney and Gemma Catney.
      p. cm. -- (International population studies)
   Includes bibliographical references and index.
   ISBN 978-1-4094-3188-6 (hbk)   1. Migration, Internal--Europe.
   2. Minorities--Europe. 3. Immigrants--Europe. 4. Europe--Emigration
   and immigration. I. Catney, Gemma. II. Title.
   HB2041.A3F56 2012
   304.8089'0094--dc23

                                                              2012025198

Contents

List of Figures

List of Maps

List of Tables

Notes on Contributors

Albert Sabater is Juan de la Cierva Research Fellow at the Centre for Demographic Studies at the University Autonoma of Barcelona. His current research focuses on analyses of internal and international population migration and Demography's contribution to debates on segregation and diversity. Other research interests include population projections and methods for producing consistent population time series for small areas and population subgroups.

Amir Hefetz was trained in economics, development and urban planning. He is currently a PhD candidate at the Public Policy Program of the School of Political Sciences, University of Haifa, where he studies public service delivery alternatives. He has published several articles in public administration journals.

Andreu Domingo is Deputy Director of the Centre for Demographic Studies at the University Autonoma of Barcelona, where he has been a researcher since 1984. His main research areas include: Demography, international migration, foreign-born populations, marriage, family and kinship. He is also the Director of the Study Group for Demography and Migrations at the Centre for Demographic Studies.

Daniel Czamanski is a Professor of urban economics, head of the Complex City research lab and Associate Dean at the Faculty of Architecture and Town Planning, Technion – Israel Institute of Technology. Czamanski serves at Springer as editor of a new book series entitled *Cities and Nature* and as member of the editorial board of *Computers, Environment and Urban Systems*.

David Manley is a Lecturer in Quantitative Geography at the University of Bristol. He is interested in understanding how inequalities are developed and reinforced in British society. He has published on residential mobility, ethnic minority segregation, the impact of neighbourhood context on individuals, neighbourhood regeneration and is currently co-editing a series of books on neighbourhood effects. David is convinced that quantitative methodologies should be at the centre of critical geography discourses.

Didier Willaert (Master in Geography, Vrije Universiteit Brussel) is a senior researcher at the Interface Demography of the VUB. His main fields of experience are population geography (especially geographic mobility/migration), Geographic Information Systems (especially cartography) and Belgian census data. He has worked on several policy-oriented research projects and has also been involved with the data management of the Belgian census of 2001 and the preparation of the ongoing census (2011).

Gemma Catney is a Population Geographer with research interests in internal migration with an ethnic group dimension, and in residential segregation and integration in Britain. In particular, she is interested in changes to population concentrations in recent decades, at a fine geographical scale. 'Religious' segregation in Northern Ireland is another focus of her work. In exploring these contemporary issues she has developed expertise in the analysis of individual and area level Census data, making use of longitudinal and advanced spatial methods.

George Kandylis is a Research Fellow at the National Centre for Social Research, Athens. He has mainly worked in research on contemporary immigration to Greece and its impact on the formation of Greek cities. His areas of interest include social segregation, social cohesion, nationalism and spatial mobility in Greek cities.

Helga de Valk (PhD. Utrecht University, the Netherlands 2006) is senior researcher at the Netherlands Interdisciplinary Demographic Institute (the Hague) and Professor at Interface Demography, Free University Brussels. Her research focuses on the transition to adulthood of immigrant youth, educational inequality, union and family formation, the second generation and intergenerational relationships in immigrant families. She currently mainly works on her ERC starting grant "Families of migrant origin: a life course perspective".

Ibrahim Sirkeci is Professor of Transnational Studies and Marketing at Regent's College London and Director of the Regent's Centre for Transnational Studies (RCTS). He holds a PhD from the University of Sheffield and a BA from Bilkent University. His research focuses on transnational mobility, conflict, human insecurity, remittances, segregation, segmentation, marketing of business schools, and transnational mobile consumers. His books include *The Cultures of Migration: The Global Nature of Contemporary Movement* (2011, University of Texas Press). He is co-editor of *Migration Letters* and a member of the IUSSP expert panel on international migration.

Jeffrey H. Cohen is Associate Professor of Anthropology at The Ohio State University. He received his Ph.D. from Indiana University. He is co-editor of the *Migration Letters* journal and is an officer for the Society of Anthropological Sciences. His research focuses on migration, economic development, and food safety/nutrition. His books include *Cultures of Migration: The Global Nature of Contemporary Movement* (2011, University of Texas Press).

Joaquin Recaño-Valverde is a geographer and demographer and currently senior lecturer in the Geography Department of the Autonomous University of Barcelona and a research fellow in the Centre for Demographic Studies (CED). He has a PhD in Geography (University of Barcelona). His substantive research interests in Demography include internal and international migration and demographical analysis. His major current project (2011–2013) is *Economical Cycle Inflexion and Transformations of Migration Movements in Spain* (CSO2010-19177).

John Stillwell is Professor of Migration and Regional Development in the School of Geography and the University of Leeds. His research interests include internal migration in different parts of the world and immigration to the UK.

Jordi Bayona is Juan de la Cierva Research Fellow at the Department of Human Geography at the University of Barcelona. His current research activities focus on urban demography and include analysis of residential migration, segregation, housing and social exclusion of population subgroups in large metropolitan areas in Spain.

Jorge Malheiros is a Professor at the Institute of Geography and Spatial Planning at the University of Lisbon. His research is in the domains of population studies, international migration and social exclusion. He is the SOPEMI (Observatory of International Migration of OECD) correspondent for Portugal and co-ordinator of one the clusters of the EU funded IMISCOE excellence network. He is also facilitator of the Portuguese Thematic Network on Social and Professional Integration of Immigrants, Ethnic Minorities and Asylum Seekers of the EU EQUAL Programme.

Michael Windzio is Professor of Sociology and Director of the Institute for Empirical and Applied Sociology (EMPAS) at the University of Bremen. His research interests include Migration and Urban Studies, Quantitative Methods, Network Analysis, Education and Social Structure, Sociology of Organisations and Delinquency. Recently, he has published on organisational ecology of immigrant employment in *Social Science Research* and on social networks and integration of immigrant children in *Sociology*.

Neriman Can is an expert demographer working on migration statistics at Turkish Statistical Institute, Ankara, Turkey. She is a graduate of Middle East Technical University and Hacettepe University Institute of Population Studies. She has been part of institutional statistical improvement programmes and projects supported by European and national funding agencies.

Nir Cohen is a Lecturer in the Department of Geography and Environment at Bar Ilan University. Trained in political sciences, international relations and geography, his main research interests include migration policy, state-diaspora relations, citizenship theory and ethnic politics in Israel.

Nissa Finney is Hallsworth Research Fellow at CCSR (the Cathie Marsh Centre for Census and Survey Research) in the School of Social Sciences at the University of Manchester. Nissa's background is in Geography and her research interests include internal migration, immigrant/ethnic integration, attitudes towards migrants and minorities and mixed methods research.

Roger Andersson is Professor of Social and Economic Geography at the Institute for Housing and Urban Research (IBF) and Department of Social and Economic Geography, Uppsala University. His current research is focused on four substantive areas in the housing and urban development field: Residential segregation, intra-urban migration, urban policy, especially area-based approaches, and economic and social integration of immigrants.

Sako Musterd is Professor of Urban Geography at the Department of Geography, Planning and International Development Studies, University of Amsterdam. His current research activities are connected to residential mobility, segregation, integration, neighbourhood effects, gentrification, and urbanisation processes. He is a member of the editorial/management/advisory boards of the journals *Urban Geography*, *Urban Studies*, *International Journal of Urban and Regional Research*, and *Housing Studies*.

Sarah McNulty graduated from the School of Geography at the University of Leeds in July 2011 with degree in BA Geography. Her research interests are in education, ethnicity and migration.

Sergi Vidal is post-doctoral researcher at Population Europe (Max Planck Institute for Demographic Research) and Research Associate at the Institute for Empirical and Applied Sociology (EMPAS, University of Bremen). His research interests include residential mobility and internal migration, life course, and methods for longitudinal data analysis.

Thomas Maloutas is Professor of Social Geography at the Harokopion University, Athens and the Director of the Institute of Urban and Rural Sociology at the National Centre for Social Research. He has a long record in urban social research and has been invited professor in French, US, UK and Spanish universities. His research is focused on urban social change, segregation, immigration and welfare regimes.

Verónica de Miguel-Luken is a statistician and demographer and a member of the Migration Research Team (Autonomous University of Barcelona) and the group Networks and Social Structures (University of Málaga). She has a PhD in Human Geography (Autonomous University of Barcelona) and an MSc in Social Research Methods and Statistics (University of Manchester). Her substantive research interests include internal migration movements and the effect of social networks on different aspects of migration processes.

Wouter van Gent is a Researcher in Urban Geography and Assistant Professor in Urban Studies at the Amsterdam Institute for Social Science Research, University of Amsterdam, the Netherlands. His current research interests are related to residential mobility, segregation, gentrification and displacement, urban policy and renewal, housing markets and institutions, and the urban geography of radical right-wing populist support.

Preface

By Philip Rees, Series Editor, International Population Studies Series

Series Preface

The International Population Studies series aims to publish the very best contributions to recent knowledge about population themes and techniques. The series comprises a mix of research monographs, advanced text books and edited collections, covering work on substantive topics and research methods. Some of the books in the series report on research findings particular to one country but many of the contributions use an international canvas for their analysis. International in scope, the books in the series cover topics such as migration and mobility, advanced population projection techniques, microsimulation modeling, life course analysis, demographic estimation methods and relationship statistics. Already published are books on *Educational Opportunities* (about access to higher education), *Demography at the Edge* (about populations in remote locations) and *Geographies of Ageing* (how the almost universal demographic process is working out in different places). Key to the series is the presentation of knowledge founded on social science analysis of hard demographic facts based on censuses, surveys, vital and migration statistics. All books in the series are subject to review.

Book Preface

In Europe we live an increasingly diverse society. The European project, that is, the European Union (EU) and its treaties guaranteeing freedom of labour migration and travel, means that minority populations of European origin within European countries have increased as the EU has expanded and will continue to grow in future. Demographic ageing across all of Europe has also meant opportunities for people to migrate to Europe from beyond its borders. These opportunities may have shrunk in the economic difficulties of 2008–2012 but there will be a recovery in time. But the movement of international migrants does not stop once they cross the frontiers of European national states; they also migrate within their chosen destination countries as internal migrants. The characteristics, directions and implications of the internal migration of minorities have not received the attention they deserve as parts of one of the most important processes changing European societies. This book on Minority Internal Migration in Europe fills that gap by bringing together high quality research on the phenomenon in thirteen European countries. So that experiences in the different countries can be compared more directly the book's editors, Nissa Finney and Gemma Catney, asked the contributing authors to adopt a common chapter structure and narrative style. The editors provide in their introductory and concluding chapters extremely useful syntheses of findings across those 13 countries. The message of the book's chapters is a positive one that identifies internal migration as leading, in general, to greater spatial integration. Of course, there are important differences in national experiences and these experiences are filtered through use of different data sets and methods. However, for scholars of minority populations this is the book to read in order to learn about the dynamics of relocation of those minorities which will influence the future shape of our ethnically diverse societies.

Editors' Acknowledgements

We thank the contributors to this volume who have engaged so enthusiastically in this project. It was excellent to have the opportunity to discuss this work at the *Minority Internal Migration in Europe Conference* held in Manchester, UK, in September 2011 and we are grateful to all who attended for their participation. The meeting was made possible by financial support from the Economic and Social Research Council (ESRC) Understanding Population Trends and Processes (UPTAP) programme, CCSR at The University of Manchester, the Population Geography Research Group of the Royal Geographical Society (with the IBG) and the journal *Population, Space and Place*.

We are grateful to Ashgate for their support for this project and for their unfailing efficiency in the production of this book. We also thank Professor Phil Rees who initially encouraged us to compile this collection and provided most helpful comments on the manuscript.

Nissa Finney, Manchester
Gemma Catney, Liverpool

Chapter 1

Minority Internal Migration in Europe: Key Issues and Contexts

Gemma Catney and Nissa Finney

Introduction

Migration within a nation state is a powerful force, redistributing the population and altering the demographic, social and economic composition of regions, cities and neighbourhoods. This internal migration may take place between provinces and across many miles, or within relatively small areas – a district, a town, a street. It is also a selective event, dependent on multiple (and inter-related) individual, household, spatial and temporal circumstances. As transnational migration increases in frequency and becomes more diverse in its composition, so too internal migration demonstrates a great deal of variation in terms of its characteristics, volume and impacts for origins, destinations and individuals. National population changes are shaped by global trends; thus, a contemporary concern common for many countries is how to understand and respond to population change at the local level. One main component of this change which has received increasing attention in much of the 'western' world in recent years is increasing ethnic diversity, brought about by immigration and subsequent family building. Cross-border movements are often followed by successive 'internal' changes of address at a variety of spatial scales, and questions concerning the selectivity of this migration by ethnicity, the consequences of changing population structures, and the magnitude and direction of internal flows, have generated considerable attention in the interconnected spheres of media, policy, and academia.

The magnitude of transnational movements across the globe has increased markedly in recent decades, reflecting adjustments in labour market demand and supply, technological innovations in communication and travel, population growth, and changes to political contexts, borders and immigration policies. Globalisation and its associated development of new forms of transnational networks and co-operation, along with the challenges it presents for traditional meanings of territorial space (Herod 2011), means that the interactions and flows of people between and within nation states are continually growing and diversifying (albeit within a context of increasingly restrictive national borders for some regions). The formation of the European Union (EU) enabled the free movement of residents of its member states to live, work and/or study within the EU. While the economic downturn beginning in mid-2008 saw the reduction of immigration to some EU countries (and an increase in emigration), net migration still remained positive for many countries for which labour-related migration constitutes a large proportion of the migrant stream, such as the UK, Spain and Italy (Koehler et al. 2010). Larger population flows have been coupled with an increase in their complexity, in terms of the reasons and motivations for migration, and the implications of the movements, for both sending and receiving countries.

Immigration to developed countries has fuelled considerable debate in the media, policy and by academics. In particular, discussions about the volume of moving individuals and households and what this may mean (both positively and negatively) for increased supply and demand on core services, and for economic progress, are shared in the United States, Australia, and much of western Europe (Castles and Miller 2009). In addition to these issues, more socially-driven debates have also emerged, relating to the experiences which immigrants may have on arrival to their destination country, and likewise any possible impacts immigration may have on existing communities. While many immigrants make subsequent return movements to their country of origin, or emigrate elsewhere, some choose to stay, following various paths of family reunification or building, social and residential mobility, educational and career advancement, depending on differing life stages and opportunities. Political and policy approaches to the integration of immigrants and their descendents, where this is seen as a priority, vary between countries. However, to a lesser or greater extent, immigrant integration policy in western counties in the past two decades has been characterised by ideas of multiculturalism and social cohesion. These policy agendas have seen shifts in their acceptability, precedence and relevance, largely as a response to the role of international migration as a major force of population change in the 1990s and 2000s (Kalra and Kapoor 2009, Catney, Finney and Twigg 2011). Debates on diversity, cohesion and integration were heightened in their national and international importance by the global context of wars in Iraq and Afghanistan, and experiences of international terrorism in western Europe and North America in the 1990s and 2000s (Finney and Simpson 2009, Catney, Finney and Twigg 2011).

This volume focuses specifically on the European case, where ethnic minority and immigrant internal migration is an emerging field of academic interest in many countries, partly as a result of increased political interest in interethnic relations and place-based policies. Only now are theories emerging to facilitate the understanding of the processes and patterns of contemporary European population change and ethnic relations. Relatively little is known about the significance of ethnicity in internal migration processes, or how migration contributes to immigrant/ ethnic integration; it is with filling these gaps, both empirical and theoretical, that the present collection is concerned. Through understanding if (and if so, how) migration may be selective by ethnic group, the research presented in this volume may also engage with wider migration themes, to develop our understandings of contemporary issues such as, for example, lifecourse transitions (Rabe and Taylor

2010), intergenerational transfers (Mulder 2007), counterurbanisation (Champion 2001) and studentification (Smith 2005).

There are several aims of this volume. We have brought together original analyses that represent the state of current knowledge in research on minority internal migration and, in doing so, we hope to advance understandings (both theoretical and empirical) of this field, providing an evidence base which may go on to inform policy. In reviewing and contributing to what is known about minority internal migration, it is also possible to identify what gaps remain, and, in turn, set an informed research agenda for future work in this field. The international approach adopted in this book enables a recognition (and aids an understanding of) the similarities and differences between countries in Europe.

Background to the Minority Internal Migration in Europe Project

A growth of interest in ethnic minority migration and integration was apparent at several conference sessions between 2008 and 2010 in which one or both of the editors participated; namely the European Population Conference in 2008 (Barcelona); the Population Geography Research Group of the Royal Geographical Society with the Institute of British Geographers (RGS-IBG)/Understanding Population Trends and Processes (UPTAP)[1] sessions at the RGS-IBG Annual Conference in Manchester, 2009; and the Centre for Demographic Studies (CED) International Seminar in Barcelona, 2010. At these meetings it was clear that issues of minority internal migration were important in countries across Europe and also that some key elements differed between countries. These include: (1) the newness of immigration – many countries in Europe have long established minorities as well as more recent immigrant flows, whereas for others, the issue of minority integration has been brought about only recently through immigration; (2) the origins of the minorities and the nature of their migration and, related to this; (3) the relative experiences of minorities in terms of, for example, their socio-economic position, language ability, and geographical locations in and outside initial settlement areas; (4) housing markets and opportunities for minority groups; (5) the rights of immigrants/minorities to, for example, housing, health care, and employment; and (6) the policy and public responses to increasing diversity.

Given the similarities and differences across Europe in the experiences of minorities, there is a need for a more collective approach to conceptualising, measuring, analysing and ultimately understanding minority internal migration, its causes and consequences. As a response, this book draws together such themes as measuring internal migration of minorities, patterns of minority internal migration, motivations for migration and residential choice, barriers to migration and residential choice, housing markets and housing experiences of immigrants/minorities, post-immigration settlement patterns, the role of ethnic

---

1   Economic and Social Research Council (ESRC) grant RES-163-25-0028. See www.uptap.net.

clusters/gateway cities, intergenerational change and transfer of residential aspirations, lifecourse understandings of internal migration, and policies which engage with themes of migration, immigration and ethnic integration. This edited collection brings together work on these themes through empirical and theoretical contributions, focussing on new immigrants, settled immigrants and subsequent second, third and even fourth generations.

The debates raised in the conference meetings led to a series of conversations, the establishment of interlinked networks, and ultimately the collaboration for this book. It is exciting to include a range of countries from across Europe. Countries for which research is presented are Belgium, Israel, Scotland and Britain as a whole, Germany, Turkey, Italy, Spain, Portugal, The Netherlands, Greece and Sweden. While most chapters present work at the national scale, some chapters concentrate on particular cities, and those metropolises represented include Lisbon, Athens, London and Amsterdam. In conceptualising Europe for the purposes of this volume, an inclusive approach has been adopted which enables us to capture the diversity of experiences and explore the distinguishing factors of national contexts throughout the continent. Our broad encapsulation of Europe's migration system recognises the fluidity and dynamism of population movement (and political allegiances). Related to this, while this book is not about the flows *between* countries, there are many important historical, policy and socio-political interconnections across national boundaries which can be captured with this broad conceptualisation of Europe. Likewise, this approach best allows for the building of academic links between countries with diverse and similar experiences, enabling a richer body of knowledge to be developed. Of course, we will have missed some potentially important contributions. However, we hope that this volume acts as a stimulus for a growing network of researchers interested in minority internal migration.

Introduction to Key Concepts and Definitions

Two key terms which are central to this book are 'minority' and 'internal migration', and given the potential variety in their use, meaning and interpretation in other research, it is important to outline clearly what we mean by these labels. Previous research has highlighted the complexities of ethnic identity, in particular how identities vary within and between groups, evolve over time, and differ according to the purpose for which the identity is made or used (Nagel 1994, Nazroo and Karlsen 2003). 'Minority', as used in this volume of chapters, refers to any population group, broadly identifiable on the basis of ethnicity, which constitutes a considerably smaller proportion of the national population than the majority group. In all chapters, the majority group is the 'White' group (although variously labelled). In all chapters apart from that on Israel (Cohen, Hefetz and Czamanski), minority groups comprise immigrants and their descendents, although there are some differences as to how easily (if at all) immigrants and subsequent generations can be separately identified. The chapter on Israel presents the very interesting case of homeland minorities. There is a great deal of variability in how minority groups are measured, or population counts reported, for statistical purposes. For some countries, self-categorisation via questions in national censuses and other surveys allows for the measurement of minority groups, while for others, population registers serve as a means of collecting country of birth data.

Internal migration is the focus of this volume, yet this is also potentially an ambiguous term. As each chapter reveals, what is meant by internal migration for each country is broadly similar in that it refers to movement *within* the given nation state (or defined sub-national region). However, there are some important differences as to how an 'internal migrant' or a 'migration event' may be defined. To some extent this variation reflects the many different ways in which migration data are collected and recorded, measured, and analysed. A migration event may be picked up by a change of address in a population register, for example, or identified via a response to previous address questions in the country's Census.

While the meaning of 'minority' and 'internal migration' are consistent in their general overall use, the inevitable variability of these terms in each chapter is clarified through a specific definition of how data on minorities and migration events are measured, collected and reported. As is discussed in more depth in the final chapter of the volume, there are as many ways of collecting, measuring, reporting, and categorising data on migrants (internal and international), minorities and migration events as there are countries. This 'diversity of measuring diversity' is exciting and full of opportunity for those concerned with understanding how our knowledge of minority migration may be enhanced.

Themes of Minority Internal Migration

Several core themes run throughout this volume, explored by all or most of the chapters. These common threads reinforce the contribution which can be made by thinking in both a collective and comparative sense about minority internal migration across Europe. They also raise challenges about what the future direction for studies of internal migration with an ethnic dimension should be in Europe and elsewhere. Thus, these core themes will be revisited in the final chapter of this book, which considers what should be prioritised in setting an agenda for future minority internal migration research.

Theories of Spatial/Residential Integration: What Happens After Immigration?

Few academic endeavours into the social or spatial 'assimilation' of immigrants (and subsequent generations) are without reference to the original works of the Chicago School of Sociology, now nearly a century old (Park, Burgess and McKenzie 1925). Since then, a great deal of research has been concerned with the processes which occur as time since immigration increases. These processes are social and spatial; both, in essence, relate to how far immigrant communities become more similar to the 'host' society over time, in terms of income, labour market position, and types of housing. Theories concerned with spatial processes of integration are related to these social processes, but also consider the geographical position of individuals and households, viewing internal migration from original settlement or 'gateway' areas into the suburbs or beyond as an indicator of integration and desegregation. 'Assimilation' and related theories of integration have engendered much debate over their decades of attention, particularly in US-based research, and they are not without contention. While some of this work has empirically tested when, how and where integration occurs, or the various paths to 'assimilation' which may be followed (Gans 1997, Boal 1999, Neckerman, Carter and Lee 1999, South, Crowder and Chavez 2005), others have criticised the 'one size fits all' assumption that over time population subgroups become more alike in their demographic, social, cultural, economic and spatial characteristics and locations (Wright, Ellis and Parks 2005). Likewise, the implications underlying some research on these themes, that movement away from residential clusters is to be favoured, has been challenged in work which highlights the positive aspects of co-ethnic concentrations. Remaining in co-ethnic clusters may be an outcome of choice for the preservation of cultural practices, the retention of a critical mass for key religious establishments, the retention of language, and the reduction of social and cultural isolation (Peach 1996, 2000, 2009). The challenge in research on minority internal migration should perhaps be concerned with how important or relevant spatial and social processes of integration are, and if (and if so how) future research should focus on testing them, for different places, subgroups and at different time points, and how they may relate to other outcomes, such as social exclusion.

The chapter by de Valk and Willaert engages with the spatial assimilation debate, considering how dispersal (in the form of suburbanisation) may be more likely for the second and third generation than for first generation immigrants in Belgium. Sabater, Bayona and Domingo also deal explicitly with these themes in their measurement of the (de-)segregation of immigrant groups in Spain. The degree to which immigrants in Portugal are contributing to the suburbanisation process is a concern in the work presented by Malheiros. Kandylis and Maloutas analyse the initial settlement and subsequent relocation patterns of immigrants in Athens, posing several important challenges to interpretations of dispersal from the inner city and what this mobility may mean for migrants. The chapter by Stillwell and McNulty explores how far immigrants to London select destinations with high concentrations of 'established' migrants of the same ethnic group, where established support networks exist, or migrate to 'new' destinations. An overt concern in the research on Amsterdam by Musterd and van Gent is with the spatial assimilation of immigrants. They ask to what extent first- and second-generation immigrants and the 'native' Dutch reside in what they term 'concentration neighbourhoods', and explore their mobility patterns within Amsterdam. The extent to which second generation immigrant movers differ from first generation and native Dutch movers is also examined, in particular to see if suburbanisaton and counterurbanisation processes are shared for all groups, explicitly testing if the second generation reach higher levels of assimilation.

Differences Between Minority Generations

The second recurring theme extends beyond the processes by which immigrants may increasingly share characteristics with the rest of the population, to the issue of the reduction in demographic/social/economic differences between *generations* of minority groups. Recent immigrant groups may have certain characteristics which would be expected to decline over time, such as, for example, higher fertility rates and larger household sizes, due to their young age profiles and the role of family reunification (Coleman 1994). How far, though, might we expect differences to reduce not only with increasing time as immigrants settle, but as second and third generations emerge? Of interest in several of the chapters is the comparison of first and second generation minority groups, asking whether convergence with the majority population's experience of migration and residential location can be identified, and whether this implies integration.

Of course, given some cultural norms and preferences which would not be expected to change in passing years, it would be unwise to assume certain socio-economic or spatial 'benchmarks' by which integration may be judged (higher employment rates, greater participation in professional occupations, increased home ownership, or relocations from the inner city to the suburbs, for example). However, such comparisons may prove useful in testing for discrimination; for example, where, *ceteris paribus*, housing careers and labour market participation should be similar for all ethnic groups.

Vidal and Windzio explore differences in mobility between first generation immigrants to Germany, the second generation and German 'natives', emphasising the importance of taking account of time since settlement and generational 'status'. In addition to their interests in the spatial assimilation of immigrants in Amsterdam over time, the housing market characteristics of first and second generation immigrants and the native Dutch are explored by Musterd and van Gent. Differences in migration propensities between immigrants and native Swedes are the focus of the chapter by Andersson; key research questions in this chapter include whether observed differences hold for inter- and intra-labour market moves, and whether the demographic composition of groups helps to explain any variation. Andersson emphasises the role of time since immigration in affecting differences in migration frequencies, and the implications of this for the second generation.

While not concerned with differences between generations *per se*, many of the chapters set out to test for variations in internal migration patterns and propensities between ethnic groups or between the 'minority' and 'majority' population (such as, for example, White and non-White). Manley and Catney's research explores if the likelihood of migrating within Scotland, and the propensity to migrate to different regions, varies between ethnic groups, or if differences in internal mobility might be explained by the socio-economic and/or demographic composition of the population. This is also considered by Cohen, Hefetz and Czamanski, in their examination of migration propensities and socio-demographic profiles of Arabs compared to Jews in Israel, and of the differences in this migration within the Arab population, according to religious sub-group. Kandylis and Maloutas explore the socio-economic differences between non-movers and the internally mobile, by immigrant group, in addition to the spatial patterns of these flows, within Athens and to/from other parts of Greece. The high mobility of immigrants within Lisbon is discussed in the chapter by Malheiros. Finney tests for ethnic differences in student mobility within Great Britain, highlighting the importance of giving due attention to sub-groups of the population if we are to understand fully in what ways (and potentially why) ethnic differences in mobility exist. Demographic differences in internal migration patterns are tested for foreigners and natives in Italy, Spain and Portugal by Recano-Valverde and de Miguel-Luken. Sirkeci, Cohen and Can are able to explore the *motivations* for internal migration in Turkey, and they consider how these differ between the native- and foreign-born.

The Importance of Place

The concept of place is far from straightforward – its meaning varies for different people and in different contexts, and it is constantly changing, continually consumed and reproduced (Massey 1994). In all of the chapters, 'place' is discussed in terms of migration origins and destinations. The way in which place is labelled is, at the most basic level, common to all chapters concerned; the country of interest is a 'destination', categorised as such mainly in the context of studies which analyse immigrants directly, but also in studies which consider how far the experiences and characteristics of subsequent generations have changed since their relatives moved to that destination. Likewise, the places from which immigrants come are considered as 'origins'. For internal migrants, one's origin may be a relatively short distance from the destination – a city to suburb move, for example. These repeated terms of 'destinations' and 'origins' act as tools for understanding behavioural similarities and differences between groups and the impact(s) of immigration and subsequent internal migration (at national and local levels). In terms of destinations, some authors ask what is the role of the new residential place for migrants? What is the impact of immigration and internal mobility? The (national) origins from which migrants come are frequently used as labels or markers, the associations with the origin remaining and being used as a means to measure minorities and, in some cases where it is assumed that certain cultures, practices and preferences might be expected to be carried forward from one's origin, *explain* differences in characteristics or behaviours.

Vidal and Windzio strongly emphasise the potential importance of immigrant origin (and associated immigration history) in explaining differences in migration behaviour, including, for example, the effect of origin-specific discrimination against certain groups, or residential preferences which may be distinct to certain groups by immigrant origin. The role of country of origin in explaining differences in migration patterns, intensities and internal migrant characteristics is a concern of Recaño-Valverde and de Miguel-Luken, and, interestingly, they are also able to compare differences between immigrant destinations at the country level (Portugal, Spain and Italy).

The importance of place is emphasised at a variety of spatial scales – as highlighted earlier, for some the country is the focal point for analyses, while others are concerned with cities. Within these larger 'units' of analyses, though, many authors explore movement at a finer spatial scale. Stillwell and McNulty consider the smaller areas within London; Manley and Catney explore differing migration propensities to regions within Scotland; Sabater, Bayona and Domingo are concerned with movement between the municipalities of Madrid and Barcelona to the rest of the province; de Valk and Willaert (Belgium), Malheiros (Lisbon) and Musterd and van Gent (Amsterdam) distinguish between 'types' of region or neighbourhood; the urban (inner city), suburban and rural periphery, for example. Similarly, distance moved, as measured either directly (for example, kilometres between internal origins and destinations) or indirectly (for example, city to suburban moves versus intra-city moves) emphasises this issue of scale (for example, Manley and Catney for Scotland, Andersson for Sweden and de Valk and Willaert for Belgium).

The multiplicity of immigrant origins is highlighted in many chapters (for example, to Spain by Sabater, Bayona and Domingo, to London by Stillwell and McNulty, to Germany by Vidal and Windzio, to Sweden by Andersson and to Turkey by Sirkeci, Cohen and Can). For many countries these origins have been significant senders of migrants for a significant period of time. However, the importance of new immigrant origins (in particular in the context of a lack of data to measure these more recent flows) is discussed by, for example, Kandylis and Maloutas (for example, Afghanistan and Somalia to Athens) and Manley and Catney (for example, immigration to Scotland from EU A8 accession countries of the Czech Republic, Estonia, Hungary, Latvia, Lithuania, Poland, Slovakia and Slovenia). Sirkeci, Cohen and Can argue that the role of Turkey as a *destination* is often overlooked, as it is considered more often in terms of its role as a source of emigration to elsewhere.

The distinctiveness of place is another important theme to emerge from all of the chapters. While commonalities can be seen, and comparisons made, between chapters, it is striking that the contexts, histories and experiences may all have a differing impact on the patterns and processes of minority internal migration. Finney's analysis of student mobility within Britain highlights how the Higher Education system generates internal migration, and how due consideration of distinctive contexts such as this should be made. Manley and Catney argue that the distinctiveness of Scotland's immigration history and integration policy context mean that Britain-wide analyses may fail to sufficiently capture variation in minority internal migration. A key theme of the contribution from de Valk and Willaert relates to the impact the particular composition of a place (its housing and neighbourhood characteristics) may have on internal migration. They are also able to use data on neighbourhood evaluations to assess how far (dis)satisfaction with one's neighbourhood in terms of, for example, cleanliness, noise levels, and attractiveness of the built and natural environment, affects out-migration rates. The interplay between physical and socio-political barriers to movement, particular to the case of Israel, is addressed by Cohen, Hefetz and Czamanski in their discussion of its impact on spatially restricted mobility of Arabs in Israel. The importance of the political and policy circumstances, particular to each country, in understanding minority internal migration is apparent for all chapters. The analyses in most chapters are framed by these contexts, and the authors variously focus on immigration history and policy, housing market structures, policies which deal with cohesion and multiculturalism, and policies related to (in)equalities in labour, housing and education.

Sabater, Bayona and Domingo discuss how some immigrants initially reside in traditional settlement areas from which they may eventually disperse, while others may directly migrate to new destinations – what they suggest may be a 'bimodal' immigrant residential pattern. The concept of chain migration assumes that 'established' immigrant destinations may be attractive to subsequent immigrants, thereby reinforcing the role of particular places as receiving areas; this process is tested for London by Stillwell and McNulty. In considering why differences in migration frequencies may exist between immigrants and 'natives' in Sweden, Andersson makes the important point that attachment to place, and 'sentiments' or 'identities' with particular regions may restrict mobility, perhaps more so for natives, compared to immigrants who may not yet have developed these attachments, or whose position in the housing/labour market may be less stable given potential discrimination, financial constraints, or less developed networks. de Valk and Willaert contribute to this debate when they question if residents in an area with a high level of co-ethnics may actually be *less* likely to move, given the attraction of better developed networks and structures in that initial settlement area. Clearly, considering these different ways in which the role of migrant origins and destinations may be re-emphasised by new flows poses a challenge for researchers in their conceptualisation(s) of place. These reproductions of the meaning of place provide an interesting challenge to research on minority internal migration (see Catney, Finney and Twigg 2011).

How to Use this Book

The book is designed to be of use to both established scholars and researchers at an earlier stage of their career. The volume of chapters presents original analyses, of relevance for a wide range of disciplines including Geography, Sociology, Policy Studies, Political Science, Urban Studies, Demography and Migration Studies. In addition to minority internal migration, those with interests in ethnicity and diversity, population dynamics, immigration, segregation, and the interrelated issues of integration/multiculturalism/social cohesion should also find the book helpful. The range of (secondary) data from national Censuses and other surveys and administrative sources used in the analyses presented, and the application of (quantitative) methodologies to explore a range of novel research questions related to minority migration, will appeal to both existing academics and the next generation of researchers.

There are a number of common elements for each chapter which lend consistency to the material and enable international comparison. Each chapter provides some material on the national context for understanding minority internal migration and, where appropriate, some background on the country's immigration history, experiences of integration, and policy frameworks. A brief review of the measurement and categorisation of minorities/ethnic groups and definition(s) of internal migration is provided by all authors. These country-specific studies, set within a wider framework of shared themes, issues and debates, will enable the book to be used as an effective teaching resource for undergraduate and postgraduate taught courses, in the aforementioned disciplines.

Chapter 2 deals with the internal migration of international migrants in Belgium; Chapter 3 is concerned with the links between immigration and internal migration in London. Chapter 4 presents research on immigrant residential mobility and 'de-segregation' in Lisbon; immigrant and 'native' mobility within Amsterdam is examined in Chapter 5, and Chapter 6 deals with Arab migration in Israel. Ethnic group migration propensities and patterns in Scotland are explored in Chapter 7, Chapter 8 looks at the mobility of immigrants and ethnic minorities within Germany, and Chapter 9 examines migration of the foreign-born in Turkey. Athens is the focus of Chapter 10, where work is presented on the internal migration of immigrants and their social integration. Chapter 11 is concerned with ethnic group-specific student mobility in Britain, while Chapter 12 provides insight into the internal migration of foreigners in Italy, Spain and Portugal. The mobility patterns of ethnic minorities in Sweden are explored in Chapter 13, before an analysis of internal migration and residential distributions in Spain is presented in Chapter 14. The final chapter, 'Minority internal migration in Europe: research progress, challenges and prospects', revisits these contributions and discusses recurring themes in the book. In considering the main messages which might be drawn from the research presented, the chapter highlights the contribution of the book to our knowledge of minority internal migration in Europe, and questions what the future research agenda for research on this topic, for Europe and beyond, might be. While the chapters can be read alone, as isolated studies, the consistencies and cross-references between chapters, and the commonalities in the research questions addressed, literature discussed, types of data used, methods employed, and findings presented, mean that the book can also be used as a cohesive source for exploring themes of minority internal migration.

Conclusions

While recent years have seen an increase in transnational and internal population mobility, a diversification of the composition and direction of flows, and increased debate on the impact of migration, there remain significant gaps in our knowledge of the patterns and processes of minority migration. The shared histories, policies and debates, as well as migration experiences, between European nation states, makes a solid case for drawing together a collection of original research on themes of minority internal migration in Europe, and it is clearly an exciting time to be doing so. A number of recurring themes have been highlighted in this introductory chapter, namely the relevance and applicability of the processes which may occur after immigration and for subsequent generations in terms of integration and convergence in residential experiences, and the meaning of place in studies of minority mobility. However, in addition to these common themes, the chapters also reveal some striking differences in the migration experiences of minorities in countries across Europe.

Acknowledgments

Ian Shuttleworth is thanked for comments on an earlier draft of this chapter.

Chapter 2

Internal Mobility of International Migrants: The Case of Belgium

Helga de Valk and Didier Willaert

Introduction

Research on international migrants has traditionally focused on their international move. Studies have tried to explain why and where people move across borders based on different theoretical assumptions. This strand of research is largely separate from the study of internal mobility and migration decisions. The latter studies mainly focus on the majority population and often take a geographic starting point. The studies that have focused on the spatial mobility of international migrants are primarily done in the US and the UK (for example, Iceland and Nelson 2008, South, Crowder and Pais 2008, Simpson and Finney 2009). It is only in recent years that these studies have been extended to continental Europe (for example, Bolt, Van Kempen and Van Ham 2008, Brämå 2008, Zorlu 2008). The growing ethnic diversity of European countries makes it ever more relevant to know more about the extent to which international migrants follow the same patterns and are influenced by the same factors of internal mobility as found for majority populations.

In this contribution we expand the previous research to Belgium and aim to advance understanding of how housing conditions and neighbourhood factors relate to internal mobility. In Belgium, like in many other European countries, the share of immigrants in the population has grown in the past decades but still little is known about internal mobility of international migrants. This study adds to the literature in three ways. First of all we test whether the same or different patterns (level and direction) of mobility are found among natives and different migrant groups. Second, very often it is suggested that the perceived quality and the ethnic composition of the neighbourhood are important factors in location decision making (South and Deane 1993, Van Ham and Feijten 2008, Schaake, Burgers and Mulder 2010). We question whether neighbourhood characteristics are important for the internal mobility of international migrants and whether they are of the same importance for different origin groups. Our data are unique in the sense that we have detailed information on the evaluated and objective neighbourhood characteristics. Third, contrary to most studies that rely on survey data, our work is based on census data including the full population. This allows for a detailed breakdown of different origins and tests whether characteristics of the neighbourhood are of similar importance for different origin groups and migrant generations. This diversity will be taken into account in our work by distinguishing six different immigrant origins and the native population as well as the migrant generation. In this way our study can provide more in-depth knowledge of the similarities and differences in residential mobility by migrant origin. For this study we use linked Belgian Census data for 2001 and 2006. This design provides longitudinal data on the full population of Belgium and allows us to disentangle the mechanisms behind mobility. The first part of our work refers to mobility patterns and covers Belgium. In the second part the role of neighbourhood characteristics is further explored and we then focus on the Brussels-Capital Region (BCR) where the highest mobility as well as the most migrants are observed.

Background

Location choices of international migrants are not accidental. Immigrants often settle, at least initially, in the larger urban areas. This is the result of the available job options, cheap housing possibilities, as well as the consequence of the importance of the network of family and friends in the migration move. Studies on international migration have focused on the fact that migrants are not randomly settling in a host society. Very often migrants arrive in the larger urban areas where housing and work are more readily available (de Valk et al. 2004, Zorlu and Mulder 2008).

Research on internal mobility, nevertheless, primarily focuses on the majority group in different countries. Only more recently has attention been paid to internal moves of international migrants in Europe. Emphasis has been put on the ethnic specific patterns of mobility (Finney and Simpson 2008), the role of family ties (Zorlu 2009), segregation patterns (Bråmå 2008) and the relationship between housing/mobility and the integration of ethnic groups (Hall 2009, Bolt, Özüekren and Phillips 2010, Sinning 2011). Despite the fact that these studies have advanced our understanding of internal mobility, what is known about how factors like neighbourhood and housing characteristics may influence the mobility of minority groups in particular is still limited.

Studies of migrant populations mainly address issues of ethnic concentration. In particular in the US there is a long tradition on analysing segregation of minority groups and its effects on different outcomes in life (Clark 1998, South, Crowder and Chavez 2005, Iceland and Nelson 2008, Mendez 2009). Recent work on Europe also links segregation to patterns of integration of immigrants in the host society (Bråmå 2008, Schönwälder and Söhn 2009, Bolt, Özüekren and Philips 2010, Crul and Schneider 2010, Karsten 2011). Although levels of segregation in Europe are reported to be lower than for the US, still many scholars and policy makers are concerned about ethnic concentration and its consequences (Ireland 2008). It has therefore been questioned whether international migrants will also follow the suburbanisation path as has been frequently observed for majority groups. The native population is in many countries found to leave the inner city centres as areas outside the inner city are perceived to provide better living conditions (Nilsson 2003, Permentier, Van Ham and Bolt 2011). Residential segregation may thus be reinforced with immigrants arriving in the urban centres which majority group members are leaving (Rérat 2011). Very often these suburbanisation moves are associated with upward social mobility (Catney and Simpson 2010). The fact that housing conditions and quality of life is perceived to be better outside the inner city potentially holds for both majority and minority group populations. However, the extent to which perceptions of housing conditions influence patterns of mobility of migrants remains to be explored.

Theoretical Frame and Hypotheses

Mobility patterns are often studied by taking an economic perspective in which costs and benefits are balanced. It is suggested that mobility is determined by preferences and needs on the one hand, and opportunities and constraints on the other hand. The decision to move to another residence will be made when the disequilibrium between the current and the desired housing and neighbourhood is too large, and when the households' budget allows it (Hanushek and Quiqley 1978). The opportunities and constraints are a result of the interaction between the income of households, and the availability and accessibility of (desired) dwellings and neighbourhoods (Clark, Deurloo and Dieleman 2006, Bolt 2008). Housing needs (in terms of size, type, quality, price and tenure) and preferences may change over the life course. In this respect, life-course events – leaving home, forming a partnership, having children, getting divorced, the death of a partner, entry into the labour market, job change, or loss, the purchase of a house, etc. – often lead to an adjustment of the housing situation, and hence trigger residential mobility (South and Deane 1993, Clark and Dieleman 1996, Nilsson 2003, Zorlu 2008). These factors can be expected to be of importance for migrant and native families alike. We thus expect that differences in life course stages, experienced events, and the economic position of the household will explain different levels of mobility. Those who are younger, are single, experience family life transitions, and have a better income position are more likely to move (Hypothesis one).

Regarding the destination of the move, previous research has indicated that a move within the central city is mainly about adjusting to space (and is thus about housing consumption), while a move to the suburbs can be rather linked to tenure change (from renter to owner), to the desire for a more comfortable dwelling in a lower density and greener environment, or to neighbourhood dissatisfaction (Clark, Deurloo and Dieleman 2006, Van Criekingen 2009). Longer-distance moves, on the other hand, are more associated with the occupational career (the search for a first job, job change, retirement) than with the housing or household career. For those who already own a house it is more likely that it satisfies their need in terms of space and comfort level, and so they are less inclined to move. And if they move, it will rather be because of other external reasons. We therefore expect that the tenure status will influence the destination of the move where renters are more likely to leave the city (to suburbs and out of the urban area) while no such influence is expected for home owners (Hypothesis two). Again we expect that this hypothesis will hold for migrant and native majority groups.

Studies of the integration of minorities into the host society after migration perceive residential mobility as one of the indicators of assimilation. According to this view, immigrants who reside in the country longer will have more similar residential patterns than is the case for those who arrived only recently (Massey 1985, South, Crowder and Pais 2008). For recently arrived migrants, settlement in certain neighbourhoods is reinforced because of the available ethnic networks that will allow for finding a job, guidance in the new environment and specific ethnic goods. This ethnic capital in the community context may make immigrants who live in an area with many co-ethnics less likely to move than those in neighbourhoods with few co-ethnics (Hypothesis three). This effect is expected to be relevant for all types of moves as well as for all groups.

Although the levels of internal mobility may be affected by the ethnic composition of the neighbourhood, the direction of the move may just as well be determined by community composition. In particular, for second-generation immigrants, boundaries with the majority population are 'blurring' (Alba 2005). Among other things this is potentially reflected in residential choices and internal mobility, and may result in moving out of the ethnic concentrated urban areas (Alba and Nee 1997). Spatial dispersion may thus increase with generation and interethnic mixing with natives. Overall we therefore expect that suburbanisation is likely to be higher for second- or third-generation households than for first-generation households (Hypothesis four). Housing quality, also in terms of context (location) and neighbourhood, are key for mobility. Studies in the US suggest that appreciation of the neighbourhood is relevant for explaining the residential mobility of different origin groups (South and Deane 1993). Van Ham and Feijten (2008) reported for the Netherlands that higher shares of ethnic minorities in the neighbourhood are related to higher preferences for out-migration among the native population and to a more limited extent also for migrant groups. Both the composition (Hypotheses one and two) and the evaluated quality of the neighbourhood can be important indicators for the extent to which people attach to the neighbourhood. Those who have a negative evaluation are more triggered to potentially change this situation in order to improve their quality of life. We thus hypothesise that a negative evaluation of the neighbourhood makes an individual, either of native or immigrant origin, more likely to move (Hypothesis five).

Immigrant Populations in Belgium

Of the population in Belgium (around 11 million inhabitants) about nine per cent is of foreign nationality and when including migrants of both first and second generations around 16 per cent of the population has a foreign origin. The largest immigrant communities in Belgium are the Italian (347,000 first- and second-generation migrants in 2001), Moroccan (245,000), French (194,000), Turkish (140,000) and Dutch (140,000). The Italian community has a long history of migration in Belgium (Morelli 2004). The first contingents of Italians moved to Belgium in the 1920s to work in the heavy industries and coal mines. Immediately after the Second World War, tens of thousands of mainly Italian workers were recruited because of a strong demand for labour in the metal and mining industries, and later also in other industries as well as in the construction and service sector. In the late 1950s and the 1960s, Italians were joined by Spaniards, Greeks and Portuguese. During the same period there was also a large influx of migrants from Morocco and Turkey (Lesthaeghe 2000). This massive immigration of foreign guest workers stopped abruptly with the economic crisis in the early 1970s. From the middle of the 1970s onward, immigration continues but was restricted to family reunification. The inflow of family members is reinforced by the import of brides and grooms by the Moroccan and Turkish population, which continues until now (Lievens 2000). In the past two decades, migration to Belgium is also characterised by asylum migration (especially in the 1990s after the outbreak of the war in former Yugoslavia), and an increased influx of (Eastern) European immigrants – mostly from Poland, Romania and Bulgaria – after the enlargement of the European Union in 2004 and 2007 (de Valk, Huisman and Noam 2011).

The Belgian migrant community in this sense covers a large range of Western (European) origin groups as well as groups originating from the Mediterranean area and Africa. Also, when it comes to labour market, educational and housing position, there is a wide variation between the different immigrant origin groups. Overall, the socio-economic position of many immigrants of non-Western origin is less favourable than that of the majority group, whereas those coming from other European countries are more often in a similar or better position than the natives.

Data and Measures

This study draws on data from the 2001 Belgian census (1/10/2001), individually and anonymously linked to the National Population Register (situation on 1/1/2006) by Statistics Belgium. This allows for the measurement of internal mobility of all residents of Belgium. The population register data of 2006 are used to identify municipality of residence and characteristics of the private households. The residence is imputed from the National Register and corresponds to the legal address (domicile). The 589 municipalities are the smallest political and administrative units in Belgium. They correspond to the LAU-2 (Local Administrative Units) level. We restrict our analyses to persons between 18 and 64 years, and not living in institutional households or student accommodation. The explanatory variables are all measured before the move took place and are derived from the 2001 Census.

Internal mobility is identified by comparing the municipality of residence in 2001 and 2006. Those living in the same municipality at both measurement points may have moved houses within the same municipality. However, our data do not capture this, limiting our analyses to moves between municipalities. From the 10,296,350 inhabitants in the 2001 Census, almost 9,550,000 (93 per cent) were still living in Belgium in 2006. Approximately 13 per cent of these inhabitants had moved to another municipality within this period (2001–2006).

Besides levels of mobility we are also interested in the direction of the move. In order to do so we focus our analyses on the Brussels Capital Region (BCR), covering the majority of moves and a large share of migrants in Belgium. We distinguish between the central city and its suburbs using the typology of Belgian city regions (Van der Haegen 1980, Luyten and Van Hecke 2007). The central city consists of the historical core, the 19th-century expansion of the city and the other densely populated urban districts. The central city covers the 19 municipalities of the BCR. In the multivariate analysis, the BCR is divided into a group of 'inner city' municipalities,[1] which broadly corresponds to the expansion of Brussels before 1914, and into an 'outer city' group[2] that was urbanised thereafter. The suburbs include the rest of the morphological agglomeration (the built-up area adjacent to the central city), the 'banlieue' (the urban fringe with a predominantly rural appearance but functionally urban) and the commuter zone (municipalities with 15 per cent or more of the labour force commuting to the urban agglomeration).

In the multivariate analysis of the determinants of mobility in the Brussels urban region, we distinguish between three types of moves: (1) from the inner city to the outer city within the BCR, (2) from the BCR to its suburbs, and (3) from the BCR to the rest of Belgium. A move from the central city to the suburbs directly results in getting into a neighbourhood with more open and green space. Compared to the inner city of the BCR, the neighbourhoods in the outer city are less dense and have more green areas. Moves from inner to outer city therefore also largely correspond to a gain in the quality of the physical environment. In the Brussels area, both types of moves generally coincide with a higher socio-economic status of the neighbourhood and with better housing as well. We apply multinomial logit analyses in which the three types of moves are compared to those who do not move (reference group). We present our findings and estimate the multivariate models separately by (migrant) origin.

In order to distinguish between migrants and the native population, we combine information on country of birth, current (Census 2001) nationality and nationality at birth of the person, as this is the only available information in the data. This gives the best available proxy for migrant background and generation. Those with a foreign nationality at the time of the census and born outside of Belgium are

---

1   Anderlecht, Bruxelles, Ixelles, Molenbeek-Saint-Jean, Saint-Gilles, Saint-Josse-ten-Noode and Schaerbeek.

2   Auderghem, Berchem-Sainte-Agathe, Etterbeek, Evere, Forest, Ganshoren, Jette, Koekelberg, Uccle, Watermael-Boitsfort, Woluwe-Saint-Lambert and Woluwe-Saint-Pierre.

defined as first-generation migrants. Those who are born in Belgium but had a foreign nationality at birth are classified as second generation (irrespective of their current nationality). Finally, those with a Belgian nationality at the Census and at birth are included as natives.

Migrant origins are categorised into six groups based on country of origin. Turkish and Moroccan migrants are distinguished separately due to the numerical size of the groups as well as their specific position in society, and religious background. Previous studies have pointed to the differences between the two groups (Lesthaeghe 2000), warranting studying them separately. Migrants coming from other non-Western origins are numerically too small to distinguish separately and are thus grouped as other non-Western. Migrants from Italy, Spain, Portugal and Greece are grouped as Southern European whereas migrants originating from Eastern Europe (with a more recent migration history) are also distinguished separately. Finally, migrants coming from other Western origins (including other European countries, North America, Japan, Australia and New Zealand) are taken together as a last category given the similarity in migration background and socio-economic status.

In line with the theoretical assumptions, our analyses include a range of demographic, socio-economic, housing and neighbourhood variables. All explanatory variables are measured in 2001 before a potential move. *Age* of the individual is measured in full years and divided into 4 age groups: 18–24, 25–34, 35–49 and 50–64 year olds. *Type of household* at the moment of the 2001 Census distinguishes between four different states (single, one parent household, couples without children and couples with children (reference group)). Furthermore, we include the household transitions experienced within the five-year period, in order to link demographic transitions in the life course to mobility patterns. Transitions between household types are defined by four dummy variables which correspond to a gain/loss of a partner or children. *Immigrant generation* is defined using nationality and country of birth. We use a six-category classification to identify the level of homogamy in the households; this classification of first-, second- and third-generation-plus immigrants is based on Ellis and Wright (2005). Third-generation-plus migrants correspond here to natives. Socio-economic status is measured by *educational level* (higher education vs. no higher education indicating less than college) and by type of *income* in the household in which those with replacement income[3] (reference category) are compared to those households where at least one person earns an income via paid labour, and households where two full-time incomes through labour market participation are earned.

In our study we are mainly interested in the role of housing and neighbourhood in affecting mobility patterns among different origin groups. *Housing characteristics* are captured by a combination of tenure status (renter vs. owner) and housing quality. The quality of the dwelling is based on the degree to which

---

3   Replacement income refers to those who are on social benefits and do not earn an income via paid work.

the person indicates large repairs are needed, the size in square meters, the number of bedrooms, and the presence/absence of a toilet, bathroom, kitchen, and double glazing (Vanneste, Thomas and Goossens 2007). The original five categories (ranging from insufficient to excellent) were recoded to three: low, medium and high quality of the dwelling. Two *neighbourhood characteristics* are included: the *self-reported appreciation of the neighbourhood*, and the *share of the own ethnic group* in the neighbourhood. Appreciation of the neighbourhood covers five items directly related to the evaluation of the immediate surroundings of the dwelling and the facilities available in the neighbourhood. It was asked what the respondent thought of: the beauty of the buildings in the immediate surroundings of the dwelling, the cleanliness, the air quality (pollution), the quietness, and the amount of green space in the neighbourhood. Answers on these questions were given on a three point Likert scale on each of the items, ranging from negative to neutral to positive appreciation. A total score was calculated by assigning one point to a positive evaluation and subtracting one point in the case of a negative evaluation. Three dummy variables were constructed for scores between -5 to -2 (negative appreciation of the neighbourhood), between -1 and +1 (neutral appreciation), and between +2 and +5 (positive appreciation). The *concentration of the own ethnic group* in the neighbourhood is calculated by dividing the 724 statistical sectors of the Brussels-Capital Region into quartiles with populations of equal size after sorting them first by increasing percentage of the own ethnic group in the total population (origin group specific). The three dummy variables correspond to the first, the two middle, and the fourth quartile.

A description of the independent variables by origin group is provided in Table 2.1. Immigrants are in general younger than the majority group: natives are more likely to belong to the older age groups whereas in particular the non-Western groups (including Turks and Moroccans) as well as the Eastern Europeans are young populations. This is also reflected in the household composition where migrant groups in general and those of Turkish and Moroccan and non-Western origin in particular are more likely to be in a couple with children. When it comes to the ethnic composition of the household, we find that among all origin groups the majority live in households with immigrants (first-generation only). Interethnic households with one partner from Belgium and a first- or a second- generation immigrant (immigrant and second- or third-generation categories) are relatively common for Western, Eastern European and other non-Western origin migrants and least common for the Turkish and Moroccan group. All migrant groups are less educated than the Belgian majority group with the exception of the Western migrants who are more likely to have higher education. Among all groups, the majority of households has to rely on one income from paid work. Two full time incomes are more common among the Belgians and the Southern European immigrants and least likely for the Turkish and Moroccan groups. Comparing the characteristics between the different origin groups there appears a dichotomy between those of Turkish and Moroccan origin and those of other migrant origin. The demographic and socio-economic characteristics of the Western origin group

Table 2.1   Description of the independent variables in the multivariate analyses by migrant origin, Brussels-Capital Region, 2001 (per cent)

	Native Belgian	Western	Southern European	Eastern European	Moroccan	Turkish	Other non-Western
Age (in years)							
18–24	3.2	2.9	3.0	4.3	7.8	9.2	4.1
25–34	23.5	27.5	25.2	32.4	35.5	36.4	31.4
35–49	39.3	40.3	44.4	38.3	38.9	37.9	46.6
50–64	33.9	29.2	27.4	25.0	17.8	16.5	18.0
Family life status/transition							
Single to couple without children, or one parent to couple with children	5.1	5.7	3.9	4.5	3.0	2.5	5.0
Household without children to household with children	8.4	10.0	8.2	8.7	11.2	7.2	11.3
Household with children to household without children	5.6	4.3	5.7	4.7	2.2	4.1	3.1
Couple to single/one parent	6.1	5.9	6.1	7.4	6.9	7.5	7.7
Single	22.9	24.2	16.8	13.7	9.0	7.3	15.2
One parent household	5.9	5.0	5.9	5.7	5.2	5.2	7.4
Couple without children	17.4	14.2	12.6	13.6	4.3	5.3	7.6
Couple with children	28.5	30.8	40.9	41.7	58.1	60.9	42.6
Migrant status and generation							
Third-generation-plus-only household	89.7	–	–	–	–	–	–
Second-generation/third-generation-plus household	2.8	6.3	7.3	3.0	1.6	1.0	2.4
Second-generation-only household	–	7.4	14.9	5.6	9.1	6.2	4.6

Table 2.1   *Concluded*

	Native Belgian	Western	Southern European	Eastern European	Moroccan	Turkish	Other non-Western
Immigrant/third generation-plus	7.5	17.4	9.2	15.2	5.1	2.9	16.8
Immigrant/second-generation	–	3.9	8.5	6.0	16.0	13.9	4.1
Immigrant-only household	–	65.0	60.1	70.2	68.1	76.0	72.1
Educational level							
Higher education	46.3	58.7	25.7	35.8	13.8	7.7	42.2
No higher education	53.7	41.3	74.3	64.2	86.2	92.3	57.8
Income and source							
Two full-time incomes	29.0	24.0	27.2	18.7	13.7	15.7	18.2
At least one labour income	53.1	62.1	54.9	59.7	56.0	54.4	59.5
Only replacement income(s)	17.9	13.9	17.9	21.6	30.3	29.9	22.3
Housing characteristics							
Renter, low quality housing	7.2	6.9	9.6	9.6	16.0	9.0	14.0
Renter, medium quality housing	30.9	32.4	33.5	41.0	40.8	28.5	42.9
Renter, high quality housing	11.8	21.2	11.3	10.2	8.1	5.1	12.8
Owner, low quality housing	2.7	1.9	3.5	3.5	4.3	7.7	2.6
Owner, medium quality housing	19.6	11.8	20.3	18.7	18.0	32.5	13.8
Owner, high quality housing	27.8	25.8	21.9	17.1	12.9	17.2	13.9
Neighbourhood evaluation							
Negative	26.7	25.8	31.3	30.7	40.1	43.5	29.6
Neutral	47.1	46.6	45.8	45.4	42.7	39.8	47.1
Positive	26.2	27.6	22.9	23.9	17.3	16.7	23.2
Total population	198,112	28,685	31,396	7,515	35,395	10,239	20,993

Source: Census 2001 and National Register (Statistics Belgium); authors' calculations.

and of Southern Europeans are closest to those of the natives. Eastern Europeans and the other non-Western group take an intermediate position. Of all immigrant groups, Southern Europeans have the highest proportion of second-generation migrants (22 per cent[4]).

The population of foreign origin in Belgium is largely concentrated in major cities and predominantly lives in the Brussels-Capital Region and its suburbs (Figure 2.1). Furthermore, larger shares of migrants are found in the border regions: with the Netherlands in the north, with France in the south, and with Germany and Luxemburg in the east of the country. Also the industrial basin and former coal mines close to the cities of Mons, Charleroi, Liège and Genk are home to a large share of the foreign population. In the other municipalities, the proportion of migrants in the total population is much lower, particularly in Flanders (North Belgium).

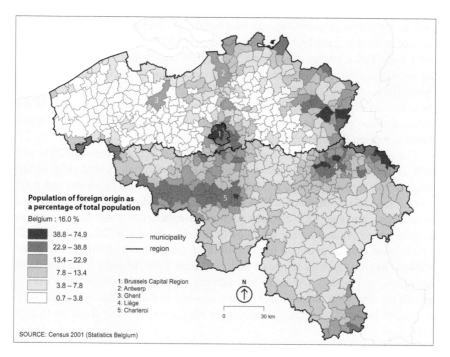

Figure 2.1   Population of foreign origin in Belgium, 2001

---

4 This figure is derived from summing the second generation and second/third generation groups.

Patterns of Internal Migration Compared

Table 2.2 shows the population distribution of native Belgians and the six migrant groups in the major regions on an urban-suburban scale (row percentages). The percentage of natives decreases sharply with the degree of urbanity. The native population is most under-represented in the Brussels-Capital Region (53 per cent of the total population). Overall, the suburbs are more populated by native Belgians and central cities have a substantial share of immigrants. The location of immigrants, though, varies by origin group (not in Table). Immigrants of Western origin are largely located in the south-eastern part of the Brussels-Capital Region (BCR) and its suburbs, and also near the border with the Netherlands, France and Germany. Southern Europeans of mainly Italian background live predominantly in the industrial areas of Mons, Charleroi, Liège and Genk (as a result of labour migration in the second half of the twentieth century), and to a lesser extent in the BCR. Eastern Europeans, Moroccans and Turks have a similar residential pattern, but are more evenly spread than Southern Europeans. The proportion of Moroccans is especially high in the inner city of the BCR and in cities like Antwerp and Liège. Turks, on the other hand, are significantly concentrated in Ghent, in the municipalities of Saint-Josse-ten-Noode and Schaerbeek (Brussels) and near Genk. The location of the other non-Western group coincides to a large extent with the more densely populated urban areas, but they are also over-represented in the suburban areas of Brussels and Liège and in several rural municipalities in the Walloon Region (South Belgium).

The geographical pattern of internal migration in Belgium between 2001 and 2006 is characterised by net losses of native Belgians in all central cities of the major urban regions, except in Ghent, and net gains in all suburban and rural areas (Table 2.3). Especially in Brussels, the urban exodus is considerable. It has to be noted, however, that also the rural areas experience a net loss of natives (mainly young adults moving for study or job reasons; not in Table). Areas with a population gain of over two per cent of the population are mainly found in the Walloon Region, such as southeast and southwest from Brussels and in the rural Ardennes. In Flanders (Antwerp and Ghent), most municipalities have only a modest positive migration balance of natives. Only in the coastal region, net gains are higher because of extensive retirement migration of native Belgians (not in Table).

One of our aims was to see to what extent similar patterns of internal migration are found for different origin groups. Our findings show a large similarity in movement out of the major central cities, and into the suburban and rural areas for each of the different migrant groups (Table 2.3, columns 2 to 7). One of the few exceptions is the net in-migration of Eastern Europeans and Moroccans in the metropolitan cities of Ghent and Charleroi. The urban exodus of the minority population is greatest in Brussels and in Liège, with even higher net losses – in relative terms – than is found for native Belgians. In the Brussels-Capital Region, however, there is a net gain in 9 of the 12 municipalities of the outer city (not in

Table). The settlement of migrant groups is most pronounced in municipalities close to the central city (whereas natives move more to rural areas further away from the city). In the suburbs, the percentage increase through internal migration is as high as 23.1 for Moroccans and 45.7 for Turks in the Brussels region, and 38.9 for Turks in the Ghent region, although one should bear in mind that these percentages are based on relatively small numbers (Table 2.2).

Table 2.2   Population distribution (per cent)[*] and total population by (migrant) origin group and urban region, Belgium, 2001

	Native Belgian	Western	Southern European	Eastern European	Moroccan	Turkish	Other non-Western	Total population (x 1,000)
Brussels – central city	52.9	9.9	9.7	2.5	13.2	3.8	8.1	974
Brussels – suburbs	89.5	4.3	2.7	0.6	0.9	0.2	1.8	1,562
Antwerp – central city	77.3	5.4	1.6	1.7	7.3	2.6	4.1	448
Antwerp – suburbs	93.2	4.1	0.4	0.4	0.7	0.4	0.8	700
Ghent – central city	86.2	2.4	0.7	0.8	1.2	5.9	2.7	225
Ghent – suburbs	96.9	1.8	0.2	0.2	0.1	0.1	0.6	348
Liège – central city	66.3	4.0	14.2	2.1	5.1	2.6	5.7	185
Liège – suburbs	77.3	3.3	14.0	1.8	0.9	1.1	1.6	546
Charleroi – central city	70.4	3.1	16.7	1.3	2.7	3.2	2.7	200
Charleroi – suburbs	79.4	2.9	13.2	1.1	0.8	1.4	1.2	312
Non-metropolitan urban regions – central city	85.9	3.7	3.6	1.0	2.1	1.6	2.2	979
Non-metropolitan urban regions – suburbs	88.9	4.8	3.6	0.6	0.5	0.8	0.8	1,246
Other (mainly rural) municipalities	89.8	4.8	2.7	0.5	0.5	0.9	0.8	2,571
Total population (x 1,000)	8,635	489	475	95	245	139	219	10,296

[*] Row percentages.

Source: Census 2001 (Statistics Belgium); authors' calculations.

Table 2.3   **Net internal migration rate per 100 group population (in 2001) by migrant group and urban region, Belgium, 2001–2006**

	Native Belgian	Western	Southern European	Eastern European	Moroccan	Turkish	Other non-Western
Brussels – central city	-3.9	-2.6	-4.2	-4.4	-3.3	-4.5	-4.1
Brussels – suburbs	0.1	2.7	8.7	10.5	23.1	45.7	0.0
Antwerp – central city	-2.0	-2.2	-2.4	-1.1	-0.6	-0.3	-0.6
Antwerp – suburbs	0.4	0.4	5.5	2.7	4.6	2.8	0.1
Ghent – central city	0.2	-0.2	-0.9	1.5	0.8	-1.0	-0.4
Ghent – suburbs	0.4	0.8	4.5	2.0	0.8	38.9	0.3
Liège – central city	-0.4	-0.5	-2.6	-2.2	-3.1	-5.0	-5.8
Liège – suburbs	0.4	0.9	0.5	0.4	8.3	5.7	0.4
Charleroi – central city	-0.8	0.5	-2.2	1.0	5.0	-0.5	2.0
Charleroi – suburbs	0.3	1.2	0.6	-1.3	3.7	-0.4	3.5
Non-metropolitan urban regions – central city	-0.4	0.9	-0.1	-0.2	-0.9	0.2	-1.6
Non-metropolitan urban regions – suburbs	0.4	0.0	0.4	0.6	4.3	0.3	2.0
Other (mainly rural) municipalities	0.6	0.4	1.3	0.9	2.6	-0.1	0.3

Note: Net migration rate uses 2001 population as the denominator.

Source: Census 2001 and National Register (Statistics Belgium); authors' calculations.

In order to get a more long-term perspective on mobility patterns we replicated our analyses also for the 1996–2001 period (not shown). The patterns found in this five-year period are overall the same as for the 2001–2006 period. For the native Belgians, the impact of suburbanisation is about the same in both five-year periods. But their movement out of the central city has substantially decreased in Ghent, Liège and Charleroi. However, one interesting difference stands: there is a striking increase in the level of urban exodus and suburbanisation of migrant groups in the Brussels urban areas. In particular, Turks, Moroccans, and Eastern Europeans have a much higher net migration out of the Brussels-Capital Region in the most recent period of 2001–2006. Furthermore, the positive net migration of people with another non-Western background into the BCR in the 1996–2001 period changed to a net outflow in the 2001–2006 period. Overall, we can conclude that for migrant groups in general and for the Turkish and Moroccan population in particular the past decade has shown an increasing level of suburbanisation in all areas of Belgium.

Internal Migration of Natives and Migrants in the Brussels Capital Region

Between 1968 and 1996, the population of the BCR declined continuously from 1,079,181 to 948,122 inhabitants, mainly as a result of consecutive waves of suburbanisation (De Lannoy et al. 2000). From 1996 onward, the combination of rising natural growth and a high positive international migration balance resulted in strong population growth (Figure 2.2). In November 2009, the population of Brussels exceeded 1.1 million inhabitants for the first time in history. Despite the growth of the BCR as a result of natural growth and international migration, an average of 12,500 inhabitants per year are lost as a result of internal migration. The geographical distribution of natives and migrant groups in the BCR varies significantly. Native Belgians predominantly live in the outer city, near the border with Flanders. The group of Western immigrants, on the other hand, is mainly concentrated in the south-eastern (most expensive) parts of the city. Turks and Moroccans live spatially segregated in the low-quality 19th-century neighbourhoods west and north of the historic city centre. Southern Europeans, Eastern Europeans, and the other non-Western migrants have more dispersed settlement patterns in mixed neighbourhoods (see, for example, Willaert and Deboosere 2005).

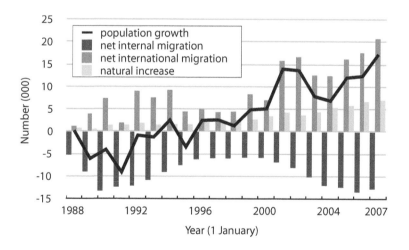

Figure 2.2   **Components of population change in the Brussels-Capital Region, 1988–2007**

Source: National Register (Statistics Belgium); authors' calculations.

In order to study the internal mobility in the BCR in more detail, we start by giving some background on moves between different areas (in terms of urbanisation). Table 2.4 presents origin-to-destination matrices for the various (migrant) origin groups. Net internal migration is calculated per 100 group population in the origin area (rows). The part right of the blank diagonal of each matrix corresponds to migration balances with districts further away from the area of departure, whereas the area to the left of the blank diagonal relates to migration balances with districts closer to the central city than the area of departure.

The matrices show a cascade of migration from more dense (the inner city) to less dense (suburban) areas, and from these suburban parts to the rest of Belgium. This 'counter-urban cascade' is very prominent for all migrant groups. The only deviation from this general pattern is a minor inward movement of Southern Europeans, Turks and the other non-Western group from outside the Brussels urban region.

On balance, all migrant groups leave the inner city of the BCR. Total net migration is also mostly negative in the outer city of the BCR, but the losses are very small because they are compensated to a large extent by a positive migration balance with the inner city. Only Turks and Moroccans have a net gain in the outer city, because the out-migration to other urban districts is not as high as the influx from the inner city. Total net migration rates in the suburban areas are always positive, except for native Belgians in the contiguous built-up area extending from the BCR. The main difference between the origin groups is that natives and migrants of Western origin most often go to the outer suburbs, whereas the other immigrants move most in the inner suburbs.

Table 2.4 Net internal migration rate per 100 group population between residential zones in the Brussels capital region (origin–destination matrix) by (migrant) origin group, 2001–2006

Native Belgian

dest. origin	1 most urban	2	3	4	5 least urban	6	total
1	–	-1.1	-1.5	-0.8	-1.2	-0.3	-4.8
2	0.9	–	-1.4	-1.1	-1.0	-0.5	-3.0
3	1.1	1.2	–	-0.9	-1.3	-0.8	-0.7
4	0.6	1.0	0.9	–	-1.4	-0.6	0.5
5	0.3	0.2	0.4	0.4	–	-0.5	0.8

Western

dest. origin	1 most urban	2	3	4	5 least urban	6	total
1	–	-2.4	-0.8	-0.7	-0.5	-0.5	-4.8
2	2.5	–	-1.3	-0.9	-0.4	-0.3	-0.4
3	1.3	2.1	–	-0.6	-0.4	-0.5	1.9
4	1.6	2.0	0.8	–	-1.2	-0.7	2.5
5	0.9	0.8	0.5	0.9	–	-0.3	2.8

All minorities

dest. origin	1 most urban	2	3	4	5 least urban	6	total
1	–	-2.1	-1.9	-0.6	-0.8	-0.4	-5.8
2	4.2	–	-1.9	-0.9	-0.6	-0.1	0.6
3	8.5	4.4	–	-0.6	-0.7	-0.2	11.5
4	3.8	2.7	0.8	–	-1.4	-0.5	5.4
5	3.0	1.3	0.6	0.9	–	-0.1	5.6

Southern European

dest. origin	1 most urban	2	3	4	5 least urban	6	total
1	–	-1.9	-2.5	-0.9	-0.8	-0.4	-6.4
2	3.1	–	-2.0	-0.8	-0.6	-0.2	-0.6
3	8.2	4.1	–	-0.8	-0.8	-0.2	10.5
4	4.3	2.5	1.1	–	-1.4	0.2	6.7
5	2.9	1.3	0.9	1.0	–	0.8	6.9

Table 2.4   Concluded

Eastern European

dest. origin	1 most urban	2	3	4	5 least urban	6	total
1	–	-2.6	-3.0	-0.6	-0.6	-0.2	-7.0
2	3.6	–	-2.7	-1.1	-0.5	0.1	-0.6
3	11.9	7.7	–	-0.5	-1.3	-0.6	17.3
4	3.7	4.4	0.6	–	-1.7	-0.8	6.2
5	1.8	1.1	0.9	0.9	–	-0.3	4.3

Turkish

dest. origin	1 most urban	2	3	4	5 least urban	6	total
1	–	-2.3	-3.1	-0.5	-0.5	-0.4	-6.7
2	15.7	–	-3.3	-1.2	-0.4	-0.3	10.6
3	71.3	11.2	–	-0.2	0.2	0.4	82.9
4	29.2	11.1	0.6	–	0.2	3.0	44.1
5	5.6	0.7	-0.1	0.0	–	0.6	6.7

Moroccan

dest. origin	1 most urban	2	3	4	5 least urban	6	total
1	–	-1.8	-1.5	-0.4	-0.7	-0.7	-5.0
2	7.5	–	-2.1	-0.6	-0.7	-0.4	3.8
3	22.6	7.5	–	-0.6	-0.9	0.1	28.8
4	10.9	3.5	0.9	–	-0.6	-1.1	13.7
5	6.1	1.5	0.5	0.2	–	-0.6	7.6

Other non-Western

dest. origin	1 most urban	2	3	4	5 least urban	6	total
1	–	-2.7	-2.2	-0.7	-1.4	0.2	-6.8
2	4.0	–	-2.4	-1.0	-1.0	0.3	-0.1
3	10.1	7.4	–	-0.4	-0.7	0.2	16.7
4	3.6	3.5	0.4	–	-2.3	-0.6	4.7
5	4.4	2.1	0.5	1.4	–	-0.3	8.0

Note: Net migration rate uses 2001 population in origin location as the denominator. Legend: 1 inner city of the BCR, 2: outer city of the BCR, 3: other municipalities of the morphological agglomeration (built-up area bordering the central city), 4: 'banlieue', 5: commuter zone, 6: other Belgian municipalities, dest.: destination – : not applicable. There are 6 destinations and 5 origins as the rows (origins) refer to the zones of the Brussels urban region, whereas the columns also include a destination outside of the Brussels urban region (thus other municipalities of Belgium).

Source: Census 2001 & National Register (Statistics Belgium); authors' calculations

Determinants of Internal Mobility

In order to evaluate the role of different factors in mobility in the BCR, separate multinomial logistic models are estimated for the native population and the six migrant groups. The results of the multinomial logistic regression analyses (living in the same municipality is the reference category) are presented as Odds Ratios of chances of moving in Table 2.5. Note that analyses are controlled for all variables specified in Table 2.1: age, family life status and transitions, educational level, and income source. Only the variables that were key for testing our hypotheses on migrant generation, housing and neighbourhood characteristics are shown.

Demographic Characteristics and Life Course Events

As expected, our analyses show that, after controlling for educational level, residential mobility varies considerably by demographic characteristics of the individual and the household as well as by life course events (Hypothesis one). Especially, young adults have a high probability of moving (not in Table; findings on demographics and life course events can be requested from the first author). At older ages, the chance of relocating is much lower with only a small increase around the ages of 55–65 when – mostly native – retirees make a (long-distance) move to the coastal municipalities or to the Ardennes region. For moves from the inner to outer city and for the long-distance migrations, the highest mobility rates are found around the age of 25. For moves from the BCR to the suburbs this is at a somewhat older age (30 years), because this type of residential move more frequently coincides with the purchase of a dwelling. These findings are overall in the same direction for all origin groups.

Our results also confirm that household transitions often trigger an internal migration move (Hypothesis one). Especially transitions that involve a gain or loss of a partner (couple formation, separation, divorce) are associated with high residential mobility since this implies a move of at least one of the partners. Also the birth of a (first) child generates a significant amount of spatial mobility among all origin groups. The household transitions are in particular relevant for inner to outer city moves in the BCR among all groups. For individuals who didn't experience a household transition, the degree of mobility varies considerably according to household type, destination and ethnic group. In line with the literature, the probability of making a suburban move is highest among couples with children. Couples without children, but especially one parent households and singles, have a much lower probability of migrating to the suburbs. This applies equally to natives and to the different migrant groups. On the other hand, singles and couples without children have a higher probability of moving within Brussels than households with children; again confirmed for all origin groups. For long distance moves, the effect of household composition is more complex. High spatial mobility is found among couples without children (natives and Southern Europeans) and among singles (Southern Europeans). With the exception of

Table 2.5 **Multinomial logit regression estimates by type of move and (migrant) origin group (Odds Ratios (exp(β)), Brussels Capital Region 2001–2006**

	Inner city to outer city within Brussels						
	Native	West	SEur	EEur	Mor	Tur	Onw
Third-generation-plus-only household	0.88***	–	–	–	–	–	–
Second-generation/third-generation-plus household	0.96	0.68***	0.84*	0.78	1.19	0.76	0.82
Second-generation-only	–	1.07	0.92	1.03	1.21***	1.23	1.02
Immigrant/third-generation-plus (Ref. natives)	1.00	0.88*	0.90	0.68***	1.17*	1.08	0.83**
Immigrant/second-generation	–	0.90	0.90	0.77	1.06	1.15	1.17
Immigrant-only household (Ref. minorities)	–	1.00	1.00	1.00	1.00	1.00	1.00
Renter, low quality housing	4.20***	4.35***	3.97***	4.09***	4.59***	3.95***	3.79***
Renter, medium quality housing	3.65***	4.10***	3.48***	3.61***	3.92***	3.13***	3.38***
Renter, high quality housing	3.57***	4.14***	3.29***	3.00***	3.47***	3.65***	2.98***
Owner, low quality housing	1.27***	1.38	1.05	0.62	1.07	1.00	1.07
Owner, medium quality housing	1.14***	1.28**	1.01	1.24	1.01	1.23	1.07
Owner, high quality housing (Ref.)	1.00	1.00	1.00	1.00	1.00	1.00	1.00
Negative appreciation of the neighbourhood	1.14***	1.44***	1.20***	1.40***	1.36***	1.11	1.22***
Neutral appreciation of the neighbourhood	1.05*	1.18***	1.03	1.20*	1.13**	0.92	1.04
Positive appreciation of the neighbourhood (Ref.)	1.00	1.00	1.00	1.00	1.00	1.00	1.00
Neighbourhood with a high concentration of the own ethnic group	0.56***	1.49***	0.93	1.18	0.68***	0.61***	1.11
Neighbourhood with an intermediate concentration of the own ethnic group	0.80***	1.18***	1.15**	1.12	0.74***	0.62***	1.23***
Neighbourhood with a low concentration of the own ethnic group (Ref.)	1.00	1.00	1.00	1.00	1.00	1.00	1.00

	Brussels to suburbs						
	Native	West	SEur	EEur	Mor	Tur	Onw
Third-generation-plus-only household	1.21***	–	–	–	–	–	–
Second-generation/third-generation-plus household	1.09	1.22*	2.03***	1.57	2.39***	1.08	1.18
Second-generation-only	–	1.48***	2.02***	1.80**	1.93***	1.31	0.95
Immigrant/third-generation-plus (Ref. natives)	1.00	1.30***	1.40***	1.26	1.91***	1.50	1.07
Immigrant/second-generation	–	1.01	1.56***	1.48*	1.43**	0.97	0.87
Immigrant-only household (Ref. minorities)	–	1.00	1.00	1.00	1.00	1.00	1.00
Renter, low quality housing	2.68***	3.49***	2.17***	2.10***	2.28***	1.36	2.95***
Renter, medium quality housing	2.74***	3.43***	2.25***	2.64***	2.49***	1.35*	2.51***
Renter, high quality housing	2.79***	3.88***	2.05***	2.99***	2.50***	1.39	3.11***
Owner, low quality housing	1.27***	1.45	0.87	1.51	0.95	1.29	1.33
Owner, medium quality housing	1.38***	1.81***	1.29**	1.25	1.44**	0.92	1.17
Owner, high quality housing (Ref.)	1.00	1.00	1.00	1.00	1.00	1.00	1.00
Negative appreciation of the neighbourhood	1.42***	1.03	1.20**	1.16	1.39***	0.92	1.00
Neutral appreciation of the neighbourhood	1.07**	0.90*	1.00	1.00	1.20*	0.97	1.03
Positive appreciation of the neighbourhood (Ref.)	1.00	1.00	1.00	1.00	1.00	1.00	1.00
Neighbourhood with a high concentration of the own ethnic group	1.08**	0.79**	0.89	0.66**	0.54***	0.60***	0.83*
Neighbourhood with an intermediate concentration of the own ethnic group	1.15***	0.88*	0.99	0.84	0.70***	0.74**	1.01
Neighbourhood with a low concentration of the own ethnic group (Ref.)	1.00	1.00	1.00	1.00	1.00	1.00	1.00

Table 2.5   *Concluded*

	Brussels to the rest of Belgium						
	Native	West	SEur	EEur	Mor	Tur	Onw
Third-generation-plus-only household	1.37***	–	–	–	–	–	–
Second-generation/third-generation-plus household	1.12	1.98***	3.76***	3.36**	2.06**	3.36*	1.74*
Second-generation-only	–	1.75***	2.46***	1.99*	1.19	0.90	1.46*
Immigrant/third-generation-plus (Ref. natives)	1.00	2.10***	2.57***	2.16**	1.94***	2.68**	1.48**
Immigrant/second-generation	–	0.81	2.50***	1.22	1.00	1.49	1.03
Immigrant-only household (Ref. minorities)	–	1.00	1.00	1.00	1.00	1.00	1.00
Renter, low quality housing	2.84***	3.51***	2.20***	2.23*	3.68***	2.36*	3.54***
Renter, medium quality housing	2.92***	2.68***	2.94***	1.78	3.01***	2.78**	3.99***
Renter, high quality housing	2.72***	2.72***	3.71***	1.69	2.67***	1.26	3.80***
Owner, low quality housing	1.38**	1.06	1.65	2.31	1.34	1.00	0.59
Owner, medium quality housing	1.36***	1.59**	1.31	0.54	1.04	1.27	1.36
Owner, high quality housing (Ref.)	1.00	1.00	1.00	1.00	1.00	1.00	1.00
Negative appreciation of the neighbourhood	1.55***	1.97***	1.50**	1.50	1.28	2.52**	1.37*
Neutral appreciation of the neighbourhood	1.12***	1.47**	1.07	1.00	1.13	2.45**	1.07
Positive appreciation of the neighbourhood (Ref.)	1.00	1.00	1.00	1.00	1.00	1.00	1.00
Neighbourhood with a high concentration of the own ethnic group	0.64***	0.48***	0.74*	1.03	1.04	0.59*	0.74*
Neighbourhood with an intermediate concentration of the own ethnic group	0.81***	0.63***	0.94	1.12	1.04	0.61*	0.95
Neighbourhood with a low concentration of the own ethnic group (Ref.)	1.00	1.00	1.00	1.00	1.00	1.00	1.00

Note: *$P < 0.05$; **$P < 0.01$; ***$P < 0.001$; – not applicable. West: other Western origin; SEur: Southern European origin; EEur: Eastern European origin; Mor: Moroccan origin; Tur: Turkish origin; Onw: other non-Western origin. Pseudo R-square [Nagelkerke]: Native: 0.25, other Western 0.24, Southern European 0.21, Eastern European 0.20, Moroccan 0.21, Turkish 0.14, other non-Western 0.19.

Source: Census 2001 & National Register (Statistics Belgium); authors' calculations.

Eastern Europeans, one-parent households have the lowest probability of moving over a longer distance.

The presence/absence and number of labour incomes in the household has a clear and substantial effect in particular on the degree of suburbanisation. For example, persons of Moroccan descent living in households with two full-time incomes or at least one labour income have respectively 4 and 2 times the odds of making a suburban move than Moroccans households with only a replacement income. This pattern is generally also observed for moves from the inner to the outer city (though not always reaching significance). Long distance moves are, on the contrary, less likely for those with two full incomes. According to our findings, the level of mobility is influenced by income status but the direction of the move is even more affected by income of the household.

Housing Conditions

As expected (Hypothesis two), the probability of moving is largely associated with tenure status: renters are much more likely to move than homeowners and even more so when their housing quality is perceived to be low (Table 2.5). Compared to homeowners in high quality housing, those who rent their house even when it is of high quality are about three to four times more likely to move from the inner to the outer city in Brussels (first column of findings in Table 2.5; ranging from 3.57 for natives, and 4.14 for migrants of Western origin to 3.47 and 3.65 for those of Moroccan and Turkish origin respectively). The analyses show that households living in a low or medium quality dwellings not only have a higher probability of moving from the inner to the outer city within Brussels, but also to the suburbs as well as to other parts of Belgium, reflecting the search for a higher quality dwelling. Renters of high quality housing seem to be somewhat more likely to move to the suburbs, which can potentially be linked to the higher probability of persons with a high socio-economic status relocating to the suburbs (see above). As hypothesised (Hypothesis two), this applies to all origin groups. For homeowners, we do not find such a clear relation with quality of housing. Only Belgian homeowners with low or medium quality housing seem to be more likely to make a move compared to those with high quality houses. For all other groups there seems to be hardly any mobility differences based on housing quality.

Neighbourhood: Composition and Evaluation

At the core of our analyses is how neighbourhood ethnic concentration might affect the mobility of particular migrants (Table 2.5, bottom two sets of rows). Our findings show that, overall, higher levels of concentration of own ethnic group in the neighbourhood reduces the likelihood of moving. The odds for inner to outer city moves from neighbourhoods with a high concentration of the own ethnic group are 0.56 for natives, and 0.68 versus 0.61 for the Moroccan and Turkish origin groups, which is in line with Hypothesis three. Furthermore, ethnic concentration also affects the direction of the move. In general, individuals who reside in neighbourhoods with a high concentration of their own group of origin have a lower probability of making a move from the inner to the outer city and to the suburbs, whereas the effects for moves to the rest of the country are less clear. Interestingly, similar patterns are found throughout for natives and the different migrant groups. Only some (significant) exceptions stand out: natives living in an area with higher shares of natives are more likely to move to suburban areas. Furthermore, other Western migrants have a higher chance of moving from the inner to the outer city within Brussels in cases where they lived in an area with more co-ethnics (odds are 1.18 for Western migrants in intermediate concentrated areas to 1.49 in areas with a high concentration of co-ethnics). The effects of ethnic composition of the neighbourhood are consistently found for a moderate level of concentration (compared to a low concentration). Even though the effects of high concentration are overall more pronounced, they indicate that ethnic concentration on a relatively small scale may already be an important determinant for moving decisions.

In general, the hypothesis based on spatial assimilation theory that suburbanisation increases with generation (Hypothesis four) is validated (first set of rows in Table 2.5). Compared to first-generation immigrant households, the degree of suburbanisation and longer distance mobility is clearly higher for households with at least a second-generation or a native member. The odds ratios for suburbanisation among the Moroccan origin group, for example, are 1.43 when a member of the household belongs to the second generation, and 2.39 for those in households with only second-generation/third-generation members. With respect to residential mobility from inner to outer city within the BCR, the assumption derived from assimilation theory only holds for Turks and Moroccans. Although the findings for the Turkish group are not reaching significance, the patterns are clearly in the same direction. The process of suburbanisation is furthermore most pronounced for the Moroccan and Southern European and other Western migrants. Additionally, our findings point to the importance of interethnic group relations: although suburbanisation is clear for the second versus the first generation, results are more pronounced for those who live together with a person of native Belgian origin.

Finally, we analyse the importance of the individual appreciation of the neighbourhood for moving (third block of rows in Table 2.5). The probability of migrating from the inner city of Brussels is clearly negatively associated with the appreciation of the neighbourhood. Those individuals who are negative about the neighbourhood quality in 2001 are much more likely to have moved from the inner city to the outer city of Brussels between 2001 and 2006; the odds range from 1.14 among natives to 1.44 and 1.40 for migrants of Western and Eastern European origin, respectively. The effect of neighbourhood appreciation is consistent for all groups. Even though the coefficient does not reach significance for the Turks, the effect for this group is also in the expected direction. A negative neighbourhood evaluation overall also seems to push people out of the city to the suburbs. Although this seems to hold for most origin groups (significant odds range from 1.20 among those of Southern European origin to 1.42 for natives), a negative association is found for the Turks (not significant) for which neighbourhood quality mainly has an effect on longer distance moves. For this latter group, our analyses suggest that bad neighbourhood quality rather pushes them further away out of the city of Brussels. For the Turkish group we namely find a highly significant effect for those who have a negative appreciation (odds ratio 2.52) and neutral appreciation (odds ratio 2.45) of the neighbourhood on moves from Brussels to other places in Belgium.

In general, the findings for those who are neutral about their neighbourhood compared to those who are positive, are in line with what is found for those who are negative about the neighbourhood, but the effects are most pronounced for the group with a negative neighbourhood evaluation. At the same time and in line with our hypothesis (Hypothesis five), the effect of neighbourhood appreciation by and large seems to operate similarly for those of different (migrant) origin.

Conclusion and Discussion

Internal mobility is often related to the integration of migrants in society and referred to as spatial assimilation. This has already been put forward in the classic work of Park and Burgess (1921). The diverse roles that housing and neighbourhood characteristics play in the mobility choices for migrants as well as natives has received only limited attention in the literature. In this paper we expanded the research on internal mobility of immigrant groups to Belgium by first of all studying the extent to which the internal migration of migrants of different origins follows the same patterns as that of the native majority group. Furthermore, our aim was to study the role of housing and neighbourhood characteristics (ethnic composition and appreciation) in affecting the level and direction of mobility. The first part of the contribution referred to Belgium, whereas the second part zoomed in on the Brussels-Capital Region, the largest urban area of the country where a substantial share of migrants live.

Immigrants in Belgium, like in many other countries, have settled predominantly in the urban centres of the country. As a result, city centres are more likely to include sizeable shares of immigrants, up to almost half of the population in Brussels. Nevertheless, our study showed that in the 2001–2006 period, levels of internal mobility among migrant groups are at least as high as found for the native population. Overall, our results also show that demographic and socio-economic characteristics are important explanatory variables for moves among all origin groups (Hypothesis one). Life course factors are relevant for migrants and natives alike. Life course events in the family domain clearly trigger moves, as has also been shown in earlier studies. Our work indicates that this is of similar importance for natives and migrants of different origin. Having children is related to a higher chance of moving from the inner to the outer city and suburbanisation moves among all groups. In this respect it is important for future planning of houses and facilities to take these patterns into account and to recognise the vast impact life course events have on mobility of households of different origins.

Among all groups and for all moves our findings show that renters are more likely to move irrespective of destination. This is contrary to our hypothesis (Hypothesis two), which suggested that renters would be more likely to leave the city for suburbs and out of the urban area. Our results suggest a general higher level of mobility among renters, irrespective of destination, while no such influence is found for homeowners. Also for the Turkish group, this factor seems to be less related to suburbanisation.

Nevertheless, suburbanisation is not only observed among the majority group, but also takes place among immigrant groups. The internal mobility of immigrants and in particular out of the city centres can be clearly observed for the 2001–2006 period. In line with spatial assimilation theory, we hypothesised that with increasing time of residence (as reflected by migrant generation), suburbanisation would increase. Our results find support for this Hypothesis (four), but also show that suburbanisation is particularly pronounced for those with a relation to the native majority population. This suggests that over generations, migrants are more likely to move to more suburban areas but that in particular relationships between migrants and natives also may be necessary for assimilation in this domain of life. Of course one has to realise that only in the past decades, the second generation has reached adulthood in larger numbers. The second generation in our study are potentially the forerunners (as the second generation is just recently coming of age) for which suburbanisation is already clearly observed. In the future, larger shares of second-generation migrants are expected to move out of the parental home and enter the housing market. It is in particular this younger generation who may move out of the urban city centre. Although this can be interpreted as a sign of ongoing spatial assimilation of the next generation (in line with our Hypothesis four), it might just as well be related to housing availability and housing costs in the urban centres which make it difficult for young people to find a home in the inner cities. Future studies should better capture the local housing market in order to grasp these different factors involved in mobility decisions. Special attention should be paid to the second generation (who are currently getting more and more numerous) in order to assess whether the patterns observed will be continued over time.

The latter point is even more important as we find in our study that the ethnic composition of the neighbourhood is of crucial importance in understanding mobility. Results suggest that more co-ethnics in the neighbourhood result in lower levels of mobility. This finding is, with few exceptions, consistent for all migrant origins. It suggests that ethnic capital and networks still play a crucial role in the location decisions of migrants in Brussels. Interestingly, it is not only high levels of concentration, but also moderate levels of ethnic concentration that keep people in the neighbourhood. To what extent this may change or remain for the second generation as well as with more interethnic relations remains to be answered. Our data did not allow us to assess this in detail. Nevertheless, it is relevant to pay more attention to ethnic composition of the neighbourhood, its changes, and the role it has for the spatial assimilation of migrants from different origin.

Dissatisfaction with the neighbourhood, in terms of cleanliness, quietness, visual attraction of the buildings, air quality and the amount of green space, is found to be a good predictor of the rate of out-migration. A less dense and more green environment is obviously what many households are looking for and contributes to their decision to move as well. Although the effect of neighbourhood satisfaction is found for all groups, it is most persistent for the natives and Western origin migrants. Those natives and Western migrants who evaluate their neighbourhood as negative are more likely to have made a move. This is somewhat less the case for the other immigrant groups. Negative evaluation particularly stimulates moves from the inner to the outer city. This may suggest that economic position and opportunities are relevant again: natives who want to leave a specific neighbourhood have more means to materialise this wish than many of the immigrant groups. One finding is striking in this respect: evaluation of the neighbourhood does not affect short and suburban moves among the Turks, but it clearly does affect the likelihood for long distance moves. Those Turks who evaluate their neighbourhood negatively are much more likely to move over a longer distance. This could again indicate that this migrant group does not perceive that better and affordable conditions are available in other areas of the city and thus decides to move further out of Brussels to achieve their preferred housing situation.

Our analyses thus show that patterns of mobility and the factors affecting these moves are overall similar for all origin groups in Belgium. They also point to the fact that it is worthwhile to take different neighbourhood characteristics into account when studying internal mobility. Ignoring this factor in studies does not do justice to the living conditions of both immigrants and natives. Finally, our analyses also point to one potentially important result of internal mobility of immigrants. Whereas first-generation immigrants mainly settled in the inner city, the residing immigrant groups in these urban centres are increasingly ageing populations as it is mainly the young adult second generation of immigrant origin leaving the city. The role of segregation and discrimination in this mobility should be further explored in future work. Nevertheless, the observed suburbanisation has potential important implications for health facilities, access to care, and suitable housing for these elderly immigrants who are not necessarily in the position of changing their housing situation. Future research should look further into the potential effects of internal mobility patterns for the individual, their families and the city. This is relevant for both scholars and policymakers in the field.

Acknowledgements

We would like to express our gratitude to the editors of this book, Gemma Catney and Nissa Finney, for providing useful and very stimulating feedback on previous versions of the manuscript. An earlier version of the paper was presented at the Minority Internal Migration in Europe Conference, Manchester 5–7 September 2011. The suggestions and comments received at that meeting were extremely valuable. Furthermore, we thank Statistics Belgium for providing the data for the empirical analyses. This work is part of the GOA project on international migration for which generous support was received from the research council of the Vrije Universiteit Brussel (GOA no 55). The contribution of Helga de Valk was funded by and carried out as part of the ERC starting grant project 'Families of Migrant origin a life course perspective' (project no. 263829).

Chapter 3

Immigration and Internal Migration of Ethnic Groups in London

John Stillwell and Sarah McNulty

Introduction

Since the early 1990s, Britain has experienced a significant increase in immigrants originating from different parts of the world which has accentuated ethnic population diversity across much of the country and fuelled debate about the implications for local labour markets (for example, Dustman, Fabbri and Preston 2005), social housing (for example, Phillips 2006a) and the use of public services as well as for community cohesion and integration (Communities and Local Government 2008). This chapter aims to examine the spatial patterns of immigration of different ethnic groups in relation to where co-ethnic populations were living in 2001 and how these populations were redistributing themselves through internal migration.

The sources of the data on immigration, population and internal migration by ethnic group used in the chapter are published and commissioned tables of counts from the 2001 Census, supplied by the Office for National Statistics (ONS). The ethnic group categories have been defined by ONS on the basis of those who arrived in the 2000–01 period and may be referred to as 'recent immigrants'. Those who arrived earlier and constitute a large proportion of the country's ethnic populations can be termed 'established immigrants'. The central aim of the chapter is to examine the destinations of recent immigrants in relation to the locations of established immigrants in London using data at the district and ward level. The process underpinning this relationship is one of chain migration (MacDonald and Macdonald 1964) in which flows of information and remittances from pioneer migrants in destination countries encourage subsequent flows of family and friends from origin countries. It is likely that whilst initial locations remain the first destinations for co-ethnic immigrants over time, there will be internal movement away as those more established ethnic group populations seek new opportunities in the housing and labour markets as they become more settled.

We therefore address two research hypotheses in the chapter: (i) recent immigrants of one ethnic group in 2000–01 chose destinations to settle where established immigrants of the same ethnic group were concentrated; and (ii) recent immigrants of one ethnic group chose to reside in those areas that were losing net migrants of the same ethnicity. Initially, we provide a brief introduction to the national context for immigration, a short review of recent research on ethnic internal migration and some theoretical considerations relating to the linkage between immigration and internal migration. We then outline the characteristics of the data used in the analysis, followed by an examination of the magnitude of ethnic populations and migrants in London *vis à vis* England and Wales. The research hypotheses are then investigated with respect to the Greater London Region since London has been the nation's major destination for both established and recent immigrants.

Context, Review and Theory

The process of human migration has been responsible for redistributing populations since the origins of mankind (King et al. 2010). International migration flows between countries have been determined by a wide-ranging set of factors. Adverse climatic or political conditions in certain parts of the world have displaced many people from their homes, creating streams of refugee migrants or migrants seeking asylum. Since the end of the Second World War, economic migrants have been drawn to countries where opportunities for work are available, in some cases on a seasonal or temporary basis, and students have been moving between countries in increasing numbers to fulfil their educational aspirations.

Britain has experienced substantial immigration flows for all the reasons listed above. Notable examples of forced migration include those leaving Uganda for Britain in the early 1970s under the Amin regime and those arriving from Hong Kong when China regained sovereignty from the British in 1997. In 1962, the Commonwealth Immigration Act restricted the freedom of passage into the UK from other parts of the Commonwealth but in more recent years, Britain has received considerable flows of asylum seekers or refugees from conflict zones such as Afghanistan, Iraq, Somalia and Zimbabwe as well as countries in Eastern Europe with oppressive regimes such as Serbia and Romania. Since 1994, the aggregate statistics on long-term migrants (LTIM) prepared by the ONS (2010) indicate a net immigration gain that reached a peak in 2004 at nearly 250,000 before dropping to 198,000 in 2009 compared with 163,000 in the previous year. In 2009, an estimated 567,000 LTIM immigrants arrived in the United Kingdom (UK), a flow similar in magnitude to the level over the previous five years. Non-British citizens accounted for 83 per cent of all immigrants; a third of these were from EU countries. Those entering with legal work permits and visas are now referred to as 'managed migration' (Home Office 2006) and the aim is to control the numbers entitled to remain in the country through a points based system whereby only skilled workers qualify to settle and acquire citizenship.

Whilst immigration has increased and the flows have become more ethnically diverse since the early 1990s, various demographic developments have taken place within Britain with respect to the ethnic populations. First, the initial ethnic minority communities in settlement cities have expanded by natural increase so that the ethnic populations defined in the Censuses of 2001 and 2011 now contain second, third and fourth generation members, many of whom have been born in Britain. The demographic dynamics of each ethnic group have resulted in distinctive population age structures.

Second, the process of ethnic mixing has accelerated. The number of mixed ethnic unions (partnering and marriage) has increased substantially, although from a small base (Coleman 2004) and this is one reason for the introduction of a 'mixed' ethnic class in the ethnic group classification used by ONS in the 2001 Census. Mixed ethnic unions provide an indication of ethnic integration at the household level and Feng et al. (2010) use micro data from the Longitudinal Study to suggest that 'out-partnering' is more likely in areas with lower concentrations of those in the same ethnic group than in areas of greater co-ethnic concentration.

Third, as well as natural change, there has been internal migration of all 'established' ethnic groups with ensuing debate about the extent of spatial ethnic mixing at the neighbourhood level (as distinct from the household level). Concerns have emerged over the extent of ethnic residential segregation in cities across Britain, with claims in 2005 by Trevor Phillips, then Chair of the Commission for Racial Equality, that the nation would experience the development of racial ghettos similar to those in the USA (Phillips 2005). The use of indices to measure segregation was pioneered in the USA by Duncan and Duncan (1955) and by Massey and Denton (1988) who subsequently introduced the idea of 'hypersegregation', the process through which black ghettos are created in large cities because of the limited educational and employment opportunities for the residents of these neighbourhoods when wealthier or white residents leave the area, with the result that the only people who are left are the poor black underclass. In the UK, studies of ethnic population distributions and the measurement of residential segregation have also been prominent (for example, Peach 1996; Ratcliffe 1996; Phillips 1998; Peloe and Rees 1999; Simpson 2004; Johnston, Poulsen and Forrest 2010). Despite the growth of ethnic minorities in England and Wales between 1991 and 2001 (Rees and Butt 2004, Sabater and Simpson 2009), the evidence suggests that ethnic residential segregation has reduced, at least for the major Black and Asian groups (Sabater 2010), although the roles of different components of population change – natural change, internal migration and international migration – are only beginning to be unravelled (Finney 2010).

The internal migration component of ethnic population change is of particular importance because of the magnitude of the flows involved and has been examined at different spatial scales by Champion (1996, 2005), Finney and Simpson (2009a, 2009b), Finney (2010), Simon (2010) and Stillwell, Hussain and Norman (2008), Stillwell (2010a, 2010b), Stillwell and Hussain (2010a, 2010b) and Catney and Simpson (2010), all of whom make use of either micro or aggregate data from the decennial census to examine variations between ethnic groups in the propensities, composition and spatial patterns of migrants and whose results suggest that internal migration is operating as a mechanism for reducing ethnic population concentration, at least for the major ethnic minority groups in the UK.

Whilst there is an extensive literature on various aspects of both international and internal migration and a growing body of work now exists on the 'ethnic dimension' of internal migration as indicated above, much less consideration has been given to the nature of the relationships between recent immigration flows, established ethnic populations and internal movements of ethnic groups in Britain, although some initial work on linkage between ethnic populations, immigration and internal migration at a district scale was reported by Stillwell and Duke-Williams (2005). Most previous work on linkage between internal and international migration comes from the USA and includes studies during the 1990s by Walker, Ellis and Barff (1992), Sassen (1994), Frey (1995, 1996), Ellis and Wright (1998) and Frey and Liaw (1998), for example. Debate in the literature centred on Frey's contention that a process of 'demographic balkanisation' was underway in the USA involving 'spatial segmentation of population by race-ethnicity, class, and age across broad regions, states, and metropolitan areas... driven by both immigration and long distance internal migration' (Frey 1996: 760). Frey's 'push' hypothesis that immigration has fuelled internal migration through 'white flight' has been contested by those (for example, Ellis and Wright 1998) who believe that outward movement is occurring which is creating 'vacancies/opportunities' that are filled by immigrants ('pull' hypothesis).

It is important to acknowledge that concepts of ethnic mixing, integration and residential segregation were the focus of classical assimilation theory emerging from the Chicago School in the USA in the 1920s (for example, Burgess (1928), reinvented as new assimilation theory in more recent times by Alba and Nee (2003)). The essence of classical theory involves ethnic minorities becoming increasingly aligned with the majority host community over time in terms of their characteristics, norms, behaviours and values. New assimilation theorists draw attention to the key role that institutions play in the integration process, whilst critics of this approach suggest that language and cultural familiarity often obstruct integration, and discrimination and institutional barriers (for example, in employment or housing) frequently block complete assimilation. In what follows, our concern is with the residential patterns of ethnic population settlement and redistribution.

2001 Census Migration and Population Data for Districts and Wards

The 2001 Census is particularly important for research on migration into and within Britain because it is reliable, comprehensive and available at different spatial levels. Districts are the local authority areas that constitute England and Wales (City of London and 32 London Boroughs; 36 Metropolitan Districts; 68 Unitary Authorities; 239 Other Local Authorities) and Scotland (32 Council Areas). Migration data are recorded from a question about where each individual resident in the country was living 12 months before the Census as well as at the time of the Census. The migration data are therefore transition counts of those in existence at two points in time rather than counts of each movement that has been made. Whilst the tables of ethnic population counts in the 2001 Census adopt an ethnicity classification comprising 16 groups, a categorisation of seven broad ethnic groups (White; Indian; Pakistani and Other South Asian (POSA); Chinese; Caribbean, African, Black British and Black Other; Mixed; and Other) was used by the ONS in the Special Migration Statistics (SMS) at district level. Only two ethnic categories (White and Non-white) were used to classify migration counts at ward level in 2001 and no ethnic breakdown was provided at output area level (Stillwell, Duke Williams and Dennett 2010). Moreover, the 2001 Census SMS and published tables provided no breakdown of the overseas origins of recent immigrants.

The migration data used in this chapter therefore come from tables commissioned from ONS in connection with an earlier research project[1] undertaken under the Understanding Population Trends and Processes (UPTAP) initiative. The data provide information on immigration and internal migration at two spatial scales (districts and wards) for the seven ethnic groups used in the SMS (listed above). Complete sets of migration data for all 16 ethnic groups were not available because of the small cell counts and the associated disclosure control issues (Duke-Williams 2010). Consequently, ONS provided two sets of migration data. First, Table CO711A is a set of district-to-district flow matrices for England and Wales for seven age groups for seven ethnic groups and Table CO711B involves counts of immigrants into the 376 districts by ethnic group (but no age disaggregation) from 56 'world regions' of previous usual residence outside the UK. Second, Table CO723 is a two-part table for which the first part contains flows from each ward to each Government Office Region (GOR) and the second part contains flows from each GOR to each ward, for the seven ethnic groups. No breakdown of immigration flows to wards by 'world region' was possible at this spatial scale, so the only data on immigration are the aggregate flows by ethnic group available from SMS Table MG103. Later in the chapter we use the wards in the London GOR, a region that contains all the 32 London boroughs and the City of London and has a boundary that is familiar to administrators and planners. Both these commissioned tables are now available for others to use via the Web-based Interface to Census Interaction Data (WICID)[2] (Stillwell, Duke-Williams and Dennett 2010).

Data on 'established immigrants' (or populations by ethnic group) were extracted using Casweb[3] from Key Statistics Table KS006 at national level and Standard Table ST101 at ward level. District populations by place of birth were obtained from Standard Table ST102 for use in the analyses that follow. An

---

1   RES-163-25-0028 funded by the Economic and Social Research Council.

2   WICID available at http://cider.census.ac.uk/ maintained by the Centre for Interaction Data Estimation and Research (CIDER).

3   Casweb available at http://casweb.mimas.ac.uk/ maintained by the Census Dissemination Unit (CDU).

initial task was to match the data on immigration counts from 56 world regions available from the commissioned Table CO711B with population counts from Table ST102 for the same seven ethnic groups and 34 places of birth around the world, including the UK. A set of 23 world regions, consistent across both data sources was derived and a database was constructed with population and immigrant counts from these world regions to 376 districts across England and Wales for seven ethnic groups.

Ethnicity is self-selected by Census respondents and the sevenfold categorisation of ethnic groups is highly restrictive given the variations that exist between sub-groups within each broad ethnic category of migrants. The White group, for example, includes both British-born Whites and those born elsewhere in the world of any nationality, whereas the non-white Other group contains an array of different ethnicities and nationalities including those non-whites born in Japan, Philippines, Republic of Korea, Malaysia and the USA. In future work, it might be possible to obtain flow data from ONS for more disaggregated ethnic groups but this would be at the expense of losing geographical detail.

International and Internal Migration

At the time of the 2001 Census, the population of England and Wales totalled just over 52 million with non-white ethnic minority populations collectively accounting for 9.7 per cent of the national population, the largest of the broadly defined minority groups being those classified as POSA. The Indian and Black populations were also each in excess of 1 million whereas the Chinese and Other populations were of much smaller size (Table 3.1, top panel). Whereas nearly 5.5 million people moved usual residence within England and Wales in the 12 months before the 2001 Census, the volume of immigrants in the same period from all parts of the world was 370,480, that is, for every one immigrant arriving at a new destination in England and Wales, nearly 15 internal migrants moved to a different place of usual residence. Columns 5–7 of Table 3.1 show the shares of population, immigration and internal migration accounted for by each ethnic group. Whereas the shares of population and internal migration are relatively consistent, the White share of immigration is just under 70 per cent whilst the non-white ethnic group with the largest share of immigrants is the Other group (6.8 per cent). In comparison with Whites, non-white recent immigrants are over-represented in relation to their population shares in England and Wales, although not for the main two ethnic minorities in London.

The lower panel of Table 3.1 illustrates the equivalent counts of population for London GOR, indicating that whilst the capital accounts for under 9 per cent of the national White population, it contains 58 per cent of the country's Black population, 40 per cent of the Other population, 31 per cent of the Asian population, and around a quarter of the Chinese and Mixed groups. The shares columns indicate the predominance of the Black group amongst the ethnic minority inhabitants as well as the immigrants to and internal migrants within London.

Table 3.1   **Population (2001) and migration counts (2000–01) by ethnic group, England and Wales and London**

Ethnic group	Established population 2001	Recent immigration 2000–01	Internal migration 2000–01	Share of population 2001	Share of immigration 2000–01	Share of internal migration 2000–01
England and Wales						
White	47,520,866	256,881	4,963,888	91.31	69.34	90.40
POSA	1,236,922	20,260	126,149	2.38	5.47	2.30
Black	1,139,572	22,151	137,888	2.19	5.98	2.51
Indian	1,036,807	20,014	100,763	1.99	5.40	1.84
Mixed	661,032	12,095	94,704	1.27	3.26	1.72
Chinese	226,948	13,635	33,343	0.44	3.68	0.61
Other	219,753	25,444	34,035	0.42	6.87	0.62
Total	52,041,900	370,480	5,490,770	100.00	100.00	100.00
London						
White	5,103,203	78,947	444,002	71.15	65.25	69.17
POSA	429,700	67,92	40,591	5.99	5.61	6.32
Black	782,849	10,558	79,804	10.92	8.73	12.43
Indian	436,993	8,688	30,573	6.09	7.18	4.76
Mixed	226,111	4,096	24,525	3.15	3.39	3.82
Chinese	80,201	3,286	8,584	1.12	2.72	1.34
Other	113,034	8,627	13,789	1.58	7.13	2.15
Total	7,172,091	120,994	641,868	100.00	100.00	100.00

Notes: POSA = Pakistani and Other South Asian.

Sources: 2001 Census Table KS006 and SMS Table M103.

A more appropriate comparison of migration propensities is obtained by computing the crude migration rates (using Census date populations) shown in Figure 3.1 in which the immigration and internal migration rates for England and Wales (including London) are juxtaposed alongside those for London alone. The histogram indicates that the rates of immigration and domestic migration are highest for the smallest ethnic minority groups – Chinese and Other. Crude rates of internal migration are greater in England and Wales than in London for all ethnic groups, and apart from Whites and Indians, immigration rates are lower for London than for the whole country. Indians have the lowest rates of internal migration, particularly in London whereas the differential between London and England and Wales for both types of migration is greatest for the Chinese, the smallest of the ethnic populations, both in London and nationally. Further discussion of the variation in internal migration rates of different ethnic groups by age can be found in Finney and Simpson (2009a) and Stillwell (2010a).

In the 1960s, immigration to England and Wales was dominated by inflows from the former colonies across the world. In the twenty-first century, international migration has become more diverse in that it involves movements of individuals originating from different parts of the world. The largest of the flows into both England and Wales and London during 2000–01 were those from the USA followed by Australia, Germany, France and South Africa, each of which sent over 20,000 to England and Wales. A full analysis of these flows by previous place of usual residence is available in Stillwell and McNulty (2011).

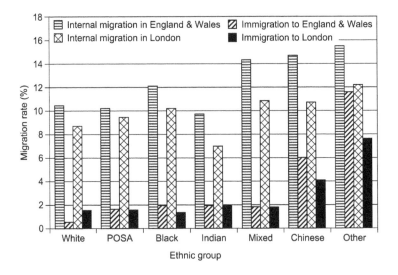

Figure 3.1   Immigration and internal migration rates by ethnic group, England and Wales and London, 2000–01

Sources: 2001 Census Table KS006 and SMS Table M103.

Established and Recent Immigrant Locations

In this section, the focus of analysis switches to the destination end of the immigrant journey and explores the geographical patterns of recent immigrant destinations within London at the district (borough) level compared with the locations of established immigrant populations for a small selection of ethnic groups. The analysis is based on the count of 'recent' immigrants by ethnic group disaggregated by place of previous usual residence outside the UK from commissioned Table CO711B, and the count of the usually resident population by ethnic group and place of birth outside the UK from Standard Table ST102 as the count of 'established' migrants.

Table ST102 indicates that almost 73 per cent of London's usually resident population were born in the UK and Figure 3.2 illustrates the shares of London's 1.9 million established migrants by world region of birth together the shares of recent immigrants by world region of previous usual residence. The European Union (EU), Oceania, USA, Other Far East, Eastern Europe and South Africa are the main regions of previous usual residence, each supplying over 5 per cent of the total immigrant flow whereas Other Africa, the EU, India, Caribbean and West Indies, Eastern Europe, the Other Far East and the Middle East are the major birthplace regions of London's established immigrants.

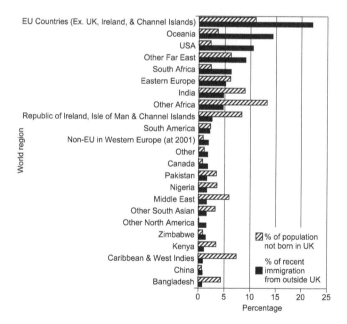

Figure 3.2   Shares of London's established and recent immigrant populations by world region, 2000–01

Sources: 2001 Census ST102; commissioned Table CO711B.

To address the question about the relationship between the spatial distributions of recent and established immigrants by ethnic group, we have selected those immigrant flows which are responsible for more than 10 per cent of the recent immigration of the ethnic group to which they belong. Location quotients (LQs) have been used to provide measures of the proportion of the established or recent immigrant counts by ethnic group in each borough, standardised by allowing for the differences in sizes between ethnic groups in each of the two populations in London. LQ values over 1 suggest over-representation or greater concentration than the average representation in London and values under 1 suggest under-representation or greater dispersal than the average representation in London.

Table 3.2 presents the results of our analysis of each of the selected ethnic groups. In the case of Whites, there are three streams of recent immigrants that each account for over 10 per cent of total White immigration to London – from the places of previous residence in the EU (29.7 per cent), Oceania (20.2 per cent) and the USA (13.3 per cent), as indicated in column 6. Columns 3 and 4 contain the equivalent numbers and percentages of the established co-ethnic populations. Indians are represented by one major stream from India whereas the POSA group has three streams – from Pakistan, Other South Asia and Bangladesh respectively – and the Chinese group has two major streams. Recent immigrants of Mixed ethnicity are drawn from the same three world regions as Whites but more come from the USA than from Oceania. Finally, over 70 per cent of the Other ethnic group come from the Other Far East world region which comprises Brunei, Myanmar, Cambodia, Hong Kong, Indonesia, Japan, Korea, Democratic People's Republic of Korea, Republic of Laos, Malaysia, Mongolia, Philippines, Singapore, Taiwan, Thailand and Vietnam.

The final column of Table 3.2 contains the coefficient of determination (R^2) that is the square of the coefficient of correlation between the location quotients for the two populations across London boroughs and indicates considerable variation in the strength of the relationship. Whereas there is relatively high correlation in London between the recent and established immigrant populations for those from Pakistan and from Bangladesh, the correlations are insignificant for the Chinese from China or from the Other Far East and for recent immigrants in the Other ethnic group that come from the Other Far East. Indians coming from India are more correlated with established co-ethnic immigrants than Blacks from Other Africa or Nigeria. Amongst the White and Mixed groups, it is those migrants from the USA that tend to move to areas where there are more established American-born immigrants, rather than those from EU countries or Oceania.

Table 3.2 **Major recent immigration streams by ethnic group in comparison with established immigrant populations for London boroughs, 2000–01**

Ethnic group	Place of birth or previous usual residence	Established immigrants		Recent immigrants		LQ comparison
		Number	% of immigrant flow	Number	% of immigrant flow	R^2
White	EU	176,972	3.47	23,425	29.67	0.24
	Oceania	50,673	0.99	15,939	20.19	0.06
	USA	28,269	0.55	10,503	13.30	0.59
Indian	India	139,253	31.87	5,404	62.36	0.56
POSA	Pakistan	60,553	14.09	1,859	27.32	0.90
	Other South Asia	47,215	10.99	1,430	21.02	0.67
	Bangladesh	81,841	19.04	940	13.82	0.88
Chinese	Other Far East	37,995	47.36	1,619	48.88	0.04
	China	11,921	14.86	828	25.00	0.00
Black	Other Africa	151,021	19.29	3,595	34.03	0.36
	Nigeria	63,515	8.11	1,818	17.21	0.35
Mixed	EU	4,285	1.90	762	18.76	0.02
	USA	1,341	0.59	488	12.01	0.47
	Oceania	1,603	0.71	438	10.78	0.14
Other	Other Far East	50,209	44.43	6,101	70.75	0.09

Notes: POSA = Pakistani and Other South Asian.

Sources: Census ST102; commissioned Table CO711.

Immigration and Internal Migration

In this section of the chapter, recent immigrant flows by ethnic group into London are examined at ward level in relation to the location of established co-ethnic group populations and to internal net migration balances. No data are available on recent immigrants by world region at the ward level so the analysis is based on total flows of recent immigrants and total established immigrants by ethnic group. The methodology adopted involves constructing a 'migration profile' for each ethnic group across the 628 wards of London based on a series of steps for each ethnic group as follows: (i) assemble the established populations for each ward and compute LQs; (ii) attach data on recent immigrants for each ward; (iii) attach data on gross inflows and outflows of internal migration and compute net balances for each ward; (iv) rank the ward data on the basis of the established population LQs from high to low; (v) divide the ranked wards into decile groups based on roughly equivalent established populations; (vi) sum the number of wards in each decile; (vii) compute the mean established population LQ, the total recent immigration and the aggregate internal net migration balances for each decile; and (viii) plot the decile information for each of the indicators on graphs and juxtapose the graphs.

LQ values of the established immigrant population of each ethnic group are computed in the first step in order to identify spatial variations of over-representation (LQ>1) and under-representation (LQ<1) across London at ward level. Counts of recent immigrants have not been subtracted from the established immigrant populations in this instance because this adjustment makes only a relatively small difference to the LQ values. The data on recent immigrants are extracted from commissioned Table CO723 in step 2 whilst the third step uses data on internal migration counts from the same table but identifies three separate counts of net migration: the net migration balance computed for each ward involving flows within London GOR (internal net migration), the net migration balance of flows between each ward and other regions in the rest of the country (external net migration) and the overall net migration balance. This disaggregation into internal and external net migration helps to expose certain key patterns of movement taking place in London (Stillwell 2010b).

In the fourth step, each ethnic group is selected and the wards are ranked in order of their established ethnic population LQ and then divided into decile groups based on roughly equivalent decile populations rather than the same number of wards. The wards in each decile are summed to give an indication of the spread of each ethnic group across London and, thereafter, the mean LQ, total recent immigration and total net migration balances for each decile for each ethnic group are computed. Plotting the decile information for each of the indicators on graphs and juxtaposing the graphs allows us to build up a migration picture/profile of each ethnic group. Thus, in the graphs which follow (Figures 3.3–3.9), the horizontal axis refers to the wards grouped into deciles according to their LQs and therefore represents the extent of population concentration from high (decile 1 on the left) to low (decile 10 on the right). The scale of the axis on each graph depicting migration reflects the count of those involved and therefore care should be taken in making direct comparisons between ethnic groups. We consider the migration profile of each ethnic group in turn.

The White population of London was spread fairly evenly over wards in 2001, with relatively little variation in the LQ values (Figure 3.3a); the number of wards per decile increases very gradually (Figure 3.3b) from one decile to the next. Nearly 79,000 White immigrants arrived in London in 2000–01 and the largest inflows were to wards of intermediate White population concentration (Figure 3.3c), that is, White immigrants were less inclined to go to areas of high White concentration which were mostly in outer London (perhaps because they could not afford the house prices) or to areas of low White concentration (perhaps because of the level of deprivation associated with these areas). As expected, the White proportion of the population and recent immigrants diminishes as the established White population concentration reduces (Figure 3.3d) – but there is a lower proportion of Whites in the recent immigrant stream than in the population as a whole in every decile. In aggregate internal net migration terms, nearly 45,000 Whites were

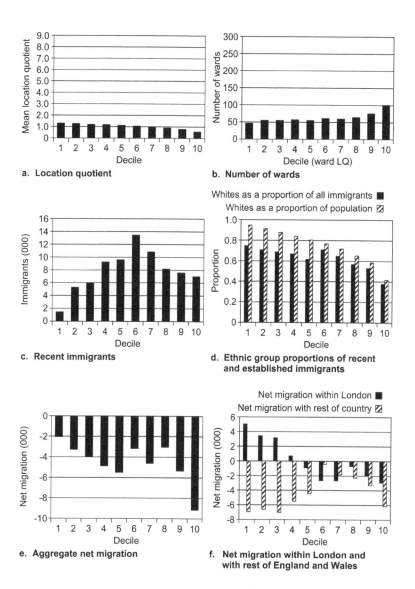

a. **Location quotient**

b. **Number of wards**

c. **Recent immigrants**

d. **Ethnic group proportions of recent and established immigrants**

e. **Aggregate net migration**

f. **Net migration within London and with rest of England and Wales**

Figure 3.3   Migration profile of White immigration and internal migration by decile of White concentration (1 indicating most concentrated) for wards in London

Sources: 2001 Census Standard Table ST101; Commissioned Table CO723.

leaving London – in greatest numbers from areas of lowest White concentration in the outer suburban areas (Figure 3.3e). Whilst this pattern is dominated by net losses from wards in all deciles to the rest of England and Wales, these aggregate net migration losses in the deciles of highest population concentration were partly offset by sizeable internal gains as residents dispersed from Inner to Outer London (Figure 3.3f).

In 2001, Indians were one of the most spatially segregated of London's ethnic populations, with concentrations of established communities in the western suburbs (Harrow, Brent, Ealing, Hounslow and Hillingdon). Consequently, there is a fairly steep decline in the population concentration of Indians across London (Figure 3.4a), with a resulting proportionate increase in the number of wards of low mean concentration, particularly in deciles 8, 9 and 10 (Figure 3.4b). Apart from decile 1, the magnitude of immigration exceeds that of net migration loss or gain for this ethnic group in other deciles and it appears that wards with the highest concentrations of established immigrants received the lowest number of recent immigrants, whereas wards with the lowest concentration of established immigrants received the most recent immigrants (Figure 3.4c). The proportions of Indians in the established and recent immigrant streams decline as the concentration of the established Indian population reduces (Figure 3.4d) so that in all but two deciles, the proportion of Indians in the recent immigrant flow is higher than the proportion of Indians in the established population. Immigration flows to wards in deciles with lower Indian LQs occurred at the same time as internal Indian migrants left areas of higher own-group concentration to move to areas of lower own-group concentration, both within and outside London (Figures 3.4e and 3.4f).

The POSA group, with concentrations in western boroughs (for example, Brent and Ealing) but also Newham, Hackney, Tower Hamlets, and Waltham Forest, exhibits a similar decline in LQ to that of Indians over the decile range (Figure 3.5a) and a similar increase in the number of wards of lower concentration in the higher deciles (Figure 3.5b). Recent POSA immigration was greatest in the deciles of intermediate and low concentration of established POSA immigrants (Figure 3.5c). It is not unexpected, therefore, that the proportion of POSA in the recent immigrant stream is much lower than the proportion of POSA in the established POSA population in the areas of highest POSA concentration (Figure 3.5d) and it is these areas that also lost migrants internally in aggregate net terms (Figure 3.5e). Whilst there were losses to the rest of England and Wales across all the deciles, POSA migrants within London were leaving areas of POSA concentration and moving to areas of lower concentration (Figure 3.5f).

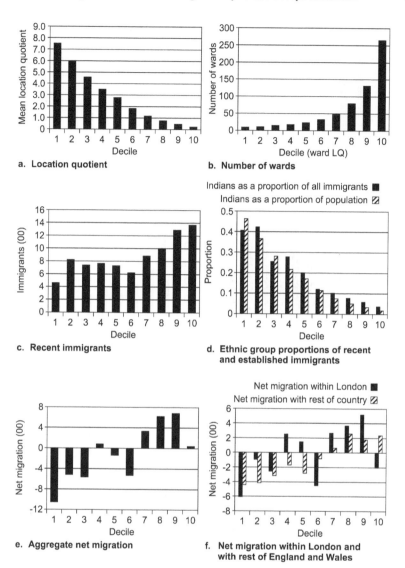

Figure 3.4   **Migration profile of Indian immigration and internal migration by decile of Indian concentration (1 indicating most concentrated) for wards in London**

Sources: 2001 Census Standard Table ST101; Commissioned Table CO723.

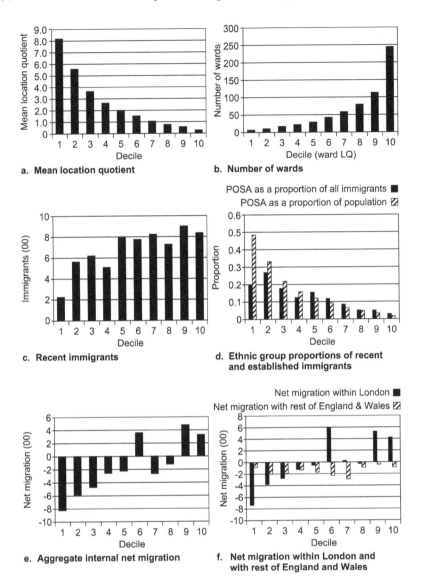

a. Mean location quotient

b. Number of wards

c. Recent immigrants

d. Ethnic group proportions of recent
   and established immigrants

e. Aggregate internal net migration

f. Net migration within London and
   with rest of England and Wales

**Figure 3.5   Migration profile of POSA immigration and internal migration
by decile of Pakistani concentration (1 indicating most
concentrated) for wards in London**

Sources: 2001 Census Standard Table ST101; Commissioned Table CO723.

The established Chinese population was much less concentrated than that of either of the two previous Asian groups in 2001 (Figure 3.6a). Chinese immigrants in 2000–01 appear to have preferred destinations where there were existing concentrations of established Chinese in London (Figure 3.6c). Moreover, recent Chinese immigrants formed a higher proportion of the immigrant stream than the proportion of Chinese in the established population of the destination areas (Figure 3.6d). The net migration balances of Chinese internal migrants, both within London and outside London, were positive for areas of more established Chinese communities and negative for those areas with low concentrations of established Chinese (Figure 3.6e and 3.6f). So, although the Chinese were a less concentrated population group, their recent migration patterns were increasing the concentration and immigration flows were much higher than net migration balances.

The Black population, by far the largest of London's minority populations, had a similar level of population concentration (Figure 3.7a) to the Chinese. However, there was a tendency for recent Black immigrants, like the two Asian groups, to favour areas with lower concentrations of established Black population (Figure 3.7c). The Black proportion of the population reduces linearly by decile, as does the Black proportion of the recent immigrant stream (Figure 3.7d). Internal migration balances indicate that Blacks are dispersing from areas of high concentration to areas of lower Black concentration within London (such as Newham, Haringey, Hackney, Lambeth, Southwark and Lewisham), and that Black people were leaving London in net terms from all areas in roughly the same numbers (Figure 3.7f), producing a gradient from net loss to net gain over the decile range for aggregate net migration (Figure 3.7e). A total of over 10,500 recent Black immigrants to London was offsetting internal net migration losses of 4,200 from London; only in the two most concentrated deciles did recent immigration not offset internal net losses.

The Mixed population, like the White population, has similar LQ scores across the decile range (Figure 3.8a) and therefore a more equal number of wards per decile (Figure 3.8b) apart from decile 10. Recent immigrant flows increase as the mean LQ per decile falls but only to decile 8 (Figure 3.8c). Wards in deciles 9 and 10 have relatively fewer recent Mixed immigrants. As the concentration of the established population of Mixed ethnicity declines, the percentage of the recent immigrant flow becomes greater than the percentage of Mixed ethnicity in the population (Figure 3.8d). The aggregate net migration flows per decile are all negative apart from in the two deciles that have the lowest concentrations of Mixed populations (Figure 3.8e). These gains were due to shifts of migrants in London from areas of higher to lower concentration whereas net losses to the rest of England and Wales were occurring across all deciles but most significantly from decile 10 (Figure 3.8f). Immigration was offsetting net internal migration losses from London.

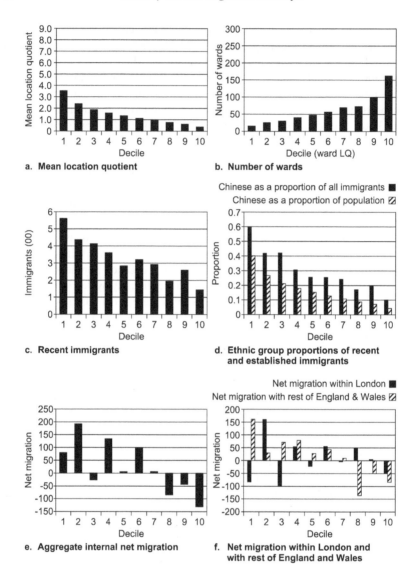

a. Mean location quotient

b. Number of wards

Chinese as a proportion of all immigrants ■
Chinese as a proportion of population ▨

c. Recent immigrants

d. Ethnic group proportions of recent
   and established immigrants

Net migration within London ■
Net migration with rest of England & Wales ▨

e. Aggregate internal net migration

f. Net migration within London and
   with rest of England and Wales

**Figure 3.6   Migration profile of Chinese immigration and internal migration
              by decile of concentration (1 indicating most concentrated) for
              wards in London**

Sources: 2001 Census Standard Table ST101; Commissioned Table CO723.

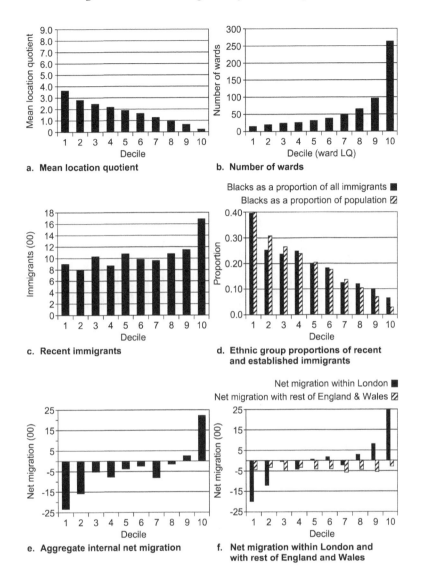

Figure 3.7   Migration profile of Black immigration and internal migration by decile of Black concentration (1 indicating most concentrated) for wards in London

Sources: 2001 Census Standard Table ST101; Commissioned Table CO723.

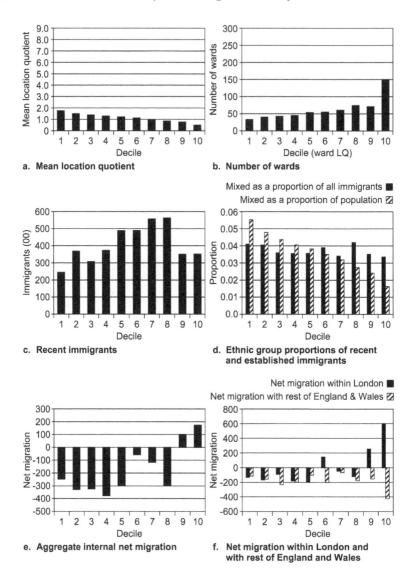

Figure 3.8   Migration profile of Mixed immigration and internal migration by decile of Mixed concentration (1 indicating most concentrated) for wards in London

Sources: 2001 Census Standard Table ST101; Commissioned Table CO723.

Whereas the mean LQ per decile declines linearly for Other non-whites (Figure 3.9a), the immigrant flows appear to be accentuating population concentration (Figure 3.10c) and there are higher percentages of Others in the immigrant flow than in the population in each decile (Figure 3.9d) – indicating that this was a much less well established population than some of the other minority ethnic groups in 2001. Recent immigration was significantly larger than the aggregate net migration balances across all the deciles (Figure 3.9e), with wards that have higher concentrations of established Other populations receiving the most immigrants. The wards in decile 1 containing the highest concentrations of Other populations gained internal migrants from within London and the rest of England and Wales. However, wards in the following four deciles lost internal migrants to other wards within London (Figure 3.9f), with net gains being particularly apparent in the wards of least Other concentration. The net balances of internal migrants with the rest of England and Wales in each decile were relatively small.

The migration profiles presented in this section convey the important message that the spatial patterns of immigration and internal migration in London during 2000–01 varied between the broad ethnic groups. In the case of the most established ethnic minorities, Asians and Blacks, recent *immigration* was occurring across the decile range but predominantly towards the areas with lower co-ethnic concentrations, whereas the recent Chinese and Other immigrants, smaller and less well established groups, were favouring areas of highest co-ethnic concentration. White immigrants, on the other hand, and those of Mixed ethnicity to a lesser extent, were choosing locations of intermediate co-ethnic concentration. In terms of *internal migration*, the spatial patterns of net migration vary according to whether the balances refer to net migration within London or between London and the rest of the country. In the case of the former, both the Asian groups, Blacks and the Mixed group were dispersing from areas of high to low co-ethnic concentration. This is the same for the Other group, except that the decile containing the highest concentration of established co-ethnic immigrants was gaining. The Chinese pattern of net migration gains and losses was rather haphazard across the decile range, whereas the pattern of White migration in 2000–01 was clearly one of migration towards areas with greater co-ethnic concentration but areas which were losing most migrants to the rest of the country. Whilst the POSA, Black and Mixed migrants were leaving London in net terms from all types of areas, those parts of London with lowest concentrations of established Indian populations were actually gaining migrants from the rest of the country whereas the Chinese were losing net migrants from areas of lowest co-ethnic concentration but gaining in areas of highest co-ethnic concentration.

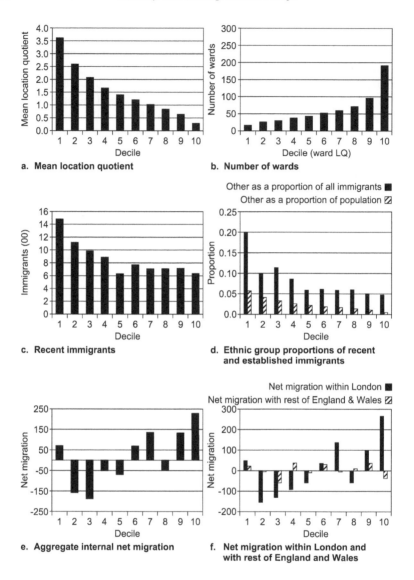

Figure 3.9   Migration profile of Other immigration and internal migration by decile of Other concentration (1 indicating most concentrated) for wards in London

Sources: 2001 Census Standard Table ST101; Commissioned Table CO723.

Summary and Conclusions

In concluding, it is appropriate to re-emphasise the restrictions that limit the analysis of secondary data on the migration of ethnic groups in the UK. Whilst the 2001 Census is an invaluable source of spatial information on ethnic populations and origin-destination migration flows, it is only a snapshot of one year and the data that are collected are hugely constrained by factors such as the categories used to classify ethnic groups, the geographical areas for which data are available and the adjustments that are made to the data in order to reduce the risk of disclosure. The classification issue is particularly frustrating because an intricate mosaic of ethnic populations and patterns of migration is conflated into counts for broad ethnic groups. Nevertheless, we have used what published and commissioned data do exist to explore the flows of 'recent' immigrants in these broad ethnic groups, to examine the movement within the country of members of these same ethnic groups, and to offer some insights into the relationship between immigration and internal migration. In doing so, we have not attempted to provide a thorough analysis of the diversity of world regions from which immigrants in different ethnic groups have originated and we refer to Stillwell and McNulty (2011) for a fuller discussion of these data.

The first of the two research hypotheses stated in the introduction to the chapter suggested that recent immigrants of one ethnic group selected destinations that were concentrations of established immigrants of the same ethnic group. This hypothesis is underpinned by the concept of chain migration and the propensity for newcomers to seek destinations where support networks are available. When investigated using data at district (borough) level for London on recent immigrants by previous usual residence and established immigrants by country of birth, this hypothesis can be accepted in the case of certain groups, such as Pakistanis, Other South Asians and Bangladeshis coming from their respective world regions of origin, but is less acceptable for other groups such as the Chinese from China or for the two predominant groups of Black migrants, those from Other Africa and from Nigeria. On the other hand, White migrants from the USA showed quite a strong tendency to choose destinations where previous immigrants born in the USA were living. This analysis was based on the assumption that the counts of populations by ethnic group and world region of birth derived from the 2001 Census would contain those 'recent' immigrants in the same ethnic group with the same world region of previous usual residence. In fact this assumption proved not to hold consistently, with some counts of recent immigrants (particularly for districts outside London) exceeding those of the established populations. In almost all cases, the differences were small and it is unlikely that the mismatch between previous residence and birthplace would have an impact on the results presented here.

The same hypothesis was also investigated at ward level in London and this more local analysis presented a somewhat different picture in the sense that 'recent' Asian immigrants were selecting wards with lower rather than higher concentrations of 'established' co-ethnic immigrants. This is also apparent for the Black group and yet Chinese and Other immigrants were moving into areas with relatively high concentrations of the members of their own group. This behaviour may be connected with the sizes of these populations and the duration of time that these ethnic communities have existed in London. The Asian and Black groups were large and well established in London in 2001 and their respective communities had reached a stage in their development in which individuals and families were dispersing through internal migration, as indicated by the net migration balances. The Chinese show the highest rates of migration of all ethnic groups and the choice by recent immigrants to concentrate in wards with higher concentrations of established Chinese may be connected with the high percentage of students in this group in London (perhaps mainly accommodated in halls of residence). This explanation is less likely to be true for the Other group, whose propensity to move to areas of higher own-group concentration is more likely to be associated with a desire to be close to family, friends, support networks, accommodation opportunities and potential sources of information about jobs.

The results of the analysis at ward level therefore indicate that the second hypothesis that recent immigrants of one ethnic group select areas that were losing net migrants of the same ethnicity cannot be accepted for all groups. As indicated above, areas with high concentrations of Asian and Black groups were experiencing dispersal through net migration to areas with lower own-group concentration as individuals and families in these ethnic groups sought to improve their living conditions. This redistribution was being reinforced by the patterns of recent immigration. On the other hand, the migration patterns of the Chinese were increasing population concentration and immigration rates were much higher than net internal migration rates. The White group was also experiencing concentration through net migration within London to the outer suburbs, where massive net outflows were taking place to the rest of England and Wales with immigrants moving to the intermediate areas. The seemingly rather paradoxical results of our analysis at district and ward level for the Chinese underline the importance of the scale used for spatial analysis of ethnic populations and their migration.

We conclude that whilst our analysis of spatial patterns of immigration and internal migration by different ethnic groups assist our understanding of how migration is contributing to population change and the debates over ethnic concentration, avoidance or flight, the aggregate nature of the data conceals information about the composition of the movements or the underpinning determinants of residential mobility in London. Within each ethnic group, migrants will be influenced by different drivers according to their circumstances and the stage in their life course, and there will be important local variations in the housing and labour markets that will create or limit opportunities to move house. The origins of the spatial patterns of many ethnic minority internal migrants have been defined by the locations of previous waves of immigrants, most of whom arrived in British cities in search of work and who settled where cheap housing was available and distance to work was minimised. We know that certain ethnic groups contain professionals who are less constrained in terms of residential location than those of lower social status. Catney and Simpson (2010) have shown how the social gradient of migrants away from settlement areas varies between ethnic groups and suggest that London's housing market may well be responsible for the shallower social gradient for non-white ethnic migrants and a reverse gradient for White migration probabilities which increase as occupational status declines. Furthermore, whilst spatial analysis of aggregate flows using the CIDER district classification (based on 2001 migration variables) indicates the importance of places with higher education establishments for attracting student migrants (Stillwell and Dennett 2012), improved understanding of ethnic differences in migration comes from analysis of the migrants in different stages of the life course, as exemplified for young adults by Finney (2011).

Finally, the analysis that has been reported is, of course, only partial and there is no element of change involved. Natural change (births-deaths) and emigration components of population change have not been considered and no attempt has been made to estimate ethnic populations at the start of the 2000–01 period for which immigration and internal migration flows have been used. Births and deaths by ethnic group, as well as start of period populations, are not available and would need to be estimated for a more comprehensive analysis of ethnic population dynamics in London over this period. Whilst this provides one avenue of further research, other work in future might usefully utilise existing data and involve: analysis at the ward level of ethnic migration by age group – particularly so as to tease out the student dimension *vis à vis* other age groups; analysis of origins and destinations of ethnic group migrants between London and the rest of the country; and analysis of the relationships between immigration and internal migration in provincial cities with concentrations of ethnic minorities. Moreover, the 2011 Census results will provide an opportunity to apply the same type of analyses reported here to data for 2010–11 and allow some comparative analysis to detect what changes have taken place both in London and across the country since 2001.

Acknowledgements

This research has been supported by grants from the Economic Research Council under the 'Understanding Population Trends and Patterns' (UPTAP) Programme (RES-163-25-0028) and the Census Programme (RES-348-25-0005). The authors are grateful for the commissioned data supplied by the Office for National Statistics and for the valuable comments on the first draft of this chapter by the book editors.

Chapter 4

Immigrants' Residential Mobility, Socio-ethnic Desegregation Trends and the Metropolises Fragmentation Thesis: The Lisbon Example

Jorge Malheiros

Introduction

Research on the residential mobility of migrants in Europe developed significantly during the 1970s and 1980s (Bonvalet, Carpenter and White 1995). The difficulties of immigrants in accessing the housing market and the effects of this process in the development of immigrants' residential concentrations, often classified as ghettos or ethnic neighbourhoods (Amersfoort 1990), has been a major research line that has developed since this period (Van Kempen and Ozkueren 1998).

Another line of research has concentrated more on the specific mobility processes of immigrants and their descendants (Finney and Simpson 2008, Stillwell, Hussain and Norman 2008, McGarrigle 2010), often stressing that the former experience higher mobility levels than nationals (Gans 1990, Bonvalet, Carpenter and White 1995, Pumarez Fernandez, Garcia Coll and Asensio Hita 2006). This has frequently been explained by the limited access of foreigners to the owner-occupied housing market that expanded all over Europe in the 1980s and 1990s. In addition, housing discrimination practices (Andersson in this book) and formal limitations in gaining access to public housing are also advanced as explanations for the higher residential mobility of foreign residents.

In addition, the particular context and specific needs of immigrants' also often require responses that involve geographical mobility. Given that labour is a major reason to move to another country, immigrants tend to geographically follow labour market opportunities in destination countries, a process that is further intensified due to their involvement in sectors characterised by high geographical mobility (e.g. public works or construction) or seasonality (e.g. agriculture, tourism). The increasing precariousness of labour market relations in contexts where irregular migration is significant is another factor to add to the puzzle of the higher mobility of immigrants. Also, personal strategies geared towards quickly improving living conditions and returning, particularly at the beginning of the migratory cycle – the period when the mobility potential is higher (Courgeau and Lelièvre 2003) – seem to lead to the devaluation of housing stability in destination countries. As a result the number of potential housing moves tends to increase (Fernandez, Garcia Coll and Asensio Hita 2006, Malheiros and Fonseca, forthcoming). In a context where a mismatch between housing market features and the housing market needs of immigrants is found, due to the concentration of offer in home ownership and a limited and expensive rental market offer, as was the case in recent years in the Iberian countries, or the absence of adequate typologies – for single male workers or extended families, for instance, immigrants' housing moves may be expected to be the rule, until labour and family stability in the destination country eventually take place.

The aforementioned elements – the link between housing careers, segregation and the geographical mobility of immigrants – have at times been contextualized in the broader framework of processes of urban change (Stillwell 2010) or urban transition. The Chicago School of Urban Sociology's thesis of invasion-succession (the settlement of new migrant groups in a neighbourhood progressively replacing the former residents), in the 1920s, is one of the processes that has been empirically tested by the analysis of the mobility and location of successive groups of immigrants (Bonvalet, Carpenter and White 1995). More recently, the issue of the gentrification and regeneration of city centres has been addressed by some researchers in connection with the opening or closing of housing market offers and the lower or higher concentration of immigrants in inner city areas (Malheiros and Vala 2004, Arbaci 2008).

As far as the process of urban fragmentation is concerned, its link to the mobility of immigrants and their spatial clustering and segregation trends are apparently less developed. The widening of the social gap in western urban milieus has caught the attention of several authors since 1990 (Mingione 1995, Marcuse and Van Kempen 2000, Wacquant 2004) who underline processes such as the progressive precariousness of labour market relations and the widening of wage differentials. In addition, the generalisation of the use of credit in the latter three decades, namely to support the promotion of expanding home ownership, has increased the indebtedness of many lower-class and even middle-class urban families, intensifying the risk of impoverishment in the present context of global economic crisis.

The geographical expression of these processes of social fragmentation in the urban territory has been addressed by Marcuse and Van Kempen (2000) or Barata Salgueiro (1997 and 2000). The latter author, based on the Lisbon case, offers a thesis of urban fragmentation that emphasises and defines the spatial dimension of the developing fractures. The fragmentation process is marked by the growth of micro-scale spatial discontinuities, not only in social terms (expansion of the urban 'enclaves' that are socially differentiated from the immediate surroundings, reflecting situations of contiguity without continuity) but also at the functional level (e.g., the spreading of large mix use complexes), with a clear expression in the urban landscape. This is generated by the promotion of polycentric urban structures within the contemporary culture of spatial planning that go hand in hand with the increasing functional and apparent social mix that, according to Barata

Salgueiro (2000), is leading to the loss of propinquity in several neighbourhoods. So, the production of more socially mixed areas within cities, which may be interpreted as a more democratic access to urban space, is in fact the result of the present stage of development of the neoliberal model that has extensively used real estate investment as a key instrument to intensify its accumulation strategies. In fact, area mix is often not followed by a strengthening of social mix and social interaction and therefore challenges rather than promotes urban cohesion (Bolt and Van Kempen 2010, Arbaci and Rae forthcoming)

If the promotion of socio-spatial and socio-ethnic mix is inherent in the development of the fragmented city, the hypothesis of a nexus between mobility, and in particular immigrants' mobility, and urban fragmentation still requires some clarification. Figure 4.1 illustrates how the key elements of contemporary urban development stimulate and are stimulated by practices and processes associated with the behaviour of immigrants as consumers and mobile citizens, contributing to the strategy of capital accumulation that took place in Portugal in the 1980s, 1990s and most of the 2000s. This strategy, that discloses socio-ethnic mixing processes (Simpson and Finney 2009) can be identified in other countries like Spain (Naredo and Montiel Márquez 2011) or even Scotland (McGarrigle 2010), is based on home ownership – supported by cheap credit expansion – and on the construction of new housing in peripheral areas. This generates fast gains and high speculation due to processes of land reclassification and increasing demand (first homes, second homes, new homes), particularly in expanding economic conjunctures. Private sector agents, namely banks and developers, and public actors, through their roles in planning (e.g. the expansion of urban perimeters and the urbanisation of the land parcels) and in market regulation (tax exemptions, credit bonuses, rent control mechanisms, etc.), facilitate the conditions that enable the success of the model.

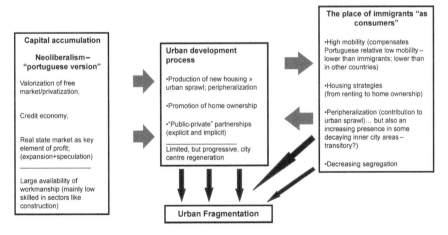

Figure 4.1   Linking the parts: capital accumulation, urban development and immigration

As far as immigrants are concerned, their simultaneous presence in the rental and home ownership markets, with progressive moves towards the latter option, and relatively high levels of geographical (and housing) mobility makes them very interesting 'clients' in the aforementioned strategy of urban development. Owing to their higher mobility, particularly in the initial stages of settlement, immigrants have contributed to the stimulation of the relatively residual Portuguese rental market. In addition, the trend towards long-term settlement in Portugal by older immigrant groups is associated with their transition from the rental market to home ownership, rendering them an interesting consumer segment. In terms of spatial patterns, which also incorporate spatially inscribed social interactions (e.g. neighbourhood relations), the result has contributed to urban fragmentation, visible in de-segregation, suburban sprawl and relatively limited neighbourhood interaction.

This chapter focuses on urban fragmentation, as an outcome of the urban development strategies that have fed the capital accumulation model between mid-1980s and late 2000s. It stresses the specific contribution of immigrants, as a mobile and interesting group of housing consumers in the Lisbon Metropolitan Area, to this fragmentation process. To achieve this, the following questions, focusing on the 1990s and the first decade of the twentieth first century, will be explored: Are immigrants actually more mobile than the autochthonous population, a situation that renders them an interesting group for the accumulation strategies associated with urban regeneration and especially urban sprawl? Is a process of peripheralisation and de-segregation associated with residential mobility in the LMA occurring among immigrant groups? How did immigrants contribute to the expansion of home ownership that took place until 2008/2009? How have immigrants contributed to the limited urban rental market?

In addition to the specific research goals, this work aims to provide a little experimental approach (despite the significant information limits) to a subject – the internal mobility of foreign immigrants in Portugal – that, so far, has been neglected by immigration research in Portugal, namely in that it has not been empirically tested.

Information Sources and Methodology

The empirical analysis that supports the approach proposed in this paper uses two basic sources: i) the database of the Non-EU Foreign Born Characterization Questionnaire of ACIDI[1] applied to a sample of 5,673 people living in Portugal in 2009/2010 (NQF) and ii) the Portuguese database of FP7 project GEITONIES (DBGEIT) that characterised 600 individuals in 2009/2010, half of them Portuguese and the other half foreigners, equally distributed in 3 multi-ethnic neighbourhoods

---

1   ACIDI stands for High-Commissariat for Immigrations and Intercultural Relations, the government body responsible for the implementation of the National integration policy.

of the Lisbon Metropolitan Area.[2] In addition, Portuguese Census data from 1991 and 2001[3] are used to analyse levels of segregation[4] and housing market features.

Despite this combination of information sources, several analytical limits prevail, namely because the aforementioned sources were not specifically designed to analyse residential trajectories. However, the inclusion of information on places of residence in these databases, such as the first place of residence in Portugal and current place of residence in the NQF, provide snapshots of specific moments of time that can be compared to provide an incomplete history. In addition, this database provides relevant complementary information about housing tenure, motives for residential choices at the municipal level,[5] and future options in this domain (intention of buying a house in Portugal and where). Unfortunately, no comparisons with native Portuguese are possible because it only covers a sample of Non-EU foreign born.

The BDGEIT contains information on the residential changes experienced by a sample of 600 foreign and national residents (aged over 18) living in the multi-ethnic neighbourhoods of Mouraria-Martim Moniz (inner city), Monte Abrãao (residential area in the Northern periphery of Lisbon) and Costa da Caparica (suburban coastal resort in LMA-South where a significant cluster of Brazilian immigrants can be found). However, in addition to the lack of information on moves prior to the age of 18, the reference times to register a residence were of

---

2   LMA is split between the North (Greater Lisbon or LMA-North) and the South (Península de Setúbal or LMA-South) banks of Tagus River, comprehending 9 municipalities in each side.

3   Although the national data collection for the 2011 Census has been finished in June, information about foreign born and foreign nationals is unavailable at the time of writing.

4   Based on the original Duncan formula, segregation indices (SI) were calculated for the main immigrant groups of the LMA, both in 1991 and in 2001, according to:

$$SI = \tfrac{1}{2} \sum_{i=1}^{n} |x_i - y_i| * 100$$

x – relationship between a specific population group x settled in a certain geographical unit i (an LMA parish, in this case) and the population of the same group x settled in the entire research area (the LMA).

y – relationship between the population of all groups settled in a certain geographical unit i (excluding group x) and the population of the same groups (again excluding group x) in the entire research area.

n – Number of geographical units comprehended in the global research area (LMA parishes, in this case).

5   The interviewees were asked to choose two motives from a set that included work, social and environmental features, housing quality and price, vicinity of family/friends, proximity to co-ethnics and quality of local shops and services.

one month in the sample neighbourhoods and 1 year outside these, which means an underestimation of the number of residential places in the latter situation. Finally, data about tenure and other features of the dwellings are absent from the questionnaire. It is important to stress that these data are used on an experimental basis and the analysis is not disaggregated by neighbourhood. In the scope of this chapter, that assumes a meso perspective, the idea is to provide elements about the mobility of immigrants in the LMA as a whole and not to particularise specific neighbourhoods.

As far as the 1991 and 2001 Censuses are concerned, only a few data on the evolution of patterns of clustering in the city and the residential conditions experienced by immigrants are used due to the intention to focus on more recent years. Information about places of residence one year and five years before the Census exists at the municipality level for 1991, 2001 and 2011, but is still unavailable for the later year, preventing the analysis of the decade that corresponds to the larger increase in the number of foreigners in Portugal.

The data used in this analysis are mostly classified by nationality. Although the utilisation of 'nationality' poses some analytical limits (e.g. discriminatory practices in access to housing are often associated with ethnic markers such as accent or colour of skin and not the formal condition of being a Portuguese citizen), it is widely used in the databases. In addition, in the Census, the use of 'place of origin' may also be misleading because it mixes up, for instance, the thousands of white Portuguese that were born in the former colonies of Africa and returned in the mid-1970s with the African population that has arrived since that time. However, because naturalisation may be seen as a sign of stabilisation in the destination country and leads to a change in formal status with impact on issues such as the access to bank loans, some NQF tables distinguish between Non-EU foreign national groups and Non-EU naturalised Portuguese.

Internal mobility measurement of Non-EU foreign born involved the construction of cross-tabulations of region of residence at the moment of arrival with region of residence at the moment when the NQF questionnaire was applied (late 2009-early 2010). With these data, a simple relative internal attraction index (AI) has been calculated for each region, dividing the relative number of immigrants nowadays (percentage of the total immigrants living in each region) by the relative number of immigrants in the moment of their arrival. In addition to these data, information about the number of job changes included in the same database has been used as a proxy for internal mobility, not only because 'work' appears as one of the two major motives for settlement choices but also because job instability is higher among immigrants than among nationals, increasing the internal mobility potential of the former. Based on this link between job mobility and geographical mobility that has been identified in previous research (Fonseca 1990), simple calculations of the average job tenure and number of jobs for the period between 1998 and 2010 (by immigrant group and by region of settlement in Portugal) have been made and are compared with global figures provided by OECD.

At the local level (LMA neighbourhoods), DBGEIT enabled the calculation of the number of internal residence changes of the Portuguese and the main foreign residents of 3 neighbourhoods in Lisbon for the periods of 2000–05 and 2005–10, distinguishing between moves inside the LMA and moves of those coming to this region. Because this sample is limited – 300 Portuguese and 300 foreigners – the results assume an experimental nature that must be developed with new information that will come, for instance, from the 2011 Census.

Housing Market, Residential Dynamics and Immigration: the Lisbon Metropolitan Area in the National Context

In the 1980s, 1990s and even early 2000s, housing production in Portugal increased significantly (IHRU 2008). This was stimulated by the arrival in the housing market of youths born in the 1960s and early 1970s, *retornados* from the ex-colonies in Africa, the return of Portuguese emigrants, especially in the 1980s, as well as the increase in general income and levels of consumption until the early 2000s. In addition, it is also the result of political options that privilege neoliberal economic orientations (reduction of direct public intervention, free market support with generalisation of individual private property) and some specific social groups and economic operators (e.g. banks). Actually, both the strategies of private financial operators and public policy in the domain of housing aimed to support the home ownership market. For instance, some of the financial applications that formerly funded the production of rental housing have been redirected towards other financial products (IHRU 2008) that have, directly or indirectly, in many cases, funded the production and acquisition of houses. The significant reductions in interest rates in the 1990s associated with aggressive bank strategies in the promotion of housing loans made these extremely attractive for the vast majority of the population. Also, this transition in housing policy was largely supported by mechanisms that privileged homeownership, such as bonuses for loans, tax exemptions for newly acquired housing or the creation of a variant of the national re-housing programme supporting the acquisition of houses in the free market (the Special Rehousing Programme PER-families). Within this framework, it is not surprising that, in the 1990s, the majority of public expenditure in the housing domain targeted, directly or indirectly, support for homeownership.

Nonetheless, the implementation of the PER in 1993, that aimed to eradicate slums in the metropolitan areas of Lisbon and Porto by the turn of the twentieth century, has led to an increase in the direct promotion of public housing (IHRU 2008). However, as this public housing initiative was almost totally targeting families registered as living in slums in 1993, any other new disadvantaged residents, such as immigrants arriving in the late 1990s and early 2000s, were completely excluded from public housing.

This apparently positive evolution hides three negative elements that will probably become even more prevalent in the future. First, access to housing is clearly more difficult for groups with low solvency levels that have arrived in the housing market more recently. This includes young people with low skills exposed to precarious labour market positions or even unemployment, and recently arrived non-EU immigrants. In the latter case, some evidence of housing discrimination and short term permanency strategies, particularly in the earlier stages of the migration cycle, make housing access even more difficult, namely in a context marked by homeownership and limited rental offer. Second, the association of generalised homeownership and economic stagnation with increasing unemployment and declining real wages since 2003/2004 is leading to an increase in housing stress. On one hand, the level of indebtedness of families has increased substantially throughout the 1990s and early 2000s, reaching an average value of 117 per cent of household income, which mostly corresponds to the value of the housing credit. On the other hand, the current economic situation is serving to reduce the capacity of many households to repay mortgage loans.

Third, this support for homeownership resulted in the reclassification of rural areas as urban land for the construction of new housing, particularly on the periphery of the Metropolitan areas, contributing to the process of suburbanization and urban sprawl. In the case of the LMA, the result has been an increasing demographic imbalance between the municipality of Lisbon, which has the oldest age indexes and experienced a significant demographic loss, only slowing in recent years (663,394 residents in 1991; 564,654 in 2011 and 545,245 in 2001), and the high growth municipalities in the periphery, with a total population of 2.3 million in 2011, corresponding to a growth rate of 22.3 per cent for the previous two decades. Decaying city centre areas of Lisbon and other old urban centres of the periphery coexist with new peripheral urbanizations some of them 'half sold' and not integrated into the metropolitan urban fabric.

The LMA is the main destination for the foreign population in Portugal. Although it constitutes just above 25 per cent of the total resident population of the country, more than 50 per cent of the foreigners live here, despite some dispersal trends that were observed between 1999/2000 and 2008/2009 (Malheiros and Fonseca 2011). The absolute and relative number of foreigners increased significantly in the early 2000s (178,000 registered foreigners living in Portugal in 1998; almost 450,000 in 2004), when a new immigration wave, dominated by Brazilians and Eastern Europeans, arrived in the country and in the LMA. This has led to a diversification of the origins of foreigners, which was dominated by Africans from the PALOP[6] since decolonisation in the mid-1970s. This jump in the number of foreigners can be associated with high economic growth, which was sustained, among other things, by asserted investment in public works and construction, largely financed by EU funds. Consumption levels also rose during the 1990s, justifying the expansion of commerce and personal services. All these

---

6   An abbreviation for Portuguese Speaking African Countries.

activities employed large numbers of immigrants who arrived in a period when official unemployment levels were reaching very low values (around 4 per cent in 2000 and 2001, according to the National Statistics Institute). In addition, the structural and generalised, though limited, upgrade of the skills of Portuguese nationals also contributed to the emergence of some labour market gaps, particularly in low skilled and socially devalued segments of construction, retail, personal and domestic services and even market agriculture.

Throughout the 1990s and early 2000s, Non-EU immigrants have become important clients of the real estate market, namely in the limited rental segment as shown in Figure 4.2. Whereas the percentage of homeownership among Portuguese, EU(15), North-American and PALOP nationals resident in the LMA increased between 1991 and 2001, reaching values over 50 per cent in all cases (more than 70 per cent in the case of Portuguese), it declined in all major groups of the post-1998/1999 immigration wave (Eastern Europeans, Brazilians, Chinese and South Asians), falling below 35 per cent. These groups, due to their recent arrival, often as irregular migrants (Malheiros and Baganha 2001), lacked the capital and the formal conditions to buy homes. In addition, an initial perspective of short to medium term settlement associated with precarious labour market insertion (short term contracts or no contracts) and involvement in sectors with high levels of geographical mobility (e.g. construction and public works) pushed these migrants into the rental housing market, both formal and informal. An analysis of tenure regimes among immigrants in 2010 confirms their significant representation in the rental market (according to the NQF survey, 72 per cent of the Non-EU immigrants were renting their homes in Portugal), particularly in LMA-

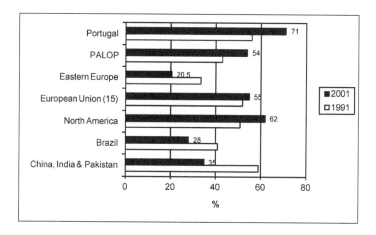

Figure 4.2   Percentage of homeownership of several immigrant groups, Lisbon Metropolitan Area, 1991 and 2001

Source: INE, 1991 and 2001 Census.

North (here, the values rise to almost 80 per cent) where house prices are higher[7] and sub-letting more frequent. Concerning LMA-South, lower house prices have contributed to an increase in the rate of homeownership among immigrants (over 40 per cent had bought their homes in 2010, according to the NQF database), particularly in the case of the longer established PALOP citizens. Despite this, informal situations are more frequent here (6.2 per cent of the cases against 1.4 per cent in the LMA-North), including undeclared renting, the occupation of illegal housing quarters, as well as slum dwelling in one of the few remaining *bidonvilles*.

The situation described has had three consequences for the immigrants, especially those that arrived in the migratory wave of the period between 1998/1999 and 2005. First, the housing conditions of the immigrants, with the partial exception of PALOP citizens, deteriorated in this period, as shown by their presence in poor quality housing and by the increase observed in the percentage of people living in over-crowded dwellings (Malheiros and Vala 2004). Second, rents paid by the recently arrived foreigners are, on average, substantially higher than the rents paid by indigenous for identical house types, contributing to the processes of overcrowding (Malheiros and Fonseca 2011). Finally, the new waves of immigrants in the 1990s became actors in the process of urban sprawl, as cheaper rents led them to the third metropolitan crown or even to peri-urban spaces. Actually, the Brazilians, Eastern Europeans and Guineans (the latest group of the PALOP immigrant workers that arrived in Portugal) show the largest increases in these kinds of spaces corresponding to their scattered patterns of settlement (Malheiros and Vala 2004). On the contrary, Chinese, Indians and Bangladeshis are feeding the rental market in some stigmatised and formerly decaying inner city areas that are now showing signs of 'marginal' gentrification (Rose 1984). The combination of these dynamics, which mean, on the one hand, the contribution of immigrants to the sprawl of the LMA, and, on the other hand, their involvement in the increasingly dynamic housing market in the decaying inner city, is part of the broader process of urban fragmentation discussed in this chapter.

Residential Mobility of Immigrants

The National and Regional Context

Lisbon is the main entry region of the country, with the main international train and bus terminals and an international airport that concentrates around 55 per cent of the passengers and commercial traffic of the country. The convergence between this mobility factor and the large concentration of the foreign population in the LMA (over 50 per cent) largely explains the fact that more than 50 per cent of the sample in the NQF declared that their first residence in Portugal was

---

7   The average housing prices in May 2011 were 1,482 euro/m2 in the Greater Lisbon and 1,204 euro/m2 in the Peninsula de Setúbal (INE 2011).

in the LMA, particularly in the Greater Lisbon (LMA-North). In the course of immigrants' time in Portugal, data point to a double geographical redistribution effect associated with the LMA-North: i) at the intra-metropolitan level, it loses resident immigrants to the LMA-South, where house prices are lower; ii) at the national level, it loses residents to all other Portuguese regions, in particular the North and the Centre. On the contrary, the Península de Setúbal (LMA-South) not only seems to benefit from the transfer of immigrants that lived in LMA-North, but also displays positive net exchanges with all the other Portuguese regions (Table 4.1). Although the re-concentration process of immigrants in the LMA, taken has a whole, does not come out from the data, the contribution of immigrants to suburban sprawl in the South Bank of Tagus is a process that must be underlined. However, it is important to recall that the database only has information for the first and the present places of residence, meaning it is impossible to track the last regional residential move that effectively took place.

In addition to the LMA-South, the autonomous Atlantic archipelagos (Madeira and the Azores) and especially the North Region are the other spaces with a positive net attraction rate (Figure 4.3). Explanations for the behaviour of these regions are still unclear and somehow unexpected, namely in the present context of economic recession that strongly hit the North Region.

Table 4.1   **Original residence places vs. residence places in the survey momentum (2009–2010) of Non-EU immigrants in Portuguese regions**

First Place of Residence in Portugal / Present place of residence	North	Centre	LMA North	LMA South	Alentejo	Algarve	Madeira	Azores	Regional total (actual residence)
North	393	209	**129**	**0**	7	21	0	0	759
Centre	33	491	**117**	**10**	7	7	1	1	667
LMA-North	**21**	**12**	**1080**	**26**	**7**	**5**	**0**	**0**	**1151**
LMA-South	**10**	**20**	**172**	**1171**	**8**	**16**	**3**	**1**	**1401**
Alentejo	1	1	**22**	**0**	143	12	1	0	180
Algarve	5	7	**27**	**0**	5	475	0	1	520
Madeira	4	8	**42**	**8**	2	6	371	3	444
Azores	10	7	**51**	**0**	0	3	6	371	448
Regional total (1st place of residence)	**477**	**755**	**1640**	**1215**	**179**	**545**	**382**	**377**	**5570**

Note: Those who did not answer to the 'first place of residence' were not included in the table.

Source: NQF Survey, 2009/2010.

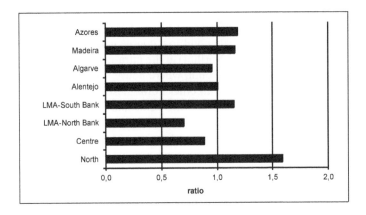

Figure 4.3   Regional attraction indices – immigrants in Portuguese regions

Note 1: Relationship between the percentage of Non-EU immigrants living in each region in the beginning of 2010 and in the moment of first residence in Portugal. Values >1 indicate relative attraction and values <1 indicate relative repulsion. Note 2: Includes foreigners and naturalised Portuguese.

Source: NQF Survey, 2009/2010.

Considering residential choices, 'proximity to family/friends' is the major motivation identified by immigrants, followed by 'work' (Table 4.2). An analysis by regions of settlement reveals a clear difference between the LMA and the rest of the country, with the partial exception of the Algarve; whereas in the first, and particularly in the LMA-South, the main reason appointed is clearly 'proximity to family/friends' (approximately 60 per cent of the respondents gave this reason), in the other – with the exception of the North of Portugal – 'work' is the principal motivation to settle (chosen by 44 per cent of those interviewed in Region Centre and Madeira and by approximately 60 per cent in Algarve and Alentejo). It is not a surprise that 'work offer' is the main reason for the internal geographical mobility of immigrants, driving them away from the LMA, where a classical combination of attraction motives – immigrants' clustering, informal housing, opportunities for irregular immigrants, diversified commercial offer – may be found. However, the underrepresentation of this motive in the case of the LMA is significant (less than 20 per cent of the respondents chose it), and may point to more definitive settlement patterns that involve buying a house and are not associated with work search moves, often of temporary nature.

Looking into residential choices by nationality of immigrants (naturalised Portuguese and foreigners from the main origins), 'work' is more relevant for the groups associated with the most recent immigration waves (Eastern Europeans, South Asians and even Chinese), whereas 'proximity to family/friends' and other motives (e.g. housing prices, natural environment) tend to be of more importance for those who have obtained Portuguese citizenship and for immigrants with a longer presence in Portugal (e.g. PALOP citizens, Brazilians). In addition, 'the presence of co-ethnics', a motive much less relevant than 'proximity to work' or 'proximity to family members' is more significant for more recent groups (Eastern Europeans, South Asians or Chinese). This is likely due to fewer social contacts and to the absence of family members and greater cultural distance.

All things considered, work motives for settlement tend to be of less importance as time in the country of destination increases and a combination of proximity to family members with a cost-benefit analysis of residential space (housing costs vs. environment/accessibility) become the key issues for choosing a place to live. Keeping in mind the over-representation of the immigrant population in the LMA and lower housing prices in the LMA-South, the re-clustering of immigrants in this area is a process that may intensify in the near future, especially if non-metropolitan unemployment rates keep increasing.

Table 4.2   Reasons for the present residential choice of Non-EU immigrants settled in Portuguese regions in 2009–2010, by legal status and main origins (per cent)

Reason for present residential choice	North	Centre	LMA-North	LMA-South	Alentejo	Algarve	Madeira	Azores	PORTUGAL
Employment	30.1	44.0	17.8	19.3	62.1	58.2	43.8	48.2	32.6
Natural environment	7.7	6.1	4.8	9.1	11.1	27.3	33.1	8.7	11.3
Social environment	7.8	3.9	4.5	6.5	2.1	4.8	10.5	4.0	5.8
Accessibility	4.9	0.6	10.2	3.1	2.1	3.1	19.2	11.3	6.4
Place of residence of co-ethnics	19.7	16.3	14.5	7.4	11.1	8.3	14.5	9.8	12.6
Proximity to family/ friends	38.6	40.1	39.6	59.8	31.1	46.6	22.4	19.6	42.0
Housing prices	26.0	5.5	27.9	15.7	3.2	3.8	16.6	10.7	16.6
Cost of living	13.9	8.5	11.6	2.5	17.4	3.6	5.1	15.3	8.5
Rehousing	0.3	0.7	3.7	1.4	0.0	0.0	0.0	0.0	1.2
Study	10.7	13.0	0.3	0.2	14.7	2.7	0.0	0.0	3.9
Others – various	19.6	20.3	11.0	12.0	18.9	8.8	9.8	5.8	13.2
N. of respondents by Region	**791**	**670**	**1176**	**1428**	**190**	**521**	**447**	**450**	**5673**

Note: The question allowed respondents to choose the 2 main reasons. Each percentage corresponds to the number of the times that each reason has been mention divided by the total of potential respondents.

Source: NQF Survey, 2009/2010.

A first clue sustaining this hypothesis is provided by the analysis of the home ownership intentions of immigrants. Looking at future intentions it may be observed that among immigrants who are not homeowners, the project of buying a dwelling is more significant for those who arrived between the early 1980s and 1997 (over 50 per cent declared that they intended to buy a house and around 40 per cent of these even said that have already made house searches). In addition to the adjustment of immigrants to the general functioning of the housing market, the over-representation of 'buying' in the aforementioned time cohort of immigrants is in line with the individual immigration cycles. Indeed, this corresponds to the consolidation of the economic and social lives of these immigrants as they begin to make decisions that 'anchor' them to the place of destination. For those who arrived in the 1970s, the interest in buying a house in Portugal is limited (approximately 60 per cent declared that they did not have such an intention); either they are not interested or do not have the right conditions to do so, or they have found other solutions such as the PER re-housing program. For those who arrived more recently (after 1997 and particularly after 2003), limited savings and the initial idea of a 'quick return', intensified by the economic crisis are clear barriers to entering the owner occupied market. As a result, just a little more than 30 per cent of the post-2003 non homeowners interviewed in NQF Survey declared the intention to buy a house in Portugal.

Going back to the regional options (Table 4.3), as expected, the LMA-South, where lower housing prices can be found,[8] is the region with the larger proportion of immigrants with the intention to buy a house in Portugal (almost 57 per cent). Actually, more than 1/5 of those interviewed living there had already acted in practical terms to realise this. In the majority of cases the relationship between those who looked for a house in the municipality where they presently live and those who looked in other places is 5 to 1. As with other situations, this is in clear contrast with the situation in the LMA-North, where the lowest intentions of homeownership can be found. Furthermore, respondents in this area were also less likely to search for a home within the municipality of residence. This seems to project a continuation of the present trends of geographical redistribution, with a growth in the settlements in the South-Bank of the LMA that involves transfers from the LMA-North following the occurrence of some social and economic stability. In addition, as there is a concentration of jobs in the LMA-North this may lead to an increase in immigrants' commuting times, unless decentralisation of employment also takes place and/or the transport system keeps improving.

In addition to the process of metropolitan recomposition, the regions where immigrants are apparently less interested in settling more permanently are the least developed and/or the more peripheral ones: Alentejo and the autonomous Atlantic archipelagos of the Azores and Madeira. This shows that an eventual contribution of immigrants to counter the low demographic dynamics (fast ageing

---

8   See note 7.

Table 4.3   Intention of buying home in Portugal by region of present settlement, Non-EU immigrant groups (per cent)

Region of present settlement Intention of buying home	North	Centre	LMA-North	LMA-South	Alentejo	Algarve	Madeira	Azores	PORTUGAL
No	37.9	39.5	46.3	29.9	41.7	40.2	50.5	44.3	39.9
Not sure	20.8	22.0	27.2	13.3	20.6	10.8	12.3	28.0	19.6
Yes, but didn't start looking	26.1	21.8	17.6	35.2	29.7	37.0	24.0	18.6	26.3
Yes, already start looking	15.2	16.7	9.0	21.6	8.0	12.0	13.2	9.2	14.3
N. of respondents	**620**	**564**	**927**	**1017**	**175**	**443**	**325**	**393**	**4464**

Note: It excludes homeowners and 'no answers'.

Source: NQF Survey, 2009/2010.

and absolute decline or very slow growth) registered in these regions[9] (Madeira is an exception because it registered a growth in the number of residents of +9.4 per cent in the 2001–11 inter-Census period) is clearly limited and insufficient.

So far, while providing evidence about the internal geographical mobility of immigrants and the reasons behind this, the analysis presented does not say much about the intensity of this mobility. As mentioned, the NQF database only provides information on the first and the actual residences and as it only targets immigrants, a direct comparison with Portuguese ethnics is lacking.

However, due to the links between job mobility and geographical mobility that have been identified in this and other research, data about job changes may function as a proxy for geographical mobility. As Figure 4.4 shows, work mobility is relatively high among immigrant workers – between 1998 and 2009/2010, the average number of jobs reached 2.8 and about a quarter of immigrants had three or more jobs in this period. This means that in this period the average job tenure of the immigrants in the NQF survey was approximately 3.9 years compared to 7.1 years for the total cohort of Portuguese workers aged between 30 and 34[10] included in the 2009 OECD Labour Force Survey database.

---

9   We are using Azores and particularly Alentejo examples, but the same would probably apply to Centre-Interior and to North-Interior. Unfortunately, the regional organization of data does not enable a sub-regional analysis of the regions of North and Centre.

10   The average age of the immigrants included in the survey was 35 years.

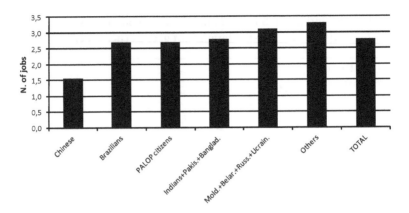

Figure 4.4   Average number of jobs between 1999 and 2009/2010, by main regions of origin

Source: NQF Survey, 2009/2010.

This confirms the higher job mobility of Non-EU foreigners as well as their low job stability reported in other research on Southern European metropolises (Pumares Férnandez, Garcia Coll and Asensio Hita (2006) for Spain, or Kandylis and Maloutas (this book) for Athens). Furthermore, it displays the adjustment of these workers to the flexible and precarious work relations that characterise the present stage of capitalism. Eastern Europeans have the highest levels of professional mobility, which is clearly related with their later arrival and the strong mismatch between skills and initial professions. Other research (Carneiro 2006) has shown that Eastern European immigrants in Portugal experience the strongest downward professional mobility, but at the same time also recover more rapidly from this situation, which is line with the evidence of higher job change.

Crossing these data with the fact that immigrants who are settled in the Algarve, Region Centre and Alentejo have the highest mobility levels (Figure 4.5), the association between high geographical mobility, high professional mobility and the need to move outside the LMA for work reasons becomes clear.

Mobility and Socio-ethnic Segregation in the Lisbon Metropolitan Area: An Experimental Micro-level Approach

So far, the internal mobility and residential options of immigrants have been dealt at the regional and municipal levels, not considering the micro-level of the urban neighbourhood. In addition, direct comparison between the behaviour of immigrants and Portuguese has, so far, been neglected due to the lack of data.

Now, with the help of the data from BDGEIT, collected in 3 multi-ethnic neighbourhoods of the LMA complemented with 2001 Census data, it is possible

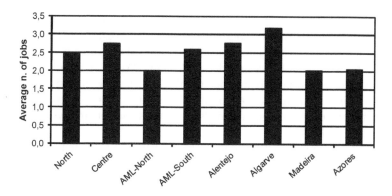

Figure 4.5   Average number of jobs between 1999 and 2009/2010, by regions of settlement (immigrants)

Source: NQF Survey, 2009/2010.

to add a few elements regarding the role of immigrants in the process of urban fragmentation, namely: i) a comparative perspective of mobility patterns of immigrants and foreigners; ii) an analysis of mobility motives at the micro-level and iii) a perspective on how segregation has evolved.

At the micro-level of the sample population interviewed in the 3 neighbourhoods, the general higher internal mobility of immigrants becomes much clearer, especially when considering the 5-year period between 2005 and 2010 (Table 4.4). Whereas only 10 per cent of the Portuguese changed residence in this time span, more than 20 per cent of the members of the 3 main foreigner groups did so. As one would expect, mobility is higher among the most recently arrived groups (Brazilians, Chinese and South Asians) – though also relevant in the case of PALOP immigrants[11] – and involves mainly intra-metropolitan movements and not intra-regional ones, which correspond to residential shifts from the other Portuguese regions to the LMA. Nevertheless, also in the latter case, immigrants' mobility is higher than the mobility of the Portuguese – and the differences are most pronounced in the 2005–10 period. This provides another clue about the possibility of immigrants re-clustering in the LMA, especially in the LMA-South, in the present context of economic crisis.

Focusing specifically on the LMA internal mobility of the immigrants, three different processes may be found. First, the PALOP immigrants, as with the Portuguese, have moved more outside the neighbourhood than inside the

---

11   Due to the size of the sample of foreigners (300 in total, 100 in each neighborhood), the interviewees were clustered in 3 broad groups – PALOP, Brazilians and Chinese and South Asians. Despite the relevance of Eastern Europeans in immigration to Portugal, the limited number of members of this group in the sample made them statistically non-significant. As a result, they were not considered in the analysis.

neighbourhood (Figure 4.6). This is the result of a longer presence but also of the options associated with a more consolidated stage of immigration, where definitive settlement leads to an expansion in local consumption that often involves buying a house. Looking into the reasons declared by the PALOP immigrants to justify the settlement in the neighbourhoods (Table 4.5), the reason 'buying a house' assumes a higher importance than for the other groups. This is consistent with the general behaviour of Portuguese and also with the settlement motives detected in the regional analysis.

Second, the Brazilians tend to move more inside the neighbourhoods than in the whole Metropolitan Area (Figure 4.6). The concentration of the arrivals in a later period in comparison to the PALOP population may explain this relative confinement to the boundaries of the neighbourhood, where knowledge and kin support is higher (one of these, Costa da Caparica in the LMA-South, corresponds to the major concentration of Brazilians of the country). However, this is insufficient to explain the finding entirely. Actually, being the most feminised and socially heterogeneous of the immigrant groups in the analysis (Peixoto and Figueiredo 2007), strategies of 'independence' and 'adventure/change of environment' emerge more strongly as motives for moving to a new home. Brazilians are also

Table 4.4   Internal changes in the place of residence of a sample of 600 people living in the multi-ethnic neighbourhoods of Martim Moniz, Monte Abraão and Costa da Caparica (LMA) – 2000– 2010 and 2005–2010 (per cent)

Internal change in place of residence	Group				
2000–2010	Total	Portuguese	PALOP citizens	Brazilians	Indians & Chinese
No change	70.5	75.9	64.3	68.0	40.0
Within LMA	26.5	21.0	33.7	28.0	55.0
Internal – coming to LMA	3.1	3.1	2.0	4.0	5.0
Total	100.0	100.0	100.0	100.0	100.0

2005–2010	Total	Portuguese	PALOP citizens	Brazilians	Indians & Chinese
No change	83.5	90.3	78.6	74.0	75.0
Within LMA	14.5	8.3	17.9	23.3	15.6
Internal – coming to LMA	2.0	1.3	3.4	2.7	9.4
Total	100.0	100.0	100.0	100.0	100.0

Source: DBGEIT (survey of 600 residents living in 3 multi-ethnic neighborhoods of the LMA).

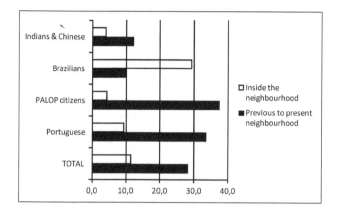

Figure 4.6   People with two or more residence changes in the LMA before moving into the present residence in the neighbourhood (%)

Source: DBGEIT (survey to 600 residents living in 3 multi-ethnic neighbourhoods of the LMA).

Table 4.5   Motives justifying the settlement in the neighbourhood of a sample of 600 people living in Martim Moniz, Monte Abraão and Costa da Caparica (LMA) (per cent)

Motive for settlement	Total	Portuguese	PALOP citizens	Brazilians	Indians & Chinese
			Group		
Accompanying/joining parents/partner	43.0	51.5	40.2	40.4	16.3
Divorce	9.0	10.6	11.1	7.1	4.1
Looking for work	24.0	7.6	16.2	47.5	77.6
Because of job / contracted by firm	14.2	13.6	13.7	13.1	18.4
Education related reasons	5.0	3.3	9.4	3.0	4.1
Trying my luck / change of environment / tourism	13.3	7.3	17.1	26.3	10.2
To improve housing conditions	28.3	27.9	23.1	40.4	18.4
Former dwelling too expensive	6.7	3.7	6.0	14.1	6.1
To buy a home	13.5	15.3	21.4	4.0	6.1
Unhappy with older neighbourhood/ village / city	2.3	3.7	1.7	1.0	0.0
To be independent	13.3	14.0	14.5	17.2	0.0
Other reasons	17.3	13.6	27.3	15.1	12.2
N. of respondents	600	301	117	99	49

Note: Percentage of people in each group that chose each option.

Source: DBGEIT (survey to 600 residents living in 3 multi-ethnic neighborhoods of the LMA).

the group that justify more their residential choices with the 'prices of houses' and the need to improve residential conditions (Table 4.5), in a context where renting is the major option. In addition to the evidence of residential discrimination that has been found in previous studies, the whole picture points to residential stress (overcrowding, discrimination, desire for 'independence') but not to high levels of neighbourhood dissatisfaction nor to a strategy aiming definitive settlement, that normally widens the geographical spectrum of a housing search.

Finally, the pattern of Chinese and South Asians is different from others in three aspects: a higher level of regional mobility, a more limited number of residential moves within the LMA and inside the neighbourhoods, and the clear supremacy of 'work' as the motive that justifies housing options. In contrast to the Brazilians, this is a highly masculinised immigrant group, that valorises ethnic clustering (South Asians were the group that valorised more the proximity of co-ethnics as a motive of residential choice) and is strongly oriented towards the origin country. Taking this into consideration, mobility is basically associated with job changes, and only when family reunion takes place, the need to move materialises, justifying the secondary importance of the motive 'improving housing conditions'. One interesting feature about the Asian group corresponds to their over-representation in the inner city, a situation that differs from Brazilians, PALOP immigrants and also Eastern Europeans, all of them displaying higher peripheralisation levels.

Considering these mobility patterns, all converge in the sense of a decrease in segregation levels, sustaining the logic of urban fragmentation. Unfortunately, there are no data that enable the computation of ethnic segregation in LMA after 2001. However, for the period 1991–2001, de-segregation was already taking place (Table 4.6) for almost all the main national groups of immigrants, something that is expected to have continued in the 2001–2011 decade.

Table 4.6   Duncan simple segregation indices: main national groups settled in the freguesias of Lisbon Metropolitan Area (1991 and 2001)

National groups	Resident population 2001	Segregation indices 1991	Segregation indices – 2001	Differences (2001–1991)
Chinese, Indians & Pakistanis	3,225	52.3	42.4	**-9.9**
Chinese	1,122	67.5	41.2	**-26.3**
Indians	1,350	54.4	54.0	**-0.4**
Pakistanis	753	71	63.8	**-7.2**
PALOP citizens	80,427	36.6	35.7	**-0.9**
Angolans	27,706	30	35.5	**5.5**
Cape Verdeas	28,702	45.6	37.4	**-8.2**
Guineans (Guinea Bissau)	13,476	51.3	45.7	**-5.6**
Mozambicans	2,758	37.6	27.2	**-10.4**
Santomese (S.Tome & Prince)	7,785	50.4	49.9	**-0.5**
North Americans	1,242	42	37.3	**-4.7**
Brazilians	16,817	32.7	27.7	**-5.0**
Eastern Europeans	7,348	59.4	28.8	**-30.6**
EU – 15 citizens (except Portugal)	12,335	38.6	39.0	**0.4**
Portuguese	2,516,812	24.8	21.5	**-3.3**

Note: Calculations use the 216 freguesias (parishes) of the Lisbon Metropolitan Area as geographical units.

Sources: INE, Census 1991 and 2001.

Final Words: Capitalist Accumulation, Urban Fragmentation and Immigrants Housing Mobility in the Lisbon Metropolitan Area: What Links?

This chapter began with the claim that contemporary urban restructuring processes taking place in Lisbon and in other metropolises, with its array of brownfield reconversion, limited city centres neighbourhood regeneration, large new housing offer, public space privatisation and urban sprawl, have been a key element for the successful reinvestment of capitalist surplus between the 1980s and the beginning of the economic crisis in 2008. This restructuring, materialised in the fragmented city pattern, involved a high stimulus to home ownership supported by the dissemination of cheap credit and the spreading of credit bonuses and tax exemptions for the households buying homes. Additionally, it has generated fast and speculative gains supported by the progressive conversion of rural areas into urban land and artificially maintained price sustainability. In order to sustain this model, demand should be continuous, directed towards home ownership and to new housing, mostly built in suburban areas.

Taking into consideration this panorama, the proposal for this chapter was to read the contribution of Non-EU immigrants, as mobile newcomers and housing consumers, to this model of capitalist accumulation based on cheap credit and urban sprawl. Because demographic dynamics of Portugal and the Lisbon Metropolitan Area have largely depended, in the 1990s and 2000s, on net migration (the contemporary immigration peak in the LMA occurred in the years of the turn of the last century), immigrants have certainly become a very interesting population segment in this domain. In order to check this supposition, three hypotheses were tested: i) the higher levels of internal and housing mobility of immigrants; ii) the progressive involvement in home ownership and in suburbanisation; and iii) the specific role of the most recent immigrant waves in animating the limited, though existing, urban rental housing market.

Although the data used present some limitations, the results obtained seem to confirm the higher mobility of immigrants. Not only do immigrants change jobs more often than the autochthonous (the average job tenure is 3.9 years for immigrants against 7.1 for the total population, according to NQF Survey of 2009–2010), but they also reveal a higher propensity to change residence (according to the BDGEIT, 20 per cent of foreigners have changed residence between 2005 and 2010, but only 10 per cent of the Portuguese did the same). Recognising that professional mobility does not equal geographical mobility, a strong link exists between the two, as previous research has shown. In the present case, work is the main reason to justify residential choices outside the LMA as well as the settlement options of the most recent immigration waves (Eastern Europeans, Chinese and South Asians).

If higher mobility seems to characterise immigrant groups in Portugal and in the LMA, we agree with other research that underlines demographic and socio-economic factors – and not ethnic factors – as mobility predictors (Finney and Simpson 2008, Manley and Catney in this book). Although these issues have not been explicitly tested in this research, immigrants' higher internal mobility is apparently connected with their high labour market participation rates and also with their labour market vulnerability. Actually, if work justifies geographical scattering throughout the country in the earlier immigration stages, the process of family and labour market consolidation leads to stronger settlement options (clearer in the case of PALOP immigrants), involving a higher trend to home ownership and residential choices based in proximity to close social networks (family, friends) and even the physical quality of the place. As far as settlement places are concerned, this is associated with the suburbanisation process, namely in LMA-South where housing prices are cheaper, resulting in benefits from relocations from both LMA-North and most of the other regions of the country. All things considered, and despite the specific labour market vulnerability of immigrants, the consolidation of their presence in Portugal and in the LMA points to options in the domain of home ownership and peripheralisation that are in line with the general behaviour of the majority of the autochthonous population.

Finally, despite the trend towards home ownership among the established immigrants, renting is still the dominant tenure among this population (72 per cent according to the NQF survey) and increases in the most recent groups, such as Eastern Europeans, South Asians and Brazilians. Being a population with a higher degree of mobility, recently arrived immigrants have become a very interesting public for the rental housing market, particularly in the city of Lisbon and in other municipalities of LMA-North Bank. In fact, they pay on average higher rents than nationals, and their geographical rotation levels and temporary presence enable interesting capital gains with limited investment in dwellings which are often annexes or clandestine homes located in peripheral spaces or decaying inner city flats. In the later case, informal contracts and subletting strategies make possible interesting gains in areas waiting for rehabilitation operations and subsequent gentrification.

If urban fragmentation is the result of the accumulation model that was implemented in the LMA in the 1980s, 1990s and most of the 2000s, then immigrants have played a role in it, contributing to gains in both the home ownership market and the rental market and participating in the suburbanisation process that has been associated with desegregation, but not necessarily to an improvement in housing conditions.

To close, it is important to remember the limits of this study. First, the data used hinder the comparison of mobility propensities of nationals and foreigners in the first decade of the twenty first century because the NQF survey is only applied to immigrants and the BDGEIT includes a very limited sample of people living in three LMA neighbourhoods. In addition, these databases do not provide full information about the housing careers of immigrants or the reasons for these moves. The Census of 2001 provides some additional elements but they do not tell anything about the changes of the last 10 years. Therefore, in order to develop and better sustain the analysis presented here, further research based on new data, starting with the elements coming from the 2011 Census, is essential to enlighten the processes of internal mobility of immigrants and their role in the urban dynamics of the LMA.

Acknowledgements

This paper was written in the framework of the projects REHURB (Rehousing and Urban Regeneration – PTDC/CS-GEO/108610/2008 of the Foundation for Science and Technology) and GEITONIES (FP7), this later one coordinated by Lucinda Fonseca at the Centre for Geographical Studies of IGOT (University of Lisbon). It also benefitted from data collected in the *Project Diagnosis and Characterization of Immigrant Population in Portuguese Municipalities* of ACIDI (Portuguese Government). The author is grateful for the permission to use the project databases of GEITONIES-Portugal and ACIDI and also wants to express his gratitude to Carla Afonso, Rui Carvalho, André Carmo and Jennifer McGarrigle for their English language corrections, their help in extracting data and their valuable comments on the first draft of this chapter. For identical reasons (plus the patience for the delays…), thanks are extensive to Nissa Finney and Gemma Catney.

Chapter 5

Residential Location and Housing Moves of Immigrants and Natives in the Amsterdam Metropolitan Context

Sako Musterd and Wouter van Gent

Introduction

The social integration of immigrants into Western European societies is often discussed in terms of immigrant geographies. To understand how immigrant, or ethnic, geographies are (re-)produced, it is important to examine migration patterns and residential mobility that follows immigration. In the end, segregation is the result of a continuous flow over time rather than a static condition of immobility (Simpson and Finney 2009; Pan Ké Shon 2010).

The objective of this chapter is to better understand social integration through 'residential choice', and spatial patterning of immigrant groups in the Amsterdam metropolitan region. This is done by comparing spatial distributions and mobility patterns of first and second generation immigrants with those of Dutch natives. In addition to immigrant concentrations, we focus on moving to and from different types of dwellings and residential environments.

Three research questions of immigrant geography will be dealt with. First, we will examine spatial assimilation in terms of concentrations: what shares of first and second generation immigrants and native Dutch reside in 'concentration neighbourhoods' and what are their mobility patterns in the metropolitan region of Amsterdam between 1999 and 2006? The second and third questions will examine assimilation in terms of housing market access and processes of (counter-) urbanisation. The second question deals with the situation of immigrants in the housing market and in the metropolitan region: what are the housing market characteristics of first and second generation immigrants and native Dutch in the Amsterdam regions in 2006? The third question will focus on mobility of urban dwellers over a seven-year period to see whether and how second generation immigrant movers differ from first generation and native Dutch movers. How do migrants and natives differ from each other in terms of suburbanisation, i.e. their moving to single-family owner-occupied dwellings in lower density residential environments outside the city, measured over the 1999–2006 period?

We differentiate between first and second generation immigrants to test the assumption that the second generation reach higher levels of assimilation.

'Canonical' statements of general assimilation theory suggest an orderly progression from first generation poverty and exclusion to the acculturation and upward social mobility of the second and following generations (Alba and Nee 1997). However, Portes (1997) doubts whether this straightforward progression applies to all immigrants equally. He cites the increased opportunities for transnational activities of immigrants, persistent discrimination of non-whites, changing labour markets which block economic progress for the second generation and trigger resentment over blocked opportunities, leading to an adversarial attitude towards mainstream society. Rather, he suggests there is 'segmented assimilation', meaning that outcomes of the second generation vary per immigrant group. Heath, Rothon, and Kilpi (2008) support this view, stating that the educational and occupational performance of second generation immigrants in Western Europe is mixed, and dependent on institutional context and on immigrant histories. Portes and Rumbaut (2005) identify four decisive factors for these outcomes: the history of the first generation; pace of acculturation among parents and children; cultural and economic barriers which the second generation may encounter; and family and community resources to overcome these barriers. Others have added that outcomes of second generation immigrants will also depend on their parents' educational, labour and housing histories and on the influence of the co-ethnic community on identity (see Waters 1994).

In subsequent sections we first address literature and theory on immigrant residential mobility and spatial outcome. We discuss the literature from three perspectives: from the perspective of 'moving to and from concentrations'; from the perspective of urbanisation and housing access; and through a focus on a more encompassing theoretical model in which various elements of different residential mobility theories have been taken together. This discussion is followed by a section on data and methodology. In the empirical section analyses are presented that parallel the three theoretical perspectives. This is followed by a brief discussion and conclusions.

Literature and Theory on Immigrant Residential Mobility and Ethnic Segregation

Moving to and from Concentrations

Several studies specifically address moving to and from ethnic concentrations in connection with spatial segregation (e.g., Bråmå 2006, Bolt, Van Kempen, and Van Ham 2008, Skifter Andersen 2010, and Pan Ké Shon 2010). Typically, these studies make statements about the moves of natives (e.g. 'white flight' and 'white avoidance') and of immigrant or ethnic groups (neighbourhood selection or self-segregation). These studies test aspects of spatial assimilation theory.

While we believe there is merit in this line of inquiry, the focus on concentrations may miss aspects of the long-term social integration process.

First, it has been shown that ethnic clusters are dynamic and volatile (Musterd and de Vos 2007). Second, immigrant mobility patterns may not always be that different from those of natives. Immigrants who have the means move out of problematic neighbourhoods about just as much or in some cases in higher numbers than natives (Simpson and Finney 2009, Pan Ké Shon 2010). Third, ethnic concentrations are not necessarily a problem (see Phillips 2007). Ethnic clustering or dispersal may be ascribed to discrimination and racism, but also to the benefits of social capital within ethnic enclaves. For some, living in an enclave concentration may be a temporary condition in the initial stages after migration that helps people find their way in society (Musterd et al. 2008, Beckers 2011), or it may offer opportunities for housing, employment or care. This is referred to as ethnic resources theory (e.g. Portes and Bach 1985). Portes and Sensenbrenner (1993) argue that living in localised ethnic communities enables individuals to draw on 'bounded solidarity' (within their own group) and 'enforceable trust' as sources of social capital. Enforceable trust refers to group members' disciplined compliance with group expectations in anticipation of greater gain ('utilities') through group membership in the future. When the outside society is perceived to be hostile to an immigrant, living in an ethnic concentration may provide individuals with altruistic support, access to economic opportunities, resources, and a sense of security in knowing code and conduct in public life (Özüekren and Van Kempen 2002, Pinkster 2009).

To be clear, these concepts of social capital do not mean that ethnic neighbourhoods are cosy and sociable. Portes and Sensenbrenner (1993) list several possible negative effects of social mechanisms that produce social capital, such as high costs of social obligations and solidarity, constraints that group norms place on action and receptivity outside the community, and levelling pressures which keep members of poorer groups in the same situation as their peers. In concordance with these ideas Musterd and colleagues (2008) found that migrants who had entered the country could benefit from residing in concentrations of co-ethnics, but only in the first few years after arrival. Those who stayed longer in these enclaves experienced negative income effects in later years.

Urbanisation and Housing Access

A complementary perspective on internal ethnic migration and geography focuses on urbanisation processes and housing market access. For the British case, Simpson and Finney (2009) propose that ethnic migration patterns are largely explained by urban de-concentration: migration from densely-built urban areas to less dense areas in the suburbs and in rural areas. This has been an established urbanisation pattern in the developed world from the 1950s onwards (Champion 1989). In contrast, urban gentrification by professionals creates an influx of natives to certain densely-built central areas, thereby pushing other households out of these areas (see also Hamnett and Butler 2010). The suggestion is that acculturation of immigrants may lead them to de-concentrate from urban areas. This would typically, but not exclusively, mean a move to single family housing as well as owner occupancy. Indeed, Hamnett and Butler (2010) found an increase in ethnic minority owner occupancy in London's suburbs between 1991 and 2001, resulting in the majority of ethnic minority households living in Outer London.

General Theory on Residential Mobility and Segregation

There are multiple approaches of the general study of residential mobility and segregation (Clapham and Kintrea 1984, Van Kempen 2002, Charles 2003). These may also provide elements for building a more general model of ethnic residential mobility with regard to different housing and residential environment situations. Moving to dwellings and neighbourhoods at an individual level may be seen as a matter of choice and constraints and different approaches stress either aspect.

Behavioural approaches link housing choices with family lifecycle and economic means (Rossi 1980, Clark, Deurloo, and Dieleman 1986, Clark 1992) and to intentions to stay in a certain place or not. This is particularly relevant when comparing first and second generation immigrants. First generation immigrants may retain a desire to return to the country of origin and view their stay in the country of immigration as temporary. This might result in minimising housing expenditures and investments (Bråmå and Andersson 2010), even though frequently the wish to return turns out to be a 'myth of return'.

Structural approaches focus on class and on the structured inequalities in the distribution and consumption of housing due to uneven access to resources (e.g., Rex and Moore 1967 and Allen 2008). Here we should also refer to discrimination. Discrimination in housing can assert itself indirectly through 'benign market forces' related to income differences, information sources, employment structures, job locations (Galster 1988), or directly through illegal regulatory discrimination related to finance and housing allocation (e.g. Aalbers 2005). Also, while immigrants may have different preferences, this does not necessarily mean that concentrations are a matter of choice. Phillips (2006a) discusses the perceived self-segregation among British Muslims and shows that members of this varied group largely have no such wishes. She asserts that the notion of self-segregation is related to the racialisation of urban space whereby groups are categorised as the homogeneous 'Others' vis-à-vis 'Whites'. These notions are reproduced by public and private housing institutions, as well as by public discourse (Phillips 2006a).

The approaches should not be seen as mutually exclusive. The housing pathways approach focuses on household's routes and housing experiences over time (see Clapham 2005). In doing so, it connects household changes, related to marriage, birth of children or divorce, and employment pathways to housing pathways. Choices are constrained by earlier decisions and existing social structures.

At this point we also want to stress the differences and similarities between different contexts and different groups. In a recent American study Sampson and Sharkey (2008) identified a structural pattern of mobility flows between residential environments, which appeared to reproduce ethnic segregation patterns. The Western European experience, however, differs from that of the US, due to different institutional arrangements in welfare, housing and integration, and differences in immigration patterns (Musterd and Ostendorf 1998, Van Kempen and Murie 2009). Moreover, US studies tend to focus on black-white race relations, while the European focus is marked by comparing immigrants with different cultural backgrounds (see Wacquant 2008). A cross-European analysis revealed that preferences to reside apart from ethnic minorities among native Europeans are fairly similar to preferences of whites in the US (Semyonov, Glikman, and Krysan 2007). However, within Europe there are differences and similarities between different welfare states and different groups. Peach (1998) explained differences with regard to housing behaviour and residential mobility in the UK in terms of culture and socio-economic integration, as well as differences in household structures between immigrant groups. He found that in contrast to Caribbean patterns, Asian own-group encapsulation remains despite upward socio-economic success. Recent research in Sweden shows that immigrants of African and Eastern European descent have a lower propensity to leave rental housing than native Swedes (Brämå and Andersson 2010). Other studies confirm patterns of ethnic sorting, and 'white avoidance': few natives move into immigrant concentrations (Brämå 2006, Bergström, Van Ham and Manley 2010). In France, a study of moving out of disadvantaged neighbourhoods showed that the trend is not 'white flight' or 'white avoidance', but rather a 'flight of all colours'. All ethnic groups move out to better residential environments when some level of socio-economic success is achieved. First generation Africans, however, find it harder to achieve this success. This suggests social and spatial assimilation rather than self-segregation (Pan Ké Shon 2010). A Danish study found that ethnic minorities tend to move to social housing and to ethnic concentrations. This was particularly the case for immigrants of Arab, Turkish and Somali origin. Ethnic minorities express preference for a neighbourhood with many of their countrymen, and proximity to family. While these findings support a degree of 'self-segregation', possibly stimulated through anti-immigrant sentiments that increased over the past decade, there was evidence of spatial assimilation as well. Those who moved away from ethnic concentrations were more likely to have a higher level of income, to be employed, to have lived in Denmark for a longer period, and to have attained Danish nationality (Skifter Andersen 2010).

Dutch researchers have observed patterns of white flight and perpetuating segregation as well. Van Ham and Clark (2009) showed that the housing stock and population composition explain most of the variation in mobility between neighbourhoods; neighbourhoods with a high mobility rate have more mobile populations (such as young people) and a housing market structure that does not inhibit this (i.e. rental dwellings rather than owner occupancy). They also show that an increase in non-Western immigrants spurs out-migration of native Dutch (see also Bolt, Van Kempen, and Van Ham 2008 and Zorlu and Latten 2009). Nonetheless, it remains to be seen whether the presence of immigrants is the reason for native Dutch to 'flee'. Feijten and Van Ham (2009) confirm that high population turnover and an increase in the proportion of non-western immigrants affect the wish to leave, while socio-economic downgrading has no effect. However, the effect of an increasing share of immigrants disappears when taking subjective notions about neighbourhood decline into account. In a study of immigrants in the Netherlands, Zorlu and Latten (2009) found no evidence for spatial assimilation across generations of Non-Western immigrants in terms of spatial proximity to natives between 2002 and 2003. Second generation immigrants from Western countries, however, are moving to neighbourhoods with high shares of natives. While immigrants are more likely to move to other neighbourhoods within the city than to move across municipal boundaries, they tend to mix with natives when they move over longer distances (Zorlu and Latten 2009).

Bolt and colleagues looked at residential mobility from and to ethnic concentrations and concluded that there is an 'ethnic specificity in the residential behaviour of households' (Bolt, Van Kempen, and Van Ham 2008: 1376). Correcting for household characteristics, they found that immigrants from non-western origins are more likely to move to concentrations and less likely to move out of them than natives. They conclude that there is a difference in preference (ibid.), although this conclusion may be premature as preferences were not linked to actual moves (Doff and Kleinhans 2011).

A study of ethnic clusters in Amsterdam between 1994 and 2004 showed that ethnic clusters are dynamic and that no immigrant group is persistently concentrated in the same concentration. All ethnic clusters of a specific ethnic group experienced a negative migration balance for that specific group. The supply of housing may explain these trends. Those who moved out of clusters moved to areas with better quality housing and higher rates of ownership. This may reflect steps towards integration and spatial assimilation (Musterd and de Vos 2007).

In the Amsterdam region between 2002 and 2003, Zorlu (2009) found evidence for counter-urbanisation among Caribbean immigrants (Surinamese and Antillean). They have a higher probability to move to the suburbs (especially to the New Town of Almere) than Moroccans and Turks. This was also found for other big Dutch cities (Bolt, Van Kempen, and Van Ham 2008). The reluctance to suburbanise among Turkish and Moroccans is partly explained by the finding that family ties have a major impact on out-migration for all second generation immigrants and is largest for Moroccans and Turks (Zorlu 2009).

Data and Methodology

This study focuses on mobility patterns in metropolitan Amsterdam (total population in 2006 of 1.4 million). The area includes rural areas, villages, and smaller cities and larger towns that function as regional (sub)centres. These centres grew substantially in the 1970s when new low density housing was constructed which facilitated the counter-urbanisation of the native middle classes. The Amsterdam region is a high-demand area, which offers a wide range of residential environments. Affordable housing can still be found in the owner occupied sector in growth areas in the regions, or in the social rental sector which is relatively large in Amsterdam. Dutch social rental housing generally is of good quality, and not stigmatised or residualised.

This study uses data from the *Sociaal Statistisch Bestand* (SSB, Social Statistical Database) of the Netherlands Bureau of Statistics. The SSB contains individual level register data on the entire population of the Netherlands, on income from work, benefits, student subsidies and pensions as well as several individual characteristics such as neighbourhood of residence, ethnicity, age, gender and household characteristics for the period 1999 to 2006. The dataset was merged with individual level housing data (on tenure) from tax records for 1999 and 2006 and from the *Woningregister* (WRG, Housing Register) for 2006 (type of dwelling, multi-family or single-family[1]). In addition, residential environments are defined by a density indicator (addresses per km², whereby 'very high urban density' is more than 2500 addresses per km²) and by a proximity variable. The proximity variable defines urban centre and urban periphery by travel time from the city centre. Through a GIS application, we defined the urban periphery as more than 20 minutes travel time from the centre by bicycle on the road network.

This paper follows the neighbourhood classification of the Netherlands Bureau of Statistics (CBS). The neighbourhood boundaries are determined by the municipalities in cooperation with CBS. The neighbourhoods are generally socially and physically homogeneous areas clearly bounded by streets, railways or waterways. The average population of a neighbourhood in the Amsterdam region was 2,765 in 2006.

The first empirical section discusses 'concentrations'. This refers to neighbourhoods with an overrepresentation of members of the own immigrant group, non-western immigrant group or first generation non-western immigrants. Overrepresentation of a group in a neighbourhood is defined as two (binominal) standard deviations above the average representation in the region (see Deurloo and Musterd 2001).[2]

First generation immigrants are individuals who were born abroad and have at least one foreign born parent. The place of birth defines group membership.

---

1   Multi-family dwellings refers to multiple housing units within the same building or in multiple buildings within the same complex (e.g. apartment buildings, tenements or flats). Single-family dwellings can be detached, semi-detached, or terraced housing with or without a private garden, typically occupied by one household per dwelling.

2   Two standard deviations above average in 1999 and 2006 is 10.8 per cent and 14.3 per cent for Moroccan, 10.2 per cent and 12.9 per cent for Turkish, 17 per cent and 20.9 per cent for Caribbean, 13.5 per cent and 19.4 per cent for 'other non-western', and 26.9 per cent and 32.2 per cent for Western. For native Dutch, two standard deviations above average surpass 100 per cent. They are set at 90 per cent. For all non-western immigrant groups the criteria for concentration are 33.6 per cent and 46.5 per cent. For first generation non-western immigrant groups they are 30.1 per cent and 37.0 per cent.

Second generation immigrants have two foreign born parents. Membership is defined by the parents' place of birth. In the case of two different origins of parents, the mother's place of birth defines the child (whether it is 'other non-western' or 'western'). Contrary to official definitions, we have labelled children with one native Dutch parent as 'native Dutch'. In our analyses, we distinguish between Moroccan, Turkish, Caribbean, western and other non-western immigrants. Caribbean immigrants are of Surinamese and Dutch Antillean descent. Most of the first generation Surinamese arrived after the decolonisation of Surinam in 1975. In Amsterdam, a considerable share was housed in newly-built estates on the south-eastern periphery. Turkish and Moroccan immigrants originally arrived in the Netherlands as 'guest workers' in the 1970s. Immigration increased in the 1980s through family reunification. Initially the workers stayed in hostels but via the private rented sector they finally gained access to social housing, frequently in post-war housing estates on the western periphery.

Findings: Moving to and from Concentrations

Table 5.1 shows the share of each immigrant group and generation that lives in three types of 'concentrations' in the Amsterdam region in 1999 and 2006. The figures reveal several differences between groups and generations. Moroccans are most likely to live in concentrations of their own group or in concentrations of (first generation) non-western immigrants. Interestingly, although national figures show a slight increase in concentrations for Moroccans in this period (Musterd and Ostendorf 2009), the share of Moroccans that lives in concentrations in the Amsterdam region has generally decreased between 1999 and 2006. Also, the shares of Caribbean and Turkish, and 'other non-western' groups living in concentrations have declined, particularly their presence in non-western concentrations. Lastly, Western immigrants do not live in own-group concentrations at all.

When we compare generations, we see that the second generation is generally less likely to live in a concentration neighbourhood than the first generation. The difference between generations is particularly large for the Caribbean group and the heterogeneous group 'other non-western immigrants'. For the latter the share of second generation living in own-group concentrations in 2006 is less than half of the first generation share. The share of second generation Caribbean living in own-group concentrations in 2006 is about two-thirds of the first generation's share. Although the share of second generation living in concentrations is also lower than that of the second generation, the generational difference is less for Turks and Moroccans.

Table 5.1 Per cent of adult group members living in 'concentrations', i.e. neighbourhoods with an overrepresentation of own-group, of Non-Western immigrants, and of first generation Non-Western immigrants per group in the Amsterdam region in 1999 and 2006

	Moroccan		Turkish		Caribbean		Other Non-Western Immigrants		Western Immigrants		Native Dutch
	1st Gen.	2nd Gen.	1st Gen.	2nd Gen.	1st Gen.	2nd Gen.	1st Gen.	2nd Gen.	1st Gen.	2nd Gen.	
2006											
Own-group concentrations*	34.5	33.2	21.1	19.2	25.8	17.0	11.4	5.0	0.0	0.0	6.7
Non-Western all immigrant concentrations	41.0	39.9	35.0	30.7	37.8	28.4	27.4	18.3	10.6	7.4	5.9
First generation all Non-Western immigrant concentrations	39.0	38.0	34.2	30.0	34.0	24.7	25.0	16.2	9.7	6.6	5.2
1999											
Own-group concentrations*	40.4	41.0	32.0	27.1	30.4	17.9	14.3	3.2	0.0	0.0	11.0
Non-Western all immigrant Concentrations	48.5	48.6	42.5	36.3	48.1	34.6	37.3	19.3	15.7	10.6	8.7
First generation all Non-Western immigrant concentrations	46.7	46.7	41.7	35.3	46.7	33.2	35.9	18.0	14.9	10.0	8.3

* Definition based on origin, not generation

Note: Percentages may not add to 100 due to rounding off to decimals.

Source: Netherlands Social Statistical Database (authors' calculations).

Table 5.2 shows the mobility patterns of the adult population from and to own-group concentrations for Moroccan, Turkish, Caribbean and native Dutch groups between 1999 and 2006. Compared with the first generation, second generation Non-Western immigrants are considerably less likely to remain in an own-group concentration for the entire period of seven years. This is particularly true for second generation Caribbeans and 'other Non-Western immigrants'. While less than the first generation, the share of second generation Moroccans living in own-group concentrations for seven years is high compared to other second generation immigrants. However, net balance figures show that more Moroccans end up in non-concentrations than in concentrations. These balance figures are the result of both mobility patterns and neighbourhood change based on a changing regional demography.

In sum, we see differences over time, between groups and between generations. First, immigrants are less likely to live in concentrations in 2006 than they were in 1999. This is the case for most immigrant groups but especially for second-generation Moroccan and all-generation Turkish immigrants whose mobility patterns showed the strongest de-concentration. First generation Turkish immigrants show the highest decrease in concentration, largely because of changing concentration neighbourhood. Second, there are differences between groups. Moroccan immigrants are most likely to live in concentrations, followed by Turkish and Caribbean, and by 'other Non-Western' and natives. Western immigrants do not live in own-group concentrations. Third, for all groups, second generation shares in concentrations were lower than first generation. De-concentration is partly the result of moving. For Turks and Moroccans, the share of those moving out of concentrations is higher for the second generation than the first generation. Second generation Caribbeans, on the other hand, are slightly more likely to move to concentration areas than the first generation.

Findings: Urbanisation and Housing Access

We are also interested in immigrant geographies, per group and between generations, in terms of housing and the residential environment.

Housing

Table 5.3 shows three housing characteristics (type, tenure and value) by group in 2006. First, in terms of type of housing, immigrants are more likely to live in multi-family dwellings than native Dutch. Moroccan and 'other Non-Western' immigrants are most likely to live in multi-family dwellings compared to 'Western', Turkish and Caribbean immigrants. This difference is also reflected in differences in tenure and residential environment (see below). There is not much difference between first and second generation Turkish, Moroccan and Caribbean immigrants. Second generation households from 'other Non-Western' and 'Western' groups,

Table 5.2 Mobility patterns of adult residents within the Amsterdam region in 1999 in a period of 7 years. Data represent distribution within groups in per cent (except net gain figures)

	Moroccan		Turkish		Caribbean		Other Non-Western Immigrants		Western Immigrants		Native Dutch
	1st Gen.	2nd Gen.	1st Gen.	2nd Gen.	1st Gen.	2nd Gen.	1st Gen.	2nd Gen.	1st Gen.	2nd Gen.	
Not moved in concentration neighbourhood	19.8	6.3	10.4	3.5	12.4	4	4.8	0.8	0	0	4.1
Moved within same concentration neighbourhood	3.0	2.8	2.6	2	3.4	1.7	2.8	0.2	0	0	0.3
Moved to another concentration neighbourhood	6.2	11.8	3.6	5.2	5.9	4.6	2.3	0.4	0	0	1.6
Total: 7 years in concentrations	**29.0**	**20.9**	**16.6**	**10.7**	**21.7**	**10.3**	**9.9**	**1.4**	**0**	**0**	**6.0**
Moved from non-concentration to concentration neighbourhood (a)	5.9	10.4	3.6	6.6	3.3	3.7	2	1.5	0	0	1
Moved from concentration to non-concentration neighbourhood (b)	8.7	18.6	8.4	12.9	7.5	7	4.5	1.7	0	0	2.3
Not moved, but neighbourhood changed to concentration (c)	0	0	1.2	0.9	0	0	0	0	0	0	0
Not moved, but neighbourhood changed to non-concentration (d)	3.1	1.7	7.2	3.6	1	0.5	0	0	0	0	3.1
Net gain to concentrations in % points (a+c-b-d)	**-5.9**	**-9.9**	**-10.8**	**-9.0**	**-5.2**	**-3.8**	**-2.5**	**-0.2**	**0**	**0**	**-4.4**
Relative gain (gain /share living in own-group concentration '99)	-0.15	-0.24	-0.34	-0.33	-0.17	-0.21	-0.17	-0.06			-0.40
Other	53.2	48.4	63.0	65.4	65.4	78.5	83.6	95.4	100	100	87.6

Note: Type of concentration is own group; definition is based on origin, not generation.

Source: Netherlands Social Statistical Database (authors' calculations).

however, show a substantial increase in single-family housing compared to the first generation. Native Dutch show the highest share in single-family housing.

Second, native Dutch display the highest share of owner occupancy, but second generation migrants are catching up. Among Non-Western immigrants, the high share of owner occupancy, often in single-family housing, among the second generation Turkish immigrants is striking. The shares in social rental housing are highest for Moroccan immigrants whereby the share of the second generation is lower than the first generation yet still higher than for other groups.

Third, native Dutch tend to live in more expensive housing (Table 5.3). The high values for first generation Western immigrants may be explained by the presence of well-paid expats in the financial and business service sector in Amsterdam. The average tax value of owner occupied dwellings owned by the (much younger) second generation tends to be equal and in some cases even higher than for the first generation. This indicates a higher level of economic integration.

Tenure transitions among movers confirm the trends above (Table 5.3). Immigrant groups are more likely to move from rental to owner occupancy than native Dutch. Exceptions are the Moroccan immigrants and first generation Western immigrants. The latter finding must be ascribed to their short-term stay. Furthermore, second generation immigrants are more likely to move into owner occupancy than first generation. Non-Western first generation immigrants, particularly Moroccans, mostly move within the rental sector. It must be mentioned that the share of owner occupancy housing in Dutch urban regions increased because of privatisations and new construction. This explains why more people change to owner occupancy than to rental housing. Also, new households are omitted in the figures; many of which are likely to start out in the rental sector.

Residential environment in the metropolitan regions

Table 5.4 shows in what types of residential environments households lived in 2006. As expected, native Dutch are overrepresented in the lower density areas outside the built-environment perimeter of the city. In the Amsterdam region, half of the native Dutch population lives in suburban areas of secondary cities, towns, villages and countryside. Non-Western immigrants are overrepresented in the more urban parts of the city (centre and urban periphery). Of all immigrants, Turkish immigrants are most likely to live in the highly urban parts in the regional towns and cities. Nevertheless, despite the lower age average, second generation immigrants are more likely to live in the lower density region than the first generation. This is particularly the case for second generation Caribbean immigrants and second generation 'Western immigrants'.

Table 5.3 Housing characteristics of households by origin in the Amsterdam region. Figures represent distribution within group in per cent

		Moroccan		Turkish		Caribbean		Other Non-Western Immigrants		Western Immigrants		Native Dutch	All
		1st Gen.	2nd Gen.	1st Gen.	2nd Gen.	1st Gen.	2nd Gen.	1st Gen.	2nd Gen.	1st Gen.	2nd Gen.		
Type of dwelling and tenure in 2006													
Single Family	Owner occupied	11.7	17.0	24.2	31.8	24.3	25.9	20.8	33.3	31.4	42.9	46.6	41.5
	Private Rent	0.4	0.7	0.5	1.5	0.5	0.7	1.2	2.1	2.1	1.2	1.3	1.2
	Social Rent	22.8	17.8	16.1	10.2	19.2	16.1	16.0	13.5	18.4	18.1	20.3	19.7
	Total Single Family	**34.9**	**35.5**	**40.8**	**43.5**	**44.0**	**42.7**	**38.0**	**48.9**	**51.9**	**62.1**	**68.2**	**62.4**
Multi-family	Owner occupied	2.8	6.4	7.2	14.4	6.2	9.6	7.5	13.7	12.8	12.0	9.7	9.5
	Private Rent	0.4	0.8	0.8	1.6	0.5	1.5	1.4	2.3	3.2	1.2	0.8	1.0
	Social Rent	62.0	57.3	51.2	40.5	49.3	46.2	53.1	35.1	32.2	24.7	21.3	27.1
	Total Multi-family	**65.1**	**64.5**	**59.2**	**56.5**	**56.0**	**57.3**	**62.0**	**51.1**	**48.1**	**37.9**	**31.8**	**37.6**
Tenure in 2006	Owner occupied	14.5	23.4	31.4	46.2	30.5	35.5	28.3	46.9	44.2	54.8	56.3	51.0
	Private Rent	0.7	1.5	1.3	3.0	1.0	2.2	2.6	4.5	5.3	2.4	2.1	2.2
	Social Rent	84.8	75.1	67.3	50.8	68.5	62.3	69.1	48.6	50.6	42.8	41.6	46.8
Tax value owner occupied dwellings in 2006 (valued in 2003, in €, x1000)													
	Single Family	204	200	214	218	216	231	245	267	291	301	278	275
	Multi-family	182	188	170	179	180	187	204	215	250	237	226	224
Transition in tenure, moving households, 1999–2006													
	Owner->Owner	2.2	2.4	6.7	11.4	11.7	12.7	9.2	19.2	24.3	32.0	36.8	31.5
	Owner > Tenant	1.4	3.1	1.8	4.9	2.2	4.5	2.6	6.0	3.5	3.9	3.8	3.6
	Tenant-> Tenant	91.0	81.9	75.7	60.3	72.0	62.0	73.9	55.0	60.1	46.8	45.9	51.2
	Tenant-> Owner	5.3	12.5	15.8	23.5	14.2	20.9	14.3	19.9	12.1	17.2	13.5	13.6

Note: Percentages may not add to 100 per cent due to rounding off to decimals.

Source: Netherlands Social Statistical Database (authors' calculations).

Table 5.4 Residential environment of all individuals by origin in the Amsterdam region in 2006. Figures represent distribution within group in per cent

	Moroccan		Turkish		Caribbean		Other Non-Western Immigrants		Western Immigrants		Native Dutch	All
	1st Gen.	2nd Gen.	1st Gen.	2nd Gen.	1st Gen.	2nd Gen.	1st Gen.	2nd Gen.	1st Gen.	2nd Gen.		
Residential environment in 2006												
City												
Centre	36.6	31.6	26.5	20.9	21.9	20.6	27.3	24.0	40.1	30.8	20.5	23.4
Periphery – high urban density	36.2	39.0	28.0	30.1	21.3	18.8	20.4	19.5	12.1	8.6	7.1	11.4
Periphery – lower densities	7.9	8.7	6.5	7.2	25.5	23.6	18.1	18.3	12.7	11.5	10.0	11.7
Metropolitan region												
High urban density	5.6	4.9	13.8	13.2	3.4	4.0	7.0	6.9	8.0	10.0	12.2	10.6
Lower densities	13.6	15.7	25.2	28.2	28.0	32.9	27.1	31.3	27.1	39.1	50.3	43.0

Note: Percentages may not add to 100 per cent due to rounding off to decimals.

Source: Netherlands Social Statistical Database (authors' calculations).

Findings: Model of Residential Mobility

The analyses above have shown that there are substantial differences between native Dutch and immigrants, between immigrant groups and between the first and the second generation. However, there are several 'intervening variables' which should be taken into consideration. Observed differences may be a reflection of life course factors (e.g. second generation immigrants are younger and may have different housing needs and preferences based on their household situation) or of different levels of economic success. Therefore, we performed multinomial logistic regression analysis to see how important immigrant group and generation are for urban dwellers in 1999 who moved after a period of seven years, controlling for household life cycle, socio-economic status, and previous position in the housing market. This allows us to better compare immigrants to natives. For the reference groups for the analyses, we have chosen moving to an owner occupied single-family dwelling and to lower density suburbs outside of the built environment of the city in 2006. These moves are typical for Dutch natives and we are interested to see the differences between first and second generation immigrants compared to the native Dutch. We only present the model outcomes for immigrants (Table 5.5 lists all variables).

As for choosing owner occupancy as the norm, we are aware that there is a policy ideology of home ownership which prescribes ownership as superior, regardless of individual household situation and employment status (Ronald 2008, Van Gent 2010). Nevertheless, households seeking more space in the Dutch housing market almost have to move to owner occupied single-family housing in lower density regions, since alternatives are limited.

Table 5.5 **Averages and frequencies for multinomial regression analyses (adults who lived in the built environment of Amsterdam in 1999 and moved house at least once in the subsequent 7 years)**

Averages

Age*	36.7
Standardised household income 1999 (natural log)*	9.74
Standardised household income 2006 (natural log)*	10.02
Development standardised income (nat. log) '99–06'*	0.080

Domain	Category	% of sample population
Origin and generation	Moroccan (1st generation)	3.9
	Moroccan (2nd generation)	0.5
	Turkish (1st generation)	3.0
	Turkish (2nd generation)	0.4
	Caribbean (1st generation)	10.9
	Caribbean (2nd generation)	2.3
	Other Non-Western (1st generation)	6.8
	Other Non-Western (2nd generation)	0.5
	Western (1st generation)	5.3
	Western (2nd generation)	6.5
	Native Dutch	59.8
Gender*	Male	48.8
Household transition*	Stable single person	20.7
	Single person > Cohabiting	16.8
	Cohabiting > single person	9.2
	Moved from parental/ custodial care	6.5
	Unknown/ other	9.6
	Stable cohabiting	37.1
Children in household*	No children	53.4
	Children left household in 2006	6.4
	Got children after 1999	21.9

Table 5.5   *Concluded*

	Underaged children in household, 99–06	18.3
Social economic transitions*	Benefits in 1999 and in 2006	12.6
	Benefits > (self-)employment	4.4
	(self-)employment > benefits	4.2
	Retired after 1999 (pension)	3.0
	Retired before 1999 (pension)	10.5
	Student > employment or benefits	2.4
	Unknown/ other	13.0
	(self-)employed in 1999 and 2006	49.8
Dwelling tenure in 1999*	Rent	84.8
Urban environment in 1999*	Centre	48.8
	Periphery - high urban density	34.8
	Periphery - lower densities	16.4

Type of dwelling and residential environment in 2006

Single-family owner-occupied dwelling in lower density region		15.6
Lower density region		6.7
High density urban periphery	Multi-family	8.7
		3.2
	Single Family	3.8
		3.3
Lower density urban periphery	Multi-family	8.5
		2.4
	Single Family	5.6
		6.5
Other*		35.7

* Model output not presented.

Source: Netherlands Social Statistical Database (authors' calculations).

In the model (Table 5.6) we combine housing type with residential environment. We are particularly interested to see if immigrants rather than suburbanise to owner occupancy, are more likely to move to single-family housing in highly urban environments or perhaps to multi-family housing in low density environments within the city. Moving to lower density neighbourhoods or to single-family housing within the city may be interpreted as a first step in counter-urbanisation. It was expected that second generation immigrants would show a higher likelihood of moving to these sorts of residential environments and/or single-family owner occupied housing.

The evidence shows that first and second generation Moroccan, Turkish, and 'other Non-Western', as well as first generation Western immigrants, are more likely than natives to move into ownership in single and multi-family housing in the higher density urban periphery rather than to suburbanise in owner occupied housing. However, compared to natives, first generation Turkish, Caribbean, and other Non-Western immigrants have a higher probability of moving into multi-family owner occupied low density urban periphery housing than moving to single-family owner occupied housing in lower density areas of the region. When we look at moving into single family owner occupied housing in the lower density urban periphery, the category most similar to our reference category, it appears that natives are more likely to move here than first generation Turkish and first generation Caribbean immigrants. Other groups show no differences.

When significant, first and second generations generally show similar trends. A notable exception is the difference in probability between first and second generation Caribbeans moving to rental housing in the lower density region. Second generation Caribbeans, just like second generation Moroccans, are more likely to do this than go into owner occupancy in the region, while first generation Caribbeans are less likely to do so. This may reflect counter-urbanisation through (social) renting.

Table 5.6 Multinomial logistic regression analysis of moving to dwellings in residential environments in 2006 (population: adult residents of city of Amsterdam in 1999; reference group is single family owner occupied dwelling in the lower density metropolitan region). Only the Odd Ratios (exp(β)) for immigrants are presented

Reference= Single-family owner-occupied dwelling in lower density region

	Lower density region	High density urban periphery				Lower density urban periphery			
		Multi-family		Single Family		Multi-family		Single Family	
	Rental	Rental	Owner	Rental	Owner	Rental	Owner	Rental	Owner
Origin and generation									
Refcat= native Dutch									
Moroccan (1st generation)	0.832	6.199***	3.075***	3.019***	2.252***	3.001***	1.226	1.007	0.445**
Moroccan (2nd generation)	2.163*	4.433***	2.742*	2.233	3.290**	3.258***	0.461	1.549	0.844
Turkish (1st generation)	0.232***	2.372***	4.040***	1.480*	4.168***	1.359	2.722***	0.384***	1.346
Turkish (2nd generation)	0.640	2.666**	4.073***	1.283	3.107**	1.664	1.763	0.632	1.066
Caribbean (1st generation)	0.615***	1.337**	1.229	0.603***	0.809	1.807***	1.752***	0.725***	0.763**
Caribbean (2nd generation)	1.736**	1.676**	1.398	0.916	1.151	2.421***	1.735**	0.936	0.974
Other non-western (1st gen.)	0.467***	2.315***	2.065***	0.646**	1.602**	2.924***	2.945***	0.560***	1.124
Other non-western (2nd gen.)	0.797	0.839	2.270*	1.063	2.503*	2.095	0.860	1.074	0.738
Western (1st generation)	0.668**	1.814***	1.609**	1.420*	1.218	1.979***	2.192***	1.049	1.238
Western (2nd generation)	0.926	0.986	1.078	0.799	1.078	1.108	0.879	1.004	1.023

*=p<0.05; **=p<0.01; ***=p<0.001 | Chi Sq. 18022.2, Df 340, Sign. .000, Nagelkerke's R² .451.

Source: Netherlands Social Statistical Database (authors' calculations).

Discussion and Conclusion

This chapter sought to analyse immigrant geographies in the metropolitan region of Amsterdam in order to gain insight into social integration, 'residential choice', and spatial patterning of different immigrant groups and generations. As expected, immigrants differ substantially from native Dutch in terms of where they live and in their mobility behaviour. However, there are also variations between immigrant groups and between generations.

There is a difference between Non-Western and Western immigrants. The latter is a rather heterogeneous group but on average more affluent than native Dutch. Consequently, they do not live in large concentrations and have ready access in the housing market with a preference for housing within the built environment of the city rather than in the surrounding region. Non-Western immigrants (Moroccan, Turkish, Caribbean and 'other'), on the other hand, are generally less affluent and more likely to live in concentrations. However, it seems that the Amsterdam city and region offer opportunities for migrants to move outside concentrations.

Opportunities also seem present in the housing market. Non-Western immigrants are still generally less likely to obtain single-family housing in low density areas outside the region than native Dutch. However, there are specific mobility and residential patterns for specific immigrant groups. Moroccans are most likely to move within the social rental sector, which means that they are overrepresented in the areas with high levels of social housing stock, usually within the city; but, second generation Moroccans also have a higher probability to move to social rented single-family housing in the urban region (suburbs) than the native populations does (compared to moving to owner occupation in a suburban area). Turkish households, particularly of the second generation, are likely to reside in owner occupied housing. Also, a large share of Turkish households resides in the highly dense areas in regional towns. Caribbean households, the oldest immigrant group, seem to prefer or end up in the lower density urban periphery, usually in multi-family housing.

So, the question is how to interpret these results in view of the spatial assimilation, integration and counter-urbanisation of Non-Western immigrants. As mentioned, all immigrants have their own dynamic compared to native Dutch urban dwellers. If we compare second generation with first generation immigrants, we see a trend of spatial assimilation and increased housing market access. In general, second generation immigrants are less represented in concentration areas, and are better able to obtain owner occupied housing which is higher in value than the first generation. Interestingly, there are no large differences in type of housing between generations. In terms of counter-urbanisation trends, while the suburban ideal of owner occupied single-family housing is less common, second generation immigrants tend not to differ significantly from the native Dutch anymore in terms of their tendency towards lower density areas. In sum, this study has found evidence for spatial assimilation over generations in terms of the housing market and residential environment. Still, many immigrants seem to be hesitant to, or deterred from, moving into typical 'white spaces' (i.e. single-family owner occupied suburbia), but they are strengthening their position in the housing market over generations by acquiring larger, more expensive (here considered as a sign of progress) owner occupied dwellings in both urban and suburban settings.

Acknowledgements

The authors wish to thank Aslan Zorlu, Sjoerd de Vos, Annalies Teernstra, Evert Verkuijlen and the editors for comments, advice and assistance.

Chapter 6

Arab Migrants in a Jewish State: Patterns, Profiles, Challenges

Nir Cohen, Amir Hefetz and Daniel Czamanski

Introduction

The past decade has seen a surge in interest in the internal migration of minority groups. Most studies suggest that while ethno-national and racial migrants display different migration behaviour to that of majorities, their socio-demographic profile is often on par with the latter (Finney and Simpson 2008, Stillwell and Hussain 2008, Catney and Simpson 2010). To date, though, most studies have been concerned with recent international in-migrants, either first or second generation, and little has been written about migration patterns among *native* minority groups. Despite their socio-cultural distinction and, quite often, unique demographic composition, internal migration of most native groups has been neglected. Such is the case of Arabs in Israel.

Considered a homeland native minority (Jamal 2009), Arabs in Israel differ considerably from the Jewish majority in virtually every aspect, including some that could potentially affect their internal mobility, such as settlement patterns and economic standing. Yet, for various reasons, research in Israel has focused exclusively upon the Jewish majority (Gonen, 1995, Newman 2000). With a few isolated and highly localised exceptions (Lipshitz 1991), Israeli migration scholars have entirely ignored internal Arab migration and therefore we currently know very little about its magnitude, characteristics or driving forces as well as migrants' socio-demographic profile.

In this chapter we begin the long overdue filling of this gap by examining patterns of internal migration among Israel's Arab population. It should be noted from the outset that our analysis here is concerned *only* with Arab citizens of Israel, namely those who reside within the pre-1967 borders (excluding the West Bank, Gaza Strip and the Golan Heights). Rather than studying post-migration geographic patterns, our main objective is to examine the propensity to migrate among Arabs in Israel and the socio-demographic profile of migrating individuals. We are primarily concerned with differences in migration both between Arabs and Jews as well as across ethno-religious groups within the Arab minority (Muslims, Christians and Druze). Using descriptive statistical methods, our chapter uncovers and attempts to explain differences (and similarities) in migration patterns and profiles within and across ethnic (sub)-groups. Following a brief theoretical prelude, we contextualise the group(s) studied, providing background information on its historical and contemporary political and socio-economic conditions. We then discuss attributes that have been potential hindering factors to Arab mobility in Israel since its independence in 1948. Finally, we analyse 1995 Housing and Population Census data at the micro scale and provide preliminary explanations for the said findings. We conclude by considering the broader implications of our findings and outlying some future directions in the study of ethnic migration in Israel.

Internal Migration and Ethno-racial Minorities

In recent years there has been a flurry of research on residential mobility and internal migration among minorities, including incoming international migrants. A growing number of studies, mostly in the UK and the US, confirm that migration patterns may vary *across* minority groups *as well as* between them and the majority group (Leon and Strachan 1993, Champion 1996, Stillwell, Hussain and Norman 2008). Differences were recorded in migration rates (Rees and Phillips 1996), distance (Simpson and Finney 2009), and eventual destinations (Stillwell and Hussain 2008). Observed differences have normally been explained by referring to either ethnic social networks or (in)-formal practices of ethno-racial discrimination in particular neighbourhoods or towns. As per the latter, studies comparing neighbourhood transitions in US cities corroborate that blacks have the lowest rates of residential mobility when compared with whites, Hispanics, and Asians (Massey and Denton 1987), even when controlling for home ownership and other socio-demographic factors (South and Deane 1993). High levels of residential segregation in specific neighbourhoods and practices of institutionalised discrimination (e.g., red-taping) have been quoted as major impediments to black mobility (South and Crowder 1998).

An alternative explanation for a lower migration level and distance among blacks and other ethno-racial minority groups centres on the role of cultural constraints faced by group members wishing to depart from neighbourhoods populated by minorities. Thus, even when neighbourhood dissatisfaction was reported, blacks were less likely to move than non-blacks (South and Deane 1993). At both the metropolitan and state levels, studies suggest that close knit social networks of co-ethnics serve as both pull (prevent out-migration) and push (play a factor in destination selection process) factors among ethno-racial minorities (Barringer Gardner and Levin 1993, Moore and Rosenberg 1995, Frey 2003). Patterns of channelised internal migration have been observed among both native-born and longer term immigrant minorities. Frey and Liaw (2005), for example, found that same-race residents were out-migration inhibitors and a strong pull factor in destination selection among African Americans as well as Asians and Hispanics. Residence in an ethnically concentrated metro area was also negatively correlated with propensity of out-migration among Hispanic (Cuban, Puerto Rican and Mexican) men (Tienda and Wilson 1992).

Simultaneously, though, research has shown that the socio-demographic profile of internal migrants remains fairly consistent. A large number of studies confirm that migrants originating in both minority and majority groups tend to be younger and possess higher levels of human capital compared with non-migrants of the same groups as well as the general population (South and Crowder 1998, Faggian, McCann and Sheppard 2006, Finney 2011). Thus, a study on ethnic migration carried out in Britain suggests that *regardless of ethnic affiliation*, those most likely to migrate were 20–29 years of age, unemployed or students, in rented accommodation, and living alone or with one other adult (Finney and Simpson 2008). And while much less work has been done outside the Anglo-American world, findings were generally in line with the above (Guest 1999). Kulu and Billari (2004), for example, report that members of ethno-linguistic minority groups tend to regionally concentrate at higher rates and migrate smaller distances than ethnic Estonians. However, they too found strong correlations between (young) age and (higher) level of education and minorities' propensity to move.

Contextualising Arabs in Israel

Now the largest ethno-national minority in the country, Arabs in Israel are those who neither fled nor were expelled from their places of residence by the Israeli military in the course of the 1948 war. It is estimated that prior to the war roughly 940,000 Arabs lived in the region that became Israel. Following independence, 160,000 of them – less than one fifth of their number before the hostilities – remained within the territory of the new state, mostly in the rural periphery (Rabinowitz 2001). While they were granted Israeli citizenship and allowed to vote for the *Knesset* (Israeli parliament), Arabs were subjected to the rule of a Military Government (*Mimshal Tzevai*), whose key objective was to spatially control them and deepen their political, economic and social segregation from the Jewish majority (Baumel 2002). In cooperation with other state agencies the Military Government concentrated Arabs in designated geographic regions (e.g., the Galilee, Eastern Negev, and the Triangle) and severely limited their physical mobility within and beyond national territory. Curtailing their everyday lives, it prevented them from acquiring state lands, relocating freely between – and establishing new – settlements, or returning to their original villages and towns (see Lustick (1980). The effects of the Military Government on the Arab population in Israel were so adverse that the period (1948–1966) came to be known as 'The Lost Years' (Ghanem 2001).

The late 1960s marked the beginning of a gradual improvement in the status of Israel's Arabs. The official abolition of the Military Government in 1966, the effect of the PLO (Palestine Liberation Organisation) on their collective identity as Palestinians, and increasing rates of urbanisation and educational attainment have propelled a noticeable social change. The next three decades have witnessed Arab political parties and NGOs turning a mélange of fragmented, mostly rural communities into a vibrant, (semi)-urban society acutely aware of its tragic past and distinct ethno-national identity. Yet, greater political, economic and cultural equality between Jews and Arabs, which many thought would be a natural outcome of the Oslo Peace Agreement (1993) between Israel and the PLO, was lingering. Notwithstanding their repeated demands for a fairer allocation of public resources,[1] and enhanced recognition of their distinct culture, Arabs in Israel remain politically under-represented and considerably segregated from the Jewish majority. More importantly, their social and economic standing is significantly inferior compared with Jews (see Table 6.1).

Arabs residing within pre-1967 Israel – that is, excluding the occupied territory of the West Bank – comprise roughly 1.53 million or 20.3 per cent of Israel's population (Central Bureau of Statistics 2010). Not surprisingly, it is not an entirely cohesive population group; in fact, it is made up of three relatively distinct ethno-religious sub-groups, namely Muslims, Christians, and Druze. Muslims are the overwhelming majority with an estimated 1.2 million (83 per cent), whereas Christians and Druze number roughly 125,000 (8.5 per cent) each. Within the Muslim sub-group, roughly 15 per cent are Bedouins, a formerly nomadic group mostly concentrated in the Southern region of the Negev, which has been gradually sedentarised over the last several decades (see Meir 1997).

Table 6.1   Jews and Arabs in Israel, selected demographic indicators (2010)

Indicator	Jews	Arabs
Annual Population Growth Rate	1.7%	2.4%
Average Household Size	3.1	4.8
Under 5 Mortality Rate (per 1,000)	2.7	7.1
Total Fertility Rate	2.9	3.7*
Life Expectancy at Birth	79.6	75.5
Gross Income per Household (NIS)	14,279	8,151

*Muslim women only.

Notes: All indicators are based on 2010 data. Annual population growth rate is calculated for 2005–2010.

Source: Central Bureau of Statistics, Various Tables, 2010.

---

    1   These demands were incorporated into a civic contract ('democratic constitution') drafted by Arab human rights groups in which they called to eliminate Israel's exclusive Jewish character and re-classify it as multicultural and bilingual state (see Ozacky-Lazar and Kabna 2008).

Officially, since the cancellation of the Military Government, Arabs are free to move and settle anywhere in the country. In practice, though, following years of discriminating planning and settlement policies (Yiftachel 2006), a disproportionately large percentage of Arabs is still concentrated in a relatively small number of mostly all-Arabs localities and even smaller number of counties. In fact, almost three quarters (74 per cent) of them presently reside in three counties (*Mechozot*) only, namely Northern (39 per cent), Jerusalem (21 per cent), and Haifa (14 per cent). Smaller concentrations exist in the Southern (13 per cent, mostly Bedouins) and Central (11 per cent, mostly non-Bedouin Muslims in the mixed cities of Lod and Ramle) counties. Currently, 76 per cent of Arabs in Israel live in one of the 122 all-Arab localities, while the other 24 per cent reside in eight mixed (Jewish/Arab) towns,[2] namely Haifa, Tel Aviv-Jaffa, Lod, Ma'alot-Tarshiha, Neve Shalom, Nazareth Illit, Akko, and Ramle (Central Bureau of Statistics 2010).

As noted earlier, Arab society has been fast urbanising since the mid-1970s. While only a handful of new urban settlements were established since, owing to especially high crude birth rate and persistent rural to urban migration, several Arab localities have reached the required population quota and been declared cities. At the present time, it is estimated that more than 90 per cent of the total Arab population in Israel live in urban or semi-urban – either all-Arab or mixed – localities, the largest of which are Nazareth (65,000), Umm al Fahm (43,000), Rahat (42,000), Tayibe (35,000) and Shefa'amr (34,000).

Data and Methodology

The data for this study are drawn from the Israeli Census of Population and Housing (1995), Individual Record Data.[3] Assuming that children are not the primary migration decision makers, we have only used records of adults who are 18 years and older in our analysis. The 1995 Census includes two questions that directly ask about migration behaviour. The first takes a temporal perspective and divides the population surveyed into four categories: those who never migrated (e.g., born in current locality of residence and have been living in the same address since), and those who either migrated from abroad, from another locality within the last five-year period (1990 onwards), or from another locality earlier (prior to 1990). The second question takes a spatial perspective, asking individuals who migrated within the last five years of their previous locality of residence. It then

---

2 It is important to note that even in large mixed cities Arabs are usually concentrated in a relatively small number of neighbourhoods (e.g., Downtown in Haifa and Jaffa in Tel-Aviv-Jaffa).

3 While the most recent Census was conducted in late 2008, its data were still unavailable for public use at the time of writing, with the exception of national aggregated data.

categorises responses along geographic scales as follows: same locality, same vicinity, same district, same region, different region, or from abroad.

These scales correspond to the various geographic tiers officially recognised by the national settlement system. In addition to Israel's 1,464 localities, there exist three other tiers, which are in ascending order: 1. (Nearest) *Vicinity or Natural Area* – a physical geographic sub-unit within which a locality is located. The Natural Areas has no administrative power and is only intended to allow specific geographic identification. 2. *District* – an administratively defined category. There are 16 districts within pre-1967 Israel, nested within the largest category of Region. 3. *Region* – the most spatially extensive and administratively powerful category in the Israeli system. There are six regions within pre-1967 Israel, including Jerusalem, Northern, Haifa, Central, Tel Aviv, and Southern.[4] To illustrate the above scalar classification system, Umm Al Fahm is an all-Arab (urban) locality, located within the Alexander Mountain natural area, belongs to the Hadera District, which is part of the Haifa Region.[5]

We constructed a migration variable that centres on the individual migration decision. The variable is composed of four categories: non-migrating individuals; individuals who migrated within the last five years, regardless of distance travelled or the administrative scale of the move (e.g., inter-district, inter-regional); individuals who migrated earlier; and individuals who migrated from abroad.

A few data limitations should be noted. First, despite salient socio-demographic differences between the general Muslim population and the Bedouin sub-group (e.g. higher total fertility rate and lower household income among the latter), the Israeli Census includes only five ethno-religious categories, namely Jewish, Muslim, Christian, Druze, and 'Others'.[6] Since Bedouins are inseparable from the larger group of Muslims, of which they are part, we were unable to perform any statistical tests to examine the extent to which they exhibit unique migration patterns. Two other limitations result from the Census' strict policy of confidentiality. First, only a survey of one fifth (20 per cent) of the total population is available for public use, and none of the records is identified. In addition, localities in which population does not exceed 2,000 are lumped into the 'small locality' category, which records are not geographically identified to prevent possible individual identification.

---

4   The Judea and Samaria Region lies east of the Green Line, and since its Arab residents are not Israeli citizens they were excluded from our analysis.

5   Source: http://www.cbs.gov.il/ishuvim/ishuvim_main.htm.

6   The formal census category is 'others and unclassified'.

Patterns of Migration Behaviour Among Arabs in Israel

Arab Migration Propensity

The 1995 Census provides evidence that 85 per cent of the Arabs in Israel are non-migrants (see Table 6.2). Only 4 per cent of Arab adults were recent migrants and an additional 9.4 per cent of the population migrated sometime in the past. As is evident from the comparative data below, Arab migration rates are significantly lower than those of the Jewish majority.[7] Comparing Muslim and Jewish males shows that the latter are seven times likelier than the former to migrate. While differences are not as large for women, it is obvious that Jewish citizens of Israel are considerably more mobile than Arabs. While a more advanced analysis is needed to determine the root causes of such differences, it is plausible that a combination of political, economic and socio-cultural factors are major constraints on Arab mobility in Israel and could thus serve as possible explanations for this finding.

Similarly to the historical experience of discriminated minorities like African-Americans in the US (Massey and Denton 1987, South and Crowder 1998), Arab relocation to and residence in mixed and/or all-Jewish localities is frowned upon and often explicitly discouraged through multiple institutional mechanisms, both public and private (see Cohen-Eliya 2003). These mechanisms, including selling

Table 6.2   Per cent in each migration category, by religious affiliation and gender (1995)

Gender	Non-Migrants		Migrated within last 5 years		Migrated before 1990		Migrated from Abroad		All, Number	
	F	M	F	M	F	M	F	M	F	M
Religious affiliation										
Jewish	33.2	38.1	12.8	12.8	42.1	38.3	12.0	10.9	237,894	220,937
Muslim	82.5	91.0	4.1	1.8	12.6	6.6	0.8	0.6	34,740	36,120
Christian	62.6	76.1	6.1	3.6	23.2	13.7	8.2	6.6	6,927	6,563
Druze	91.1	97.0	3.0	1.6	5.7	1.4	0.1	0.0	3,546	3,807
All Arab	80.1	89.4	4.3	2.0	13.7	7.2	1.8	1.4	45,213	46,490

Notes: F=FEMALE; M=MALE. Rows by gender sum up to 100 per cent, total numbers are to the right. Sample size is 550,534.

Source: Israeli Central Bureau of Statistics, 1995 Census of Housing and Population, 20 per cent of the total population of age above 18 and below 80.

---

7   Unless otherwise stated, all differences in this paper are significant.

state lands to Jews only or prioritising army veterans in the provision of lower rate mortgages, prevent *some* Arabs from leaving their birth towns and others from making additional moves in various stages of their life-cycle.[8] As well, anti-Arab migration sentiments in Israel have multiplied; in some Jewish localities, for example, a small number of local residents, clergymen and politicians, aided by out-of-town right-wing extremists, have been protesting against the gradual in-migration from nearby Arab towns.[9] While the effect of such campaigns on Arab propensity to internally migrate is still unclear, it is not implausible that the xenophobic atmosphere created deters some to-be-migrants from acting on their plan, and others to postpone it.

From an economic standpoint, the financial costs incurred by households in the process of moving are a major impediment for all Israelis, but more so for Arab households whose income – as indicated previously in Table 6.1 – is quite often lower than their co-nationals. The total cost is even higher for Arabs wishing to relocate from an all-Arab to an all-Jewish town. The latter with its improved, hence costlier, municipal services and facilities, and higher property values and taxes is invariably more expensive than the majority of peripheral, over-crowded Arab towns and villages. Overall, then, the relative and absolute potential economic burdens to be faced by most Arab households are probable migration-inhibiting factors.

Socio-cultural factors are possibly at play here as well. The role of social capital in general and social networks in particular in shaping individual and household migration decisions has been widely documented (see Portes (1998). As noted earlier, social networks are key migration enhancers (at destination) and suppressors (at origin). For Arabs, as for many other (especially discriminated) minorities, ethnic social networks make an important source of support that cushions their move into a frequently non-welcoming or hostile environment. Thus, not only would migration entail leaving behind family members, childhood friends and other members of place-of-origin-based social networks, but due to the extraordinarily segregated nature of Israel's settlement system, it would be quite difficult for at least some migrating Arab individuals to replace it with a similar co-ethnic network at destination, be it in a mixed or all-Jewish town.

Despite the small percentage of Arabs who migrate within Israel, there are significant differences in migration behaviour based on gender and the ethno-

---

8   A case in point is the long battle waged by the Ka'adan family against the Israeli Land Administration, which limits the purchase of land plots in most new settlements to Jews only.

9   See 'Right-wing activists rally to "keep Bat Yam Jewish" by kicking Arabs out'; Available online at: [http://www.haaretz.com/news/national/right-wing-activists-rally-to-keep-bat-yam-jewish-by-kicking-arabs-out-1.331682]; also 'Safed Rabbis urge Jews to refrain from renting apartments to Arabs'; Available Online at: http://www.haaretz.com/print-edition/news/safed-rabbis-urge-jews-to-refrain-from-renting-apartments-to-arabs-1.320118.

religious affiliation of migrants. As is evident from Table 6.2, Christians tend to migrate more than Muslims and Druze. This is true for both men and women, with Christian men moving twice as much as other Arab males (3.6 per cent vs. 1.8 per cent). As well, Christian women are more mobile in comparison to other minority females. A possible explanation for this pattern is that religious affiliation is a proxy for the educational and income levels of the migrants. This is made evident below where we link up ethno-religious affiliation with specific socio-demographic attributes.

Socio-demographic Profile of Arab Migrants

A large body of international research confirms that the socio-demographic profile of migrants is consistently different to that of non-migrants. The propensity to migrate is normally higher among the young, males, students, those who are childless, the employed, and those living in small households and of higher social and economic status. Age, in particular, has received much attention by migration scholars and there is clear evidence that mobility is subject to changes along the various stages of the life-cycle (Plane and Heins 2003). Regardless of ethnic affiliation, the propensity to migrate reaches its peak between the age cohorts 16–19 and 25–29 (Stillwell and Hussain 2008). Our findings, succinctly summarised in Table 6.3, are generally consistent with these findings, though as the interpretation of the results (see overleaf) suggest, some important differences do exist.

Concerning age, our results clearly indicate that Arab migrants are younger than Jewish migrants as well as their co-ethnic non-migrants. The former pattern could be partly explained by the mandatory military service for Jewish women and men – two and three years, respectively. Due to the country's compactness, soldiers in Israel rarely change their residence, which means their first move usually takes place at a much older age, normally for the purposes of college/university enrolment or taking up first professional jobs. While some minority groups do enlist in the military (all Druze men and some Bedouin men), the vast majority do not, which makes an early-age move more plausible.

Differences were also recorded between different ethno-religious sects within the general Arab population. Migrating Muslim males, for example, were four years younger than non-migrants (30 and 34, respectively), while for females the difference was seven years. Interestingly, at 31 and 32, Christian males and females were the oldest migrants. This last finding is possibly correlated with their being the most educated of the three Arab sub-groups. Stability while in college/university is expected and it is therefore reasonable that Christians – who spend longer periods at school (as evidenced by their higher educational attainments) – will migrate at a later age than others. Still, the overall pattern is consistent with the literature concerning age, life-cycle and internal migration, suggesting that similarly to minority groups elsewhere and despite objective constraints, Arab migrants are younger than non-migrants.

Table 6.3   Selected socio-demographic characteristics by migration category, religious affiliation, and gender (1995)

Migration status	Non-Migrants		Migrated within last 5 years		Migrated prior to 1990		Migrated from Abroad	
Gender	F	M	F	M	F	M	F	M
Religion	Indicator: Mean Household Income (New Israeli Sheqels (NIS), 1995)							
	Mean	Mean	Mean	Mean	Mean	Mean	Mean	Mean
Jewish	9,766	10,136	8,895	9,607	10,472	11,186	5,926	6,289
Muslim	5,340	5,767	4,329	4,870	5,571	5,644	3,145	3,002
Christian	7,396	7,816	6,238	7,189	7,911	8,557	4,674	4,901
Druze	6,485	6,650	4,560	4,730	6,624	8,017	.	.
All Arab	5,686	6,086	4,738	5,450	6,135	6,437	4,153	4,297
	Indicator: Age (years)							
	Mean	Mean	Mean	Mean	Mean	Mean	Mean	Mean
Jewish	35	34	34	35	51	50	45	43
Muslim	34	34	27	30	40	42	29	33
Christian	38	38	31	32	46	47	38	38
Druze	35	35	27	27	42	49	31	19
All Arab	35	35	28	31	42	43	35	37
	Indicator: Education (school years completed)							
	Mean	Mean	Mean	Mean	Mean	Mean	Mean	Mean
Jewish	12	11	12	12	10	11	12	12
Muslim	8	9	10	10	6	7	10	11
Christian	10	10	12	12	10	10	11	11
Druze	7	10	10	12	6	8	10	11
All Arab	8	9	10	11	7	8	11	11
	Indicator: Marital status (per cent married at Census year)							
	%	%	%	%	%	%	%	%
Jewish	57	54	67	69	69	80	59	71
Muslim	67	69	88	66	84	83	85	74
Christian	59	65	77	60	73	75	74	78
Druze	74	72	95	34	85	83	80	0
All Arab	66	69	86	62	81	81	77	77

Notes: F=FEMALE; M=MALE. About 37 per cent of the cases did not report their income. Income, age, and education variables were calculated based on the mean value of interval categories (e.g. category 1 = income 1 – 1000 → income = 500). Note 3: per cent married is the ratio between number married and total number of that item (e.g. 43 per cent of all non-migrating Jewish females are married, where marital status include: married, single, widower, divorced). Sample size is 550,534.

Source: Israeli Central Bureau of Statistics, 1995 Census of Housing and Population, 20 per cent of the total population of age above 18 and below 80.

Level of education was also a distinguishing factor between migrants and non-migrants. Regardless of ethno-religious affiliation, both male and female Arab migrants in Israel were significantly more educated than non-migrants. Muslim migrants had on average an additional year of schooling compared with non-migrants. Among Druze and Christians the gap was even larger, with migrants having two additional years of education compared with non-migrants (12 and 10, respectively). The link between education and (domestic and international) migration is well-established and has traditionally been explained by labour markets' inability to fully absorb increasing levels of educated workforce locally (see Borjas 1989). Educated individuals' desire to capitalise on their earned educational capital and translate it into economic capital is driving many to seek better employment opportunities in potential destinations, both in and out of their countries of origin. We propose that this process of self-selection is taking place among Arabs in Israel as well. Since the vast majority of Arabs still reside in the country's less-well-off, peripheral Northern and Southern counties where highly-skilled jobs are acutely scarce, a growing number among the educated choose to leave in search of employment opportunities elsewhere.

Other determinants reflecting socio-economic status were also closely related to the propensity to migrate. Migrating Arabs of all ethno-religious sub-groups had significantly lower levels of household income as compared to non-migrating individuals. Migrants' incomes were 19 per cent and 20 per cent lower than non-migrants among Muslims and Christians respectively. The greatest difference was recorded among Druze with migrating individuals earning on average 37 per cent less than non-migrants. The sharp difference between migrants' and non-migrants' incomes can be partly explained by the life-cycle approach to migration, which predicts high levels of mobility among young adults, due to different key life transitions. Leaving home to enrol in college/university, changing residence at the beginning of academic years or otherwise relocating after graduation to take up first professional jobs are typical characteristics of the lives of many young adults. At the same time, since students in Israel (as elsewhere) report lower than average incomes, the correlation between lower income and propensity to migrate is to be expected.

Our findings concerning the link between marital status and migration are somewhat at odds with the literature. While studies have often found that marriage reduces the likelihood of residential mobility, for ethnic minorities and majorities alike, our results show important variations along lines of both ethnicity and gender. Differences in marital status between migrants and non-migrants are particularly large and gendered for Arabs, where migrating women are much more likely than men (and Jewish women) to be married. This finding is possibly attributable to family relations and partnership formation among (female) migrants of non-Western origin. In some religiously observant ethnic groups, low levels of pre-marriage co-habitation have been reported, as well as lower propensity to leave the family home while obtaining a post-secondary education. This, as Finney (2011) notes, makes marriage the most viable home-leaving pathway for some minority women (e.g., South Asians in the UK).

A similar mechanism is possibly at play here, as marriage becomes an 'exit strategy' for Arab females – *but not males* – in Israel, enabling them to move away from their family homes in order to start living independently. As women in a patriarchal society, Arab women are still strongly discouraged from leaving the family home for any reason other than marriage (e.g., university enrolment or post-graduation). Such traditional conventions do not normally apply for men, who are at liberty to move prior to marriage (though expected to return once married). It should be noted that for a substantial proportion of Arab women, post-marriage migration is not simply voluntary, but rather part of patrilocalism, a cultural practice still prevalent among traditional segments of society. Expecting men to remain living in – or in very close proximity to – their parents' households after marriage, patrilocalism requires women to move with their new husbands away from their hometowns. Abu Tabih (2009), for example, reports that hundreds of recently married Arab women are uprooted from their families every year in order to join their husbands in what is often a new and unfamiliar environment. And while patrilocalism is becoming decreasingly important, especially among urban and educated families, we maintain that it remains an important factor explaining the sharp differences in migration rate between Arab men and women.

To complete our bivariate analysis, we ran four discrete choice models in which income, gender, education, age, and ethnicity were tested as possible variables explaining the decision to migrate (Table 6.4). The models were: 1) All Census population; 2) Migrants only; 3) Arabs only, migrants and non-migrants; and 4) Arab migrants only. In models 1 and 3 we compared migrants to non-migrants while models 2 and 4 compared recent (1990–1995) to earlier migrants. Models 1 and 2 test ethnicity as a migration variable in the general population (Jews=0, Arabs=1), whereas models 3 and 4 test the effect of ethnicity in the Arab population only (Christians=1, others =0). Estimates are shown alongside their marginal effects, which standardise observed effects by measuring changes in the probability of migrating as a result of change of one standard deviation unit.

The threshold propensity to migrate among the general population (model 1) is twice as high as for Arabs only (model 3).[10] This corroborates our earlier finding that Arabs in Israel are less mobile than Jews. Also, among the general migrant population (model 2), the propensity to migrate is little affected by timing (55 per cent for recent and 45 per cent for earlier migrants), whereas for Arabs, propensity to migrate is higher (84 per cent) among recent migrants when compared with earlier migrants, indicating increasing rates in the more recent past. This is not surprising given the politically motivated obstacles posed to Arab mobility by successive Israeli governments. Given, as discussed, that it is only since the 1970s – and, more forcefully, the 1980s – that migration has become more prevalent among Arabs, increasing propensity to migrate over time is expected.

---

10 These thresholds where calculated from the intercepts.

Income was also negatively correlated with migration, strengthening our earlier finding that individuals with high incomes may have little incentive to move. However, income is a relatively weak explanatory factor as its cross-model marginal effects are equal to or lower than 2.5 per cent. Similarly, although education has some effect on the decision to migrate, as more educated individuals are more likely to migrate compared with both non-migrants and earlier migrants, its marginal effect seems to be weak, reaching a high of 6 per cent in model 2 only. This suggests that a one-unit change in the standard deviation of the level of education yields only a minor increase in probability of migrating.

Age shows inconsistent effects; while models 1 and 3 show an increase in age leads to enhanced mobility (15.3 per cent and 5.8 per cent, respectively), models

Table 6.4   The impact of selected socio-economic characteristics on migration choices, 1995–1998

	Model 1 DV: Migrants vs. Non-Migrants		Model 2 DV: Recent vs. Earlier Migrants		Model 3 DV: Migrants vs. Non-Migrants		Model 4 DV: Recent vs. Earlier Migrants	
	Estimate	ME%	Estimate	ME%	Estimate	ME%	Estimate	ME%
Intercept	-1.350		.222		-2.198		1.655	
Income	-.072	-1.5%	-.060	-2.5%	-.098	-0.8%	-.094	-2.0%
Age	.050	15.3%	-.071	-3.1%	.023	5.8%	-.099	-7.6%
Education	.030	2.3%	.125	6.0%	.017	0.7%	.071	2.5%
Ethnicity	-2.129	-17.6%	-.240	-6.0%	1.033	13.8%	-.307	-4.6%
Gender (Male=1)	-.253	-3.8%	.108	2.7%	-.920	-5.8%	-.171	-2.4%
Properties:								
N	351,407		189,708		63,913		9,311	
n_1	189,708		45,583		9,311		2,028	
$X^2_{(5)}$	79,806		29,663		3,468		1,580	

Notes: DV (Dependent variable): Models 1 and 3: all migrants (=1), non-migrants (=0); Models 2 and 4: recent migrants (=1), earlier migrants (=0). ME% (Marginal Effects): The effect on the probability to migrate if the variable value varies from the mean by one unit of standard deviation. N is the number of individuals in the model. n_1 is the number of individuals in the tested group, chi-square is a measurement for goodness of fit, but since we use a large sample, this is always significant, or the null-hypothesis was rejected in all 4 models, and so is for the effect of each variable: The null hypothesis that the estimate equals zero was rejected for all variables in all models (p<0.05).

Ethnicity: Arabs =1 (models 1,2); Christians = 1 (models 3,4).

Source: Israeli Central Bureau of Statistics, 1995 National Census of Housing and Population, 20 per cent of the total population of age above 18 and below 80.

2 and 4 exhibit a decrease in the probability of migrating associated with age (-3.1 per cent and -7.6 per cent). This suggests that recent migrants are younger than earlier migrants, which is to be expected given that migration is increasingly becoming a viable option among younger age cohorts as part of rising mobility overall. Interestingly, the effect is stronger among Arabs (model 4), suggesting that recent migrating individuals tend to be younger than previous generations of Arab migrants. This may be explained by both the recent nature of Arab migration alluded to before, and the fact that it is gradually becoming a more popular strategy among young(er) individuals. Physical mobility, legally prohibited and socially unaccepted in older Arab generations, has recently become a common strategy for young men and women wishing to improve their social and economic standings.

Ethnicity, in contrast, had an important marginal effect on the decision to migrate. Not only did Arabs show a much lower propensity to migrate than Jews (-17.6 per cent), corroborating our findings above, but in-group variations re-confirm the results of our initial analysis concerning the enhanced mobility of Christian Arabs (13.8 per cent). These strong effects suggest that of all variables examined ethnicity remains key in explaining migration decisions propensities. It seems, however, that over time Christians are losing their 'mobility edge' over other Arabs, as is evident by their lower propensity to migrate among recent Arab migrants (-4.6 per cent). This recent closing of the mobility gap may be explained by the enhanced mobility among other Arab groups as a result of increasing rates of education and income in the last three decades. Alternatively, accelerated waves of Christian Arab emigration from Israel since the 1990s, and particularly during the second Intifada (2000–2004), have undoubtedly lessened their share in the general Arab population and, possibly, amongst Arab migrants.

Finally, gender also effects the decision to migrate; as shown, men in the general population were generally less migratory than women, though they have been showing tendencies towards increased mobility in the more recent past. Interestingly, among Arabs, higher migration rates among women do not seem to change over time (-2.4 per cent). In fact, entrenched patrilocalism on the one hand and young cohorts of women entering the (mostly skilled) job market in search of jobs, are likely to maintain the feminisation of Arab migration.

Discussion and Conclusions

This chapter set out to study the migration patterns of Arabs in Israel. We examined both the propensity to migrate as well as the specific socio-demographic profile of migrating individuals among Arabs' three major ethno-religious groups and compared them with the Jewish ethnic majority. Despite explicit differences between Arabs and ethnic groups in other countries that could be generally attributed to their unique social and political history as a homeland minority in an ethnic democratic state, our findings were generally in line with the literature.

We found that as a group, Arabs are migrating at a significantly lower rate than Jewish citizens of Israel. Sub-group variation notwithstanding, the low rate has been explained by drawing on a plethora of political, economic and cultural factors. In addition to the immobilising effects of the Military Government and discriminating planning and settlement practices which suppress economic development of Arab settlements, while simultaneously discouraging Arab migration to Jewish towns, the importance of ethnic social networks and the heavy financial costs associated with relocation have been quoted as possible explanations for the low migration rates among Arabs.

With notable exceptions, the dominant profile of Arab migrants in Israel resembles that of ethnic migrants in other economically developed countries. Although slightly older than ethnic migrants in other countries, Arab migrants are significantly younger than their co-ethnic non-migrants as well as Jewish migrants. Migration during one's late 20s and early 30s is usually attributed to major changes occurring along the life-cycle that require relocation, including college graduation, landing of a first job, marriage, and first home purchase. We presume that the same applies for Arabs during this period, although, with some important differences, which we discuss below.

Our findings also show links between (high) levels of education and migration. Greater mobility among the more educated is explained by their desire to maximise economic return on educational capital. Especially in less developed regions, educated individuals facing difficulties accessing suitable jobs (over-qualification) are more likely to relocate in quest for better opportunities. For educated Arabs, the majority of whom still reside in the Israeli periphery where fewer, less lucrative jobs are available, migration is a viable option. In a small country, even a short-distance move could prove beneficial for the highly educated as it allows them better access to a larger pool of jobs as well as improved pay. While the same is true for less educated Arabs, the value added from the move is undoubtedly smaller and therefore counteracts migratory tendencies. This explanation is also supported by Arab migrants' lower incomes, which suggests that they seek to capitalise on recently earned cultural capital by moving away from present communities in the rural periphery in search of opportunities elsewhere in Israel. Lesser incomes are explained by both the traditional lower wages offered for both skilled and unskilled jobs in the periphery, as well as migrants' early stage in their professional careers.

Finally, we found that across religious denominations Arab women migrate at higher rates than men. This higher level of mobility, in some cases more than double, is partly explained by the still high prevalence of patrilocalism, an entrenched cultural practice which strongly encourages Arab women to relocate upon marriage in order to live in close proximity to their new husband's extended family. An alternative explanation, which merits further investigation, is that increased female mobility is tied to the fast growing representation of women among Arab students in Israel. This may partly explain higher rates of migration among women coming from less traditional families, as they temporarily leave their home communities to enrol in urban universities and, subsequently, take up professional jobs. Taken together, these explanations shed some light on the radically changing position of women in contemporary Arab society and its impact on mobility. Whether following their husbands for family reasons or realising educational and professional goals, Arab women in Israel are clearly on the move.

The unique position of Arabs in Israel notwithstanding, our findings have implications for migration research beyond the local scene. Three are worth mentioned here in brief; first, our research shows the importance of looking beyond the overarching 'ethnic' category. Our refined consideration of ethnic sub-groups revealed important variations in migration dynamics, reflected in both rates and profiles, and subsequent studies should take these differences into account. Secondly, our research points to extremely gendered migration patterns among Arabs in Israel. Not only did we find that Arab women move around more than men, but their motives appear to be quite different at times. Although the study of gendered patterns is on the rise (see Faggian, McCann and Sheppard 2007), further analysis is needed to better understand their specific motivations and the extent to which they vary between ethnic and religious denominations. Moreover, as we pointed out briefly, within the general minority female population, class – and possibly level of religiosity – lie at the heart of cross-group observed differences. Future research should therefore pay greater attention to these critical intersectional (gender, class, and level of religiosity) dimensions. In this context, students of migration must strive for a broader socio-economic and political contextualisation of particular minority groups, and females in particular, in order to evade simplistic, indeed essentialising, theorisations, which explain their migratory routes solely in cultural terms.

Finally, while we briefly touched upon regional variations in migration behaviour, there is a need for further examination of these. It is particularly important to study the intersections between sub-groups and their mobility, both within and across various counties. Juxtaposing ethnicity and the scalar attributes of migration is important for understanding the geographic extent of ethno-migration and the extent to which it is affected by both physical (e.g., distance) as well as socio-political (e.g., tradition, majority sentiments) characteristics. There is an important need to go beyond the regional scale and analyse the type of settlements from and to which individuals move. While most migration is from the rural periphery, future research should characterise places of origin and destination to determine the extent to which other patterns become prevalent.

Chapter 7

'One Scotland'? Ethnic Minority Internal Migration in a Devolved State

David Manley and Gemma Catney

Introduction and Background

Like most other European countries, the UK has a long history of immigration from a diverse range of origins. In turn, these flows have followed a common course of family reunification and subsequent family building. Increasing diversity from these and new immigration streams is both celebrated and contested. Policy, public discourse, and media coverage of this widened cultural and social mix have become entwined with issues of inequality, integration and social cohesion (Kalra and Kapoor 2009, Catney, Finney and Twigg 2011). These debates have been accompanied by substantial academic concern with the internal mobility and spatial distribution of immigrants and their descendants, exploring the characteristics, magnitude and direction of internal flows (Champion 2005, Finney and Simpson 2008, Stillwell and Hussain 2010, Simon 2011).

Despite the growth in our knowledge of the migration of ethnic minorities in Britain, and especially in England, ethnic minority groups in Scotland have been largely ignored in contemporary research (Houston 2010). On the one hand, where research has reported on Scottish ethnic minority migration, Scotland tends to be treated as a regional entity within the rest of Britain (see, for instance, Champion 1996; Finney and Simpson 2008; Rees 2007; and Stillwell and Hussain 2008). These studies ignore the heterogeneity of the Scottish population. On the other hand, papers which have tackled the question of Scottish ethnic minority migration have done so using smaller case studies, for the most part adopting either a qualitative methodology, or restricting the analysis to a small-scale city-based quantitative approach. For instance, Bowes, Dar and Sim (1997), Pacione (2005a), Houston (2010), McGarrigle (2010) and Muñoz (2010, 2011) all investigated elements of ethnic minority populations in single Scottish cities. Thus, there has to date been no national-level study investigating the migratory patterns of minorities in Scotland.

In comparison with England, the proportion of individuals identified as belonging to an ethnic minority in Scotland is relatively small. According to the 2001 Census, 8 per cent of the population in England identified themselves as belonging to the 'non-White' population. The comparable figure for Scotland was 2 per cent. While Scotland is home to a much smaller proportion of ethnic minorities than England, these are numerically not inconsiderable with a population of over 100,000.[1] Most immigration in the UK has traditionally been to southern England, especially London, and the urban conurbations in the north of England. However, Census data show that immigration to Scotland, particularly the main urban areas of Glasgow and Edinburgh, increased between 1991 and 2001 (Dorling and Thomas 2004). As in Wales and the South West of England there are also minority groups associated with language. Perhaps the best known of these are the Gaelic speakers concentrated in Eilean Sair (in the Western Isles). Although they represent a minority group we do not focus on language minorities in this chapter.

There are many similarities in the histories of immigration to the rest of the United Kingdom and Scotland. In Glasgow, the majority of the non-White population is comprised of the Pakistani group many of whom came to the city to work in the textile and clothing trade. On the East coast of Scotland, many emigrants entered the city of Dundee during the 1950s and 1960s, as a response to the associated labour requirements of the Jute trade (Jones and Davenport 1972). However, there are also some notable differences: there is a distinct Protestant-Catholic divide, especially in Western Scotland (notably in Glasgow), resulting predominantly from the Irish immigrations between the 1798 Rebellion and the 1845 famine (Pacione 2005b). Many of the immigrants entered what is now known as the Central belt, including the City of Glasgow, Clydeside and also Perthshire, pulled by the attraction of employment opportunities in the new industries associated with railway locomotive manufacture, ship building, engineering and mining (Lawton 1959). The influx of Irish into Scotland, especially Glasgow, created large religious communities which can still be found in some parts of contemporary Scotland. Later waves of immigration into Glasgow, as well as other Scottish cities, have been fuelled by demand from the service and transport industries (see Muñoz 2011). More recently, all major cities of Scotland have experienced growth in migration from East Asia, fuelled by the increase in international moves by students into British Universities (for evidence for Dundee, see Houston 2010).

It is clear therefore, that Scotland is not a homogenous area and it is argued here that it should not be treated merely as a region within Great Britain (comparable to, for example, the North East or South West of England). There are some major differences between the rest of the United Kingdom and Scotland. Politically, Scotland has had a degree of devolution since the Act of Union in 1707 and has maintained education and legal systems that are separate to the rest of the United Kingdom. Until 1998 the Westminster Government recognised Scotland's special status with the UK through provision of a Scottish Executive and Secretary of State of Scotland. After the formation of the UK Labour Government in 1997, there was an increase in devolved powers, with the Scottish Executive assuming legislative powers on 1st July 1999, after the first elections to the Scottish parliament of the same year. After the 2007 Scottish Parliament elections when

---

1   The exact figure is 101,747, and can be derived from Table 7.1.

the Scottish National Party formed an administration the name was symbolically changed from the Scottish Executive to the Scottish Parliament, although in practice there were no changes in status relative to Westminster. Whilst many issues around foreign and high-level economic policies remain the preserve of the central Westminster Government, the Scottish Government (which superseded the Scottish Executive) has been able to chart a distinctly different course in relation to domestic policies for ethnic minority integration. Indeed, the ethnic minority experience in Scotland contrasts in many ways with that of ethnic minorities in the other parts of the UK, and especially compared with England. Despite some racially-motivated crime in Scotland (Kelly 2000) there is little history of ethnic related riots (as experienced in Northern English towns in the 1990s, or towns in Southern England in the 1980s). A recent poll of public attitudes to immigration found that respondents from Scotland were less in favour of reduced immigration to the UK than their English and Welsh counterparts (The Migration Observatory 2010). The Scottish Government has been proactive in the promotion of strategies to create an 'inclusive' and 'multicultural' society, and to minimise the potential for racially-orientated tensions. Under the banner 'One Scotland', the Scottish Government has launched successive attempts to raise awareness of issues of diversity and ethnic equality within the population through large-scale initiatives aimed at addressing racism and racist attitudes, both within the residential and employment spheres. The 'One Scotland' campaign became part of 'Scotland Against Racism' during 2008 and the Scottish Government has aimed to promote ethnic minority inclusion across the country for all ethnic groups, including those arriving more recently from the A8 countries (A8 accession countries of the Czech Republic, Estonia, Hungary, Latvia, Lithuania, Poland, Slovakia and Slovenia). Given this specific context, the need for work on internal migration of minorities with a Scottish focus is clear.

This chapter engages with the existing Britain-wide literature on minority internal migration to explore two key questions which to date remain unanswered for the Scottish case. Previous research for Britain as a whole has suggested that ethnicity is not, in itself, a determinant of migration propensities, but rather the 'normal' characteristics of migrants cited repeatedly in the literature (young, more affluent, better educated, renters, for example), are determinants of migration for the *whole* population (Finney and Simpson 2008). Thus, the socio-economic and demographic profile of an individual, rather than their ethnic group, can explain migration propensities for Britain (ibid.). However, do these findings still hold if we examine the constituent parts of Britain? We propose that given differences in immigration histories, political and policy contexts, and non-White group population size, it is insufficient to explore Britain as a whole. Examining the characteristics of migrants in Scotland alone, we ask, does ethnicity matter in explaining mobility propensities in Scotland? Understanding better the Scottish case will, in turn, contribute to our knowledge of how far patterns and processes observed for a set of regions may apply when we consider these regions individually.

Our second question aims to contribute to the existing literature on the geography of migration destinations, to explore how the relative popularity of destinations within Scotland varies. Despite accusations that Britain has experienced an increase in ethnic segregation, with the evolution of so-called 'parallel lives' (Cantle 2001), it has been shown that deconcentration from minority clusters is commonplace; dispersal, in the form of counterurbanisation, has been demonstrated regardless of ethnic group (Simpson and Finney 2009), for those with the economic means for this form of spatial mobility (Catney and Simpson 2010). Migration from non-White concentrations (Simon 2011), along with shared housing aspirations and migration motivations (Phillips 2006b), have been features of ethnic minority internal mobility in recent years. Therefore, while the *characteristics* of those who make a residential move have been shown to be common for ethnic groups, so too have the *spatial patterns* of migration. The broad national level trend of population decentralisation from large cities, and associated counterurbanisation, is also known to be occurring in Scotland (Stockdale, Findlay and Short 2000). These processes are similar regardless of ethnic group in England and Wales, with some exceptions worth noting. Simpson and Finney (2009) found common patterns of counterurbanisation for all ethnic groups in Britain apart from the Chinese, who were moving instead towards concentrations of their own group in urban areas. In addition to these group-specific experiences, certain place-specific experiences (with an ethnic dimension) have been noted by some, in particular the distinct migration system of London (Catney and Simpson 2010, Stillwell and Hussain 2010). Little is known about the nature of these processes and experiences in Scotland (an exception is the work of Muñoz (2011) for Indian populations in Glasgow and Dundee), an omission given that place-specific experiences have been evidenced elsewhere. The spatial patterns of ethnic minority internal mobility in Scotland may well differ from the rest of the UK given its more recent history of immigration. As spatial assimilation theory suggests, increased time since immigration is likely to be coupled with residential patterns more akin to the 'native' population (as discussed in the US context by South, Crowder and Chavez 2005). Where the minority population may be more recently established, the traditional settlement areas (namely, urban centres with superior housing and labour opportunities, and established support networks) may be more attractive. Here, we ask if there are differences by ethnic group in migration to large cities in Scotland and more peripheral (suburban) locales, testing the likelihood of migration to the various regions of Scotland.

To provide context for this analysis, it is necessary to consider the demographic and socio-economic composition of ethnic groups within Scotland. The next section of the chapter details the data and methods used to explore the composition of ethnic groups in Scotland, ethnic group change over time, and the mobility propensities of each group. Following this, the results are split into three major sections, dealing with (1) the proportions of ethnic groups in Scotland, 1991–2001; (2) the socio-economic and demographic characteristics of ethnic groups in Scotland, and (3) the internal mobility of ethnic groups, considering both migration propensities and variations in destination choice.

Data and Methods

Ideally, we would answer questions about migration using longitudinal data at the individual-level, which would allow the same individuals to be analysed at multiple time points. However, whilst there are a number of potentially useful datasets covering the Scottish population, none are ideal for the proposed analysis. For instance, the British Household Panel Survey (BHPS, now superseded by Understanding Society) and the Labour Force Survey (LFS) contain sufficient information to model ethnic minority groups and mobility outcomes across the whole of the United Kingdom. However, when focusing on Scottish ethnic minorities alone the numbers are too small in both these datasets for reliable statistical modeling.[2] In the UK, the largest data source (in terms of both population and geographic coverage) is the decennial population Census. There are a number of outputs from the Census that could be utilised. An individual and longitudinal Census-based resource is the Scottish Longitudinal Study (SLS), which contains a 5.3 per cent sample of the Scottish population, i.e., around 270,000 individuals. However, this also suffers from small number problems when ethnicity and mobility are combined in the outcomes of a set of models. Given our specific interest in both detailed ethnic group breakdowns and geography, the SLS does not present a suitable alternative to standard cross-tabulated (area-level) data from the Census. The analysis that we present is implemented using both individual and ecological cross-sectional data from the 1991 and 2001 Censuses of Population, and whilst we acknowledge that data drawn from 2001 are unlikely to represent accurately the current migration situation in Scotland (in particular given the increased flows from the new European Union A8 countries), they do, nevertheless, provide a comprehensive data source from which to start to understand internal ethnic minority migration in Scotland.

The first substantive section of the chapter deals with the proportions of ethnic minority groups in Scotland. Counts of ethnic group were taken from UK Census Tables SAS06 and KS006 for 1991 and 2001 respectively. We restrict our analysis to the 1991 and 2001 Censuses as only District level estimates of ethnic group size are available for 1981 (Rees and Phillips 1996); there are no comparable lower level ethnic group data from the 1981 Census, when only country of birth was asked. In 1991 and 2001, however, a question on ethnic group was included, which consisted of a tick box option with pre-determined ethnic group categories, or, alternatively, a write-in option for the respondent if it was felt that the categories did not represent their ethnic group. Between 1991 and 2001 the question relating

---

2   This is despite attempts within the BHPS to use booster samples of specific groups within the population, including ethnic minorities, to overcome small number problems.

to ethnic group changed, resulting in inconsistent classifications between the two Censuses. To overcome this problem we employ an ethnic group classification based on Simpson and Akinwale (2007) which allows comparison over time; the resultant ethnic groups from this classification are White, Indian, Pakistani, Bangladeshi, Caribbean, African and Chinese. In addition, we use an eighth residual category, Other. It is important to note that the 2001 Scottish Census had a different ethnic group question to that of the 2001 England and Wales Census, whereby the decomposed mixed categories (White and Black-Caribbean, White and Black-African, White and Asian and Other Mixed) were omitted and replaced with the categories 'Black Scottish and any other Black' along with 'Any Mixed Background'. In addition, for brevity in the first analysis we concentrate on 'visible' ethnic minority groups. As such, the potentially heterogeneous 'White' group which contains the Scots, English, Irish and all other individuals identified as White in one category is left unexplored. In the second analysis below we acknowledge the heterogeneous nature of this category and use a secondary set of indicators based on country of birth to distinguish between the various groups.

The second analysis section explores the composition of ethnic groups in Scotland by providing summary statistics of ethnic minorities in Scotland across a range of socio-economic and demographic variables. This section concentrates only on the most recent available data (2001 Census), making use of theme Table CT007. The table provides only country-level information, with a more limited ethnic group breakdown than that which is available for standard counts of the population by ethnic group: namely only White, Indian, Pakistani and Other South Asian, Chinese, and Other.

Individual-level cross-sectional data from the 2001 Population Census are available through the Sample of Anonymised Records (SARS) and the Small Area Microdata (SAMs). The SARs have more detailed ethnic group breakdowns but relatively little geographical information, while SAMs have better low-level geocoding although there is a loss in information about ethnic groups. As we are explicitly interested in mobility as well as ethnic grouping we chose the SAMs for the individual-level modelling of migration propensities component of the analysis as it offers the best compromise between ethnic group detail and geographical resolution. The SAMs provides a 5 per cent sample of the Scottish population, including information on ethnic group, along with some geographic coding for migration and other individual characteristics. We use five ethnic groups in the modelling: White (White Scottish, Other White British, White Irish, White Other), Indian (Indian), Other Asians (Pakistani and Other South Asian), African and Caribbean (Caribbean, African, Black Scottish, Other Black, any Mixed or Other), and Chinese (Chinese). A migration event is indentified in the SAMs if the respondent's address one year prior to Census day was different from their address at Census enumeration (completed via a change of address question on the Census form).

We adopt a twin strategy to model the mobility behaviour of ethnic minorities: firstly, to overcome the problem that the SAMs does not have low-level neighbourhood geocoding we model the likelihood that individuals will migrate over either short- or long-distances. We use long- and short-distances as a proxy for different types of mobility, an approach common in the migration literature (Boyle 1995, Boyle et al. 2001, Nivalainen 2004). Short moves are likely to be associated with residential motivations; for example, to improve individual circumstances through accessing better educational opportunities for children, or changing housing conditions by moving from an inner city area to the suburbs. Long-distance moves tend to be labour market-orientated and associated with gaining employment or career advancement (Boyle et al. 2001, Nivalainen 2004). Previous studies of internal migration in Britain have used 30 or 50 kilometres (km) as a cut-off point between long and short moves (see Boyle et al. 2003 and van Ham, Mulder and Hooimeijer 2001). However, SAMs data only provide discrete categories for distance moved: 0–4km, 5–19km, 20–99km and 100km or over. Clearly, a move of 99km cannot be considered short-distance. As commuting distances are relatively short across Scotland, especially in comparison with England, we adopt 19km or less as a cut off point for short-distance moves. All moves of 20km or greater are treated as long-distance moves.

A second stage of modelling further explores migration in Scotland to identify the locational choices of migrants; rather than stratifying by short- and long-distance moves to explore variation in migration propensities, this next stage of modelling examines differences in migration to (and within) selected regions of Scotland. All individuals in these models made a move in the year prior to Census enumeration; non-migrants are excluded, as are those who migrated but left Scotland. Possible destinations for migrants are: the major cities of Edinburgh and Glasgow, selected given the dominance of minority groups in these cities; Aberdeen City and Dundee City, included to allow moves to the smaller Scottish Cities to be evident; the Central Belt of Scotland, a region within commutable distance to Edinburgh and/or Glasgow and home to suburban 'sprawl' of the two cities; and the Rest of Scotland. The final category of 'Rest of Scotland' serves as an 'other' category which subsumes any remaining within-Scotland moves, but which does not suffer from small numbers; this category is the reference category for these geographically-based models. It is not possible to include an origin control variable in the modelling or to produce an origin and destination matrix for the movers, given that the origin of migration is only available indirectly (through either 'staying within the Local Authority' or moving 'between Local Authorities'). Whilst an origin control would be desirable, small numbers of ethnic minorities in the more rural parts of Scotland force a compromise.

Selected socio-economic and demographic characteristics are included in latter versions of the models. It should be noted that the characteristics we account for are partly dictated by what information is gathered and made available from the Census. We make use of the standard proxies for socio-economic status and demographic and household contexts applied commonly in other UK-based studies. We follow standard classifications for employment types, tenure and household characteristics. However, as the UK Census did not directly ask questions relating to income, we make use of the common proxy for income, the National Statistics Socio-economic Classifications (NS-SeC). To avoid problems from small numbers we collapse the standard nine category measure into four groups: Managerial and Professional (Large employers and higher managerial occupations and Higher professional occupations), Intermediate and Self-employed (Lower managerial and Professional occupations, Intermediate occupations, and Small employers and Own account workers), Intermediate and Lower (Lower supervisory and technical occupations, Semi-routine, and Routine occupations) and Never worked and Long-term unemployed.

Proportions of Ethnic Groups in Scotland, 1991–2001

We start by describing change in ethnic minority populations in Scotland between the 1991 and 2001 Censuses. The first four columns of Table 7.1 show the absolute change (2001 counts minus 1991 counts) and proportional change (2001 percentage minus 1991 percentage) and ratios of absolute change (2001 counts over 1991 counts) and proportional change (2001 percentages over 1991 percentages) for each ethnic group. All minority ethnic groups have increased their proportional share, although it should be noted that, compared to the White group, minority groups remained relatively small in Scotland. While the proportions of non-White populations are small at both time points, examining the ratios of change does provide some insight into the relative growth of non-White populations in Scotland. Despite a slight proportional loss, in absolute terms the White group gained population, like all other groups. The largest proportional gains in the overall share of the population were by the Other group, followed by Pakistanis. (Note, however, that the subgroups which compose the Other group changed between 1991 and 2001, which could make assessment of changes in this group problematic; Simpson and Akinwale (2007) describe the impact of changes to the Other category in England and Wales). The Pakistani group was the largest non-White ethnic group at both time points (not shown in Table 7.1). The ratio of percentage change shows the Caribbean, African and Other groups to be large gainers relative to the other ethnic groups, with each of these groups nearly doubling their populations between 1991 and 2001. The final two columns of Table 7.1 contrast the ethnic minority composition of Scotland with Great Britain as a whole. The largest ethnic group in Scotland is the Pakistani population, twice the size of the next two largest groups, the Chinese and Indian. In contrast with all of Great Britain, the African and Caribbean groups only make up 0.10 per cent and 0.04 per cent of the total population respectively.

Table 7.1   Comparing the ethnic minority composition of Scotland with Great Britain and absolute and proportional change in ethnic group populations in Scotland only, 1991–2001

Ethnic Group	Absolute change	Ratio absolute change	Proportional change 1991 to 2001	Ratio % change	Scottish (% 2001)	Great Britain (% 2001)
White	23,868	1.00	-0.75	0.99	97.99	92.13
Pakistani	10,557	1.50	0.20	1.48	0.63	1.34
Chinese	5,817	1.55	0.11	1.54	0.32	0.48
Indian	5,003	1.50	0.10	1.48	0.30	1.82
African	2,327	1.83	0.05	1.81	0.10	1.20
Caribbean	820	1.86	0.02	1.83	0.04	0.89
Bangladeshi	858	1.76	0.02	1.74	0.04	0.51
Other	13,614	1.85	0.26	1.83	0.49	1.63
Total Population	-	-	-	-	**5,062,011**	**58,789,174**

Notes: Categories adapted to follow ethnic group classifications used below, following Simpson and Akinwale (2007).

Source: 1991 and 2001 Censuses, Tables SAS06 and KS006. Authors' own calculations.

Characteristics of Minorities in Scotland

This section explores the composition of all major ethnic groups in Scotland to assess similarities and differences between groups across multiple aspects of life. The descriptive information discussed here sets the context for the main analysis sections, which deal with models of migration propensities and spatial mobility patterns, by ethnic group. Table 7.2 shows the proportion of each ethnic group (with more limited ethnic group categories, as discussed in the Data and methods section) by a selection of socio-economic and demographic indicators.

Table 7.2 shows that, of the ethnic minority groups, the Pakistani and Other South Asian group have the largest proportion of their population born in Scotland (nearly 45 per cent). In contrast, the Indian group has the largest percentage born in the rest of the UK of any group, which we suggest reflects migration to Scotland from England. Those in the Chinese ethnic group have the largest proportion born outside the UK.

Table 7.2   Socio-economic and demographic composition of ethnic groups in Scotland in 2001

Ethnic Group		All	White	Indian	Other Asians	Chinese	Other
	Total	5,062,011	4,960,334	15,037	39,970	16,310	30,360
Country of birth	Scotland	87.13	88.14	33.76	44.18	29.75	35.42
	Rest of UK	9.09	9.06	14.92	10.60	4.38	11.36
	Outside UK	3.78	2.80	51.32	45.22	65.87	53.21
Economic activity	Employed	60.61	60.82	57.58	44.57	51.29	50.05
	Unemployed	4.36	4.33	4.88	6.28	4.77	7.18
	Economically inactive	35.03	34.85	37.54	49.15	43.94	42.77
NS-SeC	Higher and Professional	32.76	32.82	41.52	19.63	24.96	39.03
	Intermediate and Self-employed	20.51	20.46	23.71	26.21	26.61	16.79
	Lower and Routine	41.07	41.32	20.81	25.25	36.42	29.29
	Never worked and Long-term unemployed	5.67	5.40	13.95	28.91	12.01	14.89
Qualifications	None	33.23	33.26	24.13	39.58	37.93	22.99
	Level 1	24.69	24.85	14.84	18.59	13.03	15.22
	Level 2	15.65	15.66	14.80	14.07	15.99	16.94
	Level 3	6.95	6.95	5.95	7.73	5.91	7.49
	Level 4	19.47	19.28	40.28	20.03	27.15	37.35

Notes: (i) Economic activity rates are calculated as a proportion of those aged 16–74. (ii) NS-SeC rates are calculated as a proportion of the total population, excluding those in the 'Not classified' category, who are individuals who have not been coded and cannot be allocated to an NS-SeC category. Not classified also includes all full-time students. (iii) Qualifications rates are calculated only for the population aged 16–74. None: those people without formal qualifications. Level 1: 'O' Grade, Standard Grade, Intermediate 1, Intermediate 2, City and Guilds Craft, SVQ level 1 or 2 or equivalent. Level 2: Higher Grade, CSYS, ONC, OND, City and Guilds Advanced Craft, RSA, Advanced Diploma, SVQ level 3 or equivalent. Level 3: HND, HNC, RSA Higher Diploma, SVQ level 4 or 5 or equivalent. Level 4: First degree, Higher degree, Professional Qualification.

Source: 2001 Censuses, Table CT007. Authors' own calculations.

Economic activity, in addition to the income proxy of the National Statistics Socio-economic Classifications (NS-SeC) variable, gives some indication of the relative socio-economic status of the ethnic minority groups. In terms of the economic activity indicator, we can see that the highest unemployment rates are for Pakistani and Other South Asians and Others, at 6.3 per cent and 7.2 per cent respectively. There are comparable employment rates between the White and Indian groups. Considering NS-SeC, the Indian group has the largest proportion of any group, including White, in Higher and Professional occupations. The White group has the largest proportion in Lower and Routine forms of employment. Pakistanis and Other South Asians have the smallest proportion in the Higher and Professional forms of employment.

The final characteristic important to our understanding of the composition of ethnic minority groups in Scotland relates to the levels of educational qualifications achieved by individuals. Reflecting the grouping of the education variable in the SAMs, we have five categories of education, ranging from individuals without formal qualifications to individuals who have gained at least one higher education qualification. The descriptive information for qualifications shows that the Indian population has the largest proportion with the highest level of qualifications (with over 40 per cent of the population possessing a first degree or higher). This is considerably greater than the just under 20 per cent observed for the full population, whilst the diverse group of Pakistani and Other South Asian are the least well-qualified, with nearly 40 per cent having no qualifications as recognised in this classification. As would be expected, the educational distribution is very similar to the NS-SeC distribution, suggesting that occupational achievement is related to educational achievement at the aggregate level in Scotland and is not dependent on ethnic background (a finding consistent with the individual longitudinal analysis of van Ham et al. 2010).

The Internal Mobility of Ethnic Groups in Scotland

Migration Propensities

To begin the analysis of mobility patterns in Scotland we look at the proportion of migrants for all ethnic groups identified above. We divide the analysis into short- (19km or less) and long- (20km or more) distance moves given that, as noted above, the motivations behind these actions are very different (van Ham 2002). Table 7.3 shows the percentages of distance moved by ethnic group.

Table 7.3   Distance moved within Scotland, 2000–2001, by ethnic group (per cent)

Distance	White	Indian	Other Asians	Caribbean	Chinese	Total
No move	89.67	87.46	87.62	80.71	86.51	225,205
Short move	7.69	7.52	9.41	12.06	9.41	19,433
Long move	2.65	5.01	2.96	7.23	4.07	6,752
Total	246,593	678	1,923	1,410	786	251,390
Ratio: long/short	0.34	0.67	0.31	0.60	0.43	0.35

Notes: Rates exclude moves from outside the UK.

Source: 2001 Small Area Microdata (SAMs). Authors' own calculations.

As would be expected, the majority of the population do not move. Of those who do move, the majority move a short distance, although there are clear differences in the propensity to make long- and short-distance moves. All ethnic minority groups are more likely to make moves than the White majority. The Caribbean group has the greatest propensity to move both short- and long-distances. Ratios of the percentage of long- to short-distance moves allow us to compare the relative importance of move type between ethnic groups. The Indian group's largest ratio of long- to short-moves shows that long-distance moves are proportionately greater for Indians than for any other group; the reverse is true for the Pakistani and Other Asian group. The final section of the analysis investigates if ethnic group helps to explain differences in the propensity to move or if such variation can be accounted for by introducing other explanatory factors. As noted in the Data and methods section, the ethnic group categories available for modelling are the White majority (the reference ethnic group in all models), the Indian, the Other Asians, the African and Caribbean, and the Chinese groups.

Table 7.4 shows the odds ratios from a logistic regression predicting the likelihood that an individual will make a short-distance move (compared with not making a move at all). Model 1 controls for ethnic group alone and confirms that all minority groups are more likely to make a short-distance move than the White population, with the African and Caribbean group being the most likely. Numerous studies have demonstrated variation in the likelihood of migrating by socio-economic and demographic characteristics (for a review see Boyle, Halfacree and Robinson 1998), while Finney and Simpson (2008) have demonstrated that internal migrants in Britain share these 'typical' characteristics, regardless of ethnic group. Thus, in Model 2 we add in individual-level controls, including the socio-economic position of an individual using a collapsed version of NS-SeC. The first control in Table 7.4 is country of birth, used as a proxy to distinguish between first generation immigrants and those who have been born in the UK. Only individuals born Outside the UK or in Ireland are more likely than the Scots to make short-distance moves. The most mobile NS-SeC category is the Professional and Managerial group (the reference category), whilst the least mobile category represents those who have never worked or are long-term unemployed. For individuals making short-distance moves, the motivation is likely to relate to the opportunity to improve one's residential circumstances, and we expect that those working in occupations classified in the more professional groups are more likely to have the means with which to make such moves. Conversely, those in the lower classified occupations are less likely to have the means to move, and therefore exhibit lower levels of short-distance mobility. Controls for the main individual characteristics that are associated with migration, including level of educational attainment, age, household composition and housing tenure, are also included. In line with previous research, those individuals with higher levels of education are more likely to migrate than individuals without educational qualifications. It is notable that as education level increases, so does mobility. Unsurprisingly, and as demonstrated in other work, an increase in age is associated with a lower likelihood of migrating, those in employment have lower mobility than the unemployed and economically inactive (a group which includes retirees), while those classified in groups with child(ren) are less mobile than the individuals in households without children (for a general overview of these commonly observed trends see Boyle, Halfacree and Robinson 1998). Finally, home owners are less mobile than private renters, who are in turn more mobile than social renters (tenants living in public housing).

The most important difference between the two models is the change in significance levels associated with the five ethnic groups. Once all contextual effects are included, ethnic group ceases to be a significant factor in predicting short-distance migration. This suggests that ethnic group is not a predictor of differentials in migration propensities, but, rather, that the differences observed in Table 7.3 are the result of structural differences in the population, explained by the socio-economic and demographic characteristics of movers and non-movers.

Table 7.4 **Logistic regression models of not move versus short-distance (under 20km) moves in Scotland, 2000–01**

	Model 1	Model 2
	Odds Ratios (exp(β))	Odds Ratios (exp(β))
Ethnic Group		
White	*1.00*	*1.00*
Indian	1.11	0.79
Other Asian	1.34***	1.12
African and Caribbean	2.03***	0.90
Chinese	1.36**	0.83
Country of Birth		
Scotland		*1.00*
England		1.04
Wales		1.12
Northern Ireland		1.27**
Outside UK		1.10**
NS-SeC		
Professional and Managerial		*1.00*
Intermediate and Self-employed		0.98
Routine		0.96
Never worked and Long-term unemployed		0.79***
Age		
16–25		*1.00*
26–35		0.67***
36–45		0.30***
46–55		0.14***
56–65		0.10***
66 and over		0.09***
Qualifications		
None		*1.00*
Level 1		1.15***
Level 2		1.16***
Level 3		1.27***
Level 4		1.33***

Table 7.4   Concluded

Economic Activity		
Employed	*1.00*	
Unemployed	1.07	
Inactive	1.25***	
Household composition		
Single	*1.00*	
Single with child(ren)	0.55***	
Couple	1.09**	
Couple with child(ren)	0.45***	
Tenure		
Owner	*1.00*	
Social Renter	1.49***	
Private Renter	2.97***	
Log Likelihood	-67834.71	-34425.48
Pseudo R^2	0.00	0.10
Number of Obs.	132135	132135

Notes: Significance levels reported where *=$p<0.05$; **=$p<0.01$; ***=$p<0.001$. Qualifications rates are calculated only for the population aged 16–74. None: those people without formal qualifications. Level 1: 'O' Grade, Standard Grade, Intermediate 1, Intermediate 2, City and Guilds Craft, SVQ level 1 or 2 or equivalent. Level 2: Higher Grade, CSYS, ONC, OND, City and Guilds Advanced Craft, RSA, Advanced Diploma, SVQ level 3 or equivalent. Level 3: HND, HNC, RSA Higher Diploma, SVQ level 4 or 5 or equivalent. Level 4: First degree, Higher degree, Professional Qualification.

Source: 2001 Small Area Microdata (SAMs). Authors' own calculations.

Table 7.5 shows odds ratios for long-distance moves; here, as for short-distance moves, the likelihood of making a long-distance move is compared with not migrating at all. The pseudo R^2 value is much greater than in the previous model, suggesting that the processes leading to a long distance move is more accurate captured in the modelling. Model 3 controls for the ethnic group, and again all minority ethnic groups are more likely to move over long-distances than the White reference category. Model 4 adds the control characteristics. Individuals born outside Scotland are significantly more likely than the Scottish-born to make long-distance moves. Of all the groups, those born outside the UK are the most likely to make long-distance moves. This model suggests that first generation migrants (those born outside the UK) are more likely to make long-distance moves in Scotland; a period of initial high mobility might be expected as new arrivals 'find their feet' in terms of their housing, location and employment options. The control for socio-economic position, NS-SeC, demonstrates that the reference group, Professional and Managerial, is the most likely to make long-distance moves (as with short-distance moves, shown earlier). Educational achievement clearly demonstrates that individuals with higher levels of attainment are more mobile, with all groups being significantly more likely to make a long-distance move than individuals without qualifications. The odds ratios for the age bands, household composition and economic activity follow a similar pattern as for short-distance moves. The odds ratios for tenure show that private renters and owner occupiers are more mobile than social renters, an observation to be expected given that there tends to be limited mobility across social housing administrative boundaries (which is likely in the case of moves over larger distances) (Boyle 1995).

Table 7.5   Logistic regression models of not move versus long-distance (20km or greater) moves in Scotland, 2000–01

	Model 3	Model 4
	Odds Ratios (exp(β))	**Odds Ratios (exp(β))**
Ethnic Group (reference = White)		
White	*1.00*	*1.00*
Indian	4.86***	1.29
Other Asian	2.02***	0.75
African and Caribbean	5.74***	1.25
Chinese	3.31***	0.71
Country of Birth		
Scotland		*1.00*
England		2.56***
Wales		2.31***
Northern Ireland		2.40***
Outside UK		3.51***
NS-SeC		
Professional and Managerial		*1.00*
Intermediate and Self-employed		0.69***
Routine		0.58***
Never worked and Long-term unemployed		0.36***
Age		
16–25		*1.00*

Table 7.5   Concluded

26–35	0.49***	
36–45	0.25***	
46–55	0.15***	
56–65	0.10***	
66 and over	0.11***	
Qualifications		
None	*1.00*	
Level 1	1.26***	
Level 2	1.67***	
Level 3	1.79***	
Level 4	2.55***	
Economic Activity		
Employed	*1.00*	
Unemployed	2.11***	
Inactive	2.01***	
Household composition		
Single	*1.00*	
Single with child(ren)	0.45***	
Couple	1.18***	
Couple with child(ren)	0.53***	
Tenure		
Owner	*1.00*	
Social Renter	0.956	
Private Renter	5.60***	
Log Likelihood	-35135.52	-15561.33
Pseudo R^2	0.01	0.20
Number of Obs.	125520	125520

Notes: Significance levels reported where *=p<0.05; **=p<0.01; ***=p<0.001. Qualifications rates are calculated only for the population aged 16–74. None: those people without formal qualifications. Level 1: 'O' Grade, Standard Grade, Intermediate 1, Intermediate 2, City and Guilds Craft, SVQ level 1 or 2 or equivalent. Level 2: Higher Grade, CSYS, ONC, OND, City and Guilds Advanced Craft, RSA, Advanced Diploma, SVQ level 3 or equivalent. Level 3: HND, HNC, RSA Higher Diploma, SVQ level 4 or 5 or equivalent. Level 4: First degree, Higher degree, Professional Qualification.

Source: 2001 Small Area Microdata (SAMs). Authors' own calculations.

Ethnic group is no longer a significant factor predicting long-distance internal migration in Scotland after including socio-economic classifications and other household and individual characteristics in the model. After the addition of controls, the odds ratios are no longer significant for all groups except the Other Asians, who are now less likely than the White population to move over long-distances (note, though, that the significance level for this group has also declined).

The Geography of Migration Destinations

The second stage of modelling uses multinomial regression to explore differences in the likelihood of migrating to (and within) various parts of Scotland (namely, Edinburgh, Glasgow, Dundee, Aberdeen, the Central Belt and the 'Rest of Scotland'), to examine the relative 'pull' of different regions, and how this may vary by ethnic group. We would expect that the major employment centres of Edinburgh and Glasgow would be the most likely destinations for all minority groups, as with the White population. While not shown here given space constraints, it should be noted that the largest concentration of ethnic minorities in Scotland are in Glasgow, Edinburgh and the Central belt region, and the minority population is less widely dispersed in Scotland than in England; minority groups are found mainly in the larger urban areas.

Table 7.6 reports a complete model showing the likelihood of moving into/within one of the four major urban areas in Scotland or moving to the suburban Central Belt, against moving to what we have termed the 'Rest of Scotland'. The pseudo R^2 value is relatively low for this model, although this is not unexpected given the complex processes that are being modelled. Differently to the previous set of models, all individuals in the model have made a move. The relative risk ratio (which for interpretation purposes can be regarded as effectively the same as an odds ratio but in a multinomial setting) for ethnic group, country of birth, distance moved (either as moves within Council Areas (relatively local moves), moves between Council Areas either within Scotland or from England and Wales (medium to long-distance moves), or moves from outside the Great Britain (long-distance moves)) and NS-SeC are included in the table, but in the interests of space the results for the added controls are not; these are available from the authors on request. The controls that were included match those used in the previous models and have similar effects on the outcomes.

Unlike the models for distance of move (reported in the previous section), when geographical region is modelled ethnic group remains a significant predictor of destination, even when the other controls are included. This is the case for all ethnic groups for the destinations of Edinburgh and Glasgow, and suggests a 'Glasgow and Edinburgh effect' on the migration patterns in Scotland. As the major population centres of the country this is perhaps to be expected. Only the Indian ethnic group are significantly more likely to move to (or within) Aberdeen than the White population. The Indian and the Other South Asian groups are more likely than the reference White population to choose Dundee as a destination (as are the African and Caribbean and Chinese, although the results are not significant for these two groups). For the Central Belt, however, no minority groups (except the Indian group, although the relative risk ratios are not significant here) are more likely than the White group to select the Central Belt region. It should be remembered that given data limitations on the origins of internal migrants, these results will partly reflect the original spatial distribution of ethnic minorities (that is, the tendency for minority groups to reside in urban as opposed to non-urban locales, and the fact that many moves are short distance and within a given urban area), rather than their propensity to move into and out of the major urban centres. Nevertheless, large relative risk ratios for the ethnic minorities are still informative in that they demonstrate how urban areas are retaining minority populations.

The country of birth relative risk ratios highlight a stark difference for mobility by place of birth. The Northern Irish group are consistently more likely than the English to move into all the destination categories, and are the most likely to move into Edinburgh, Dundee and the Central Belt. In contrast, individuals born in Wales are only significantly different to the English born with regard to moves around the Central Belt. The relative risk ratios for Glasgow and Aberdeen (although not significant for this second region) suggest that these urban destinations are attractive to international migrants, with individuals born outside the UK more likely to move into the urban areas than individuals born in England.

Table 7.6   Ethnic minority migration and location in Scotland (2000–01), including all individual controls (not shown)

	Edinburgh RRR.	Glasgow RRR.	Aberdeen RRR.	Dundee RRR.	Central Belt RRR.
Ethnic Group					
White	*1.00*	*1.00*	*1.00*	*1.00*	*1.00*
Indian	6.30***	5.34**	4.85**	13.20***	2.62
Other Asian	2.91**	7.64***	0.86	7.29***	1.75*
African and Caribbean	2.09**	2.53**	1.97	2.44	1.75
Chinese	3.93***	3.87**	1.43	3.19	1.46***
Country of Birth					
Scotland	*1.00*	*1.00*	*1.00*	*1.00*	*1.00*
England	1.22**	0.52***	0.83	0.39**	0.57
Wales	0.65	0.80	1.49	0.37	1.03
Northern Ireland	3.12***	1.39	1.82*	1.51	1.10
Outside UK	2.48***	1.34*	1.76**	0.97	0.98
Distance moved					
Outside UK	*1.00*	*1.00*	*1.00*	*1.00*	*1.00*
Within Council Area	0.70**	0.82	0.58**	1.02	1.18
Between Council Areas	0.52***	0.8	0.54**	0.85	1.27
NS-SeC					
Professional and Managerial	*1.00*	*1.00*	*1.00*	*1.00*	*1.00*
Intermediate and Self-employed	0.48***	0.62***	0.51***	0.71*	0.86*
Routine	0.32***	0.50***	0.42***	0.76	0.82*
Never worked and Long-term unemployed	0.53***	1.05	0.43***	1.50	0.95
Log Likelihood	-22217.68	Pseudo R^2	0.11	Number of Obs.	15635

Notes: RRR=relative risk ratio, which can be interpreted as an odds ratio in binary logistic regression. Significance levels reported where *=$p<0.05$; **=$p<0.01$; ***=$p<0.001$.

Source: 2001 Small Area Microdata (SAMs). Authors' own calculations.

There are still clear differences between those born outside Great Britain (the first generation migrants) and the rest of the population. The distance moved proxy (within Council Area, between Council area, or from outside the UK), demonstrates that those entering Edinburgh and Glasgow are most likely to have moved from outside the UK, highlighting the attraction of these cities as international centres. There is a similar result for Aberdeen. Although a smaller city, Aberdeen has a substantial presence in the oil industry which is likely to have an international pull for labour migration. This is less true for Dundee and the Central Belt, where moves are mostly likely to have been made within the immediate Council Areas or between Council Areas.

Discussion and Conclusions

Although ethnic minorities make up a relatively small proportion of the Scottish population compared with the English population, the tendency to submerse the Scottish experience within Britain-wide studies has meant that little is known about ethnic minority migration within Scotland. This is a failing, given the rather different national contexts between Scotland and the rest of Great Britain. Differences in ethnic group characteristics and experiences between England and Wales and Scotland might be expected given their different settlement histories, government policies, and population sizes and distributions, yet this has not been tested to date.

While there is a rich and growing literature on ethnic minority internal migration and population distributions in Britain, and in particular England (Champion 2005, Finney and Simpson 2008, Simpson and Finney 2009, Catney and Simpson 2010, Stillwell and Hussain 2010, Simon 2011), only a few studies have concentrated on ethnic population sub-groups in selected parts of Scotland (Bowes, Dar and Sim 1997, Pacione 2005a, Houston 2010, Muñoz 2010). In order to fill this gap, this chapter set out to explore the ethnic group-specific internal migration propensities and geographies of mobility in Scotland. Informed by previous research for Britain as a whole, and emphasising the need to explore the case for Scotland separately, the chapter addressed two questions: namely, does ethnicity matter in explaining mobility propensities in Scotland? How does the propensity of migration to different regions vary (by ethnic group)? In turn, we also argue that, given the heterogeneity and diversity of the ethnic minority experience within Britain, studies with such a broad geographical coverage may not always be appropriate.

Although the relative size of the ethnic minority population in Scotland is lower than in England and Wales, the proportions of the groups which make up the total population are relatively similar. Of the minority groups represented in the Scottish population, the Indian group have the largest percentage born in the rest of the UK, suggesting perhaps that the group comprises second or third generation individuals who are likely to have migrated from England to Scotland. This group also has the largest proportion of any group (including White) in Higher and Professional occupations, and the greatest proportion of people most highly qualified. Although not directly tested here, this does suggest a link between occupational and educational achievement and length of time spent in the UK.

In their Britain-wide study, Finney and Simpson (2008) demonstrated the similar propensities to migrate for all ethnic groups, with differences explained by the 'usual' socio-economic and demographic determinants. Taking Scotland as a separate region, we were interested to see how far ethnic group might relate to differentials in the likelihood to move short- and long-distances. Modelling mobility propensities showed that individual and household circumstances explained differences in the likelihood of migration, whereas ethnic group was not a significant predictor of migration behaviour. Indeed, after controlling for socio-economic, demographic and household characteristics, differences between ethnic groups largely disappeared.

As with the modelling of distances moved, the introduction of controls for socio-economic and demographic characteristics of individuals showed a reduction in the significances for ethnic differences in the likelihood of migrating to Aberdeen and the Central belt and, for the Indian and Other Asian groups, Dundee. This is not the case, however, for those migrating to Edinburgh or Glasgow. Ethnic minority groups are more likely to migrate into and around the large urban areas of Scotland, and for the two major cities significant differences in migration behaviour by ethnic groups remain.

It was suggested earlier in the chapter that while dispersal from minority concentrations is occurring in the rest of Great Britain (Simpson and Finney 2009, Simon 2011), the spatial mobility of minorities in Scotland may differ, given the more recent history of immigration to Scotland. While movement to the suburbs or more rural locales may be attractive to long-established minority groups, in Scotland, where many minority communities are still fairly new, the attraction of major urban centres is likely to 'hold' minorities. There is a wide labour market literature that points to marked differences in the performance of Edinburgh and, to a lesser extent, Glasgow compared with the rest of Scotland and their function as regionally important escalators (Findlay, Short and Stockdale 2000, Findlay et al. 2008, van Ham et al. 2010). The results presented suggest that, unlike for other regions in Scotland, variation in the likelihood of migrating is not well explained by differences in the characteristics of individuals, as measured by socio-economic and employment status, education, age, household composition and housing tenure. Instead, the results imply differences in mobility for migrants to Edinburgh and Glasgow which may be attributable to one's ethnic group – that is, there may be an ethnic group 'dimension' to internal migration, when geographical region is modelled. There are several key distinctions between Edinburgh and Glasgow and the other regions in Scotland, which may help to explain this effect, beyond their labour market characteristics. The cities are likely to be attractive to newcomers and see considerable churn given, for example, the cities' statuses as the capital (Edinburgh) and largest city (Glasgow), the fact that they are home to by far the largest proportion of individuals in non-White groups, and have the longest history of both immigration and ethnic minority populations more generally. In turn, it may be expected that extended social networks, knowledge transfer on housing (see Bowes, Dar and Sim 1997) and chain migration may lead to biases in preferences for certain regions.

It should be noted, though, that despite this reduction in significance for most ethnic groups for most regions bar Edinburgh and Glasgow, there remained statistically significant effects for the Indian and Other Asian groups moving into (or within) Dundee, and for Indians to/within Aberdeen. The 'preference' for major urban centres compared with the 'Rest of Scotland' by minority groups is notable, but not surprising; it is in these areas where the greatest housing, job and educational opportunities exist, and where longer established networks of immigrants and subsequent generations, with their associated support mechanisms, are available. The Rest of Scotland category includes the Highlands and remote coastal regions of Scotland, which, given the fairly uneven geographical distribution of minority groups in Scotland, are unlikely to offer the same attractive characteristics. This is supported by the findings which suggested that Glasgow and Aberdeen were particularly attractive destinations for those born outside the UK, and that those moving to Edinburgh, Glasgow and Aberdeen are most likely to have come from outside the UK. The counterurbanisation process occurring within Scotland, as explored by Stockdale, Findlay and Short (2000), is therefore likely to be a largely 'White' phenomenon until time since immigration increases and the growth of subsequent generations occurs for minorities in Scotland. Given that we are not able to distinguish migrant origins (due to data restrictions), the dominance of minority groups in urban rather than suburban and rural areas may explain this urban 'preference', however the results do allow us to identify the attraction of urban areas for minority groups.

While it has been possible using a collection of Census data to provide some insight into minority internal migration in Scotland, we must recognise the limitations of the study. Firstly, using the cross-sectional, individual-level SAMs data to access geographical detail imposes a restriction in terms of the ethnic minority group information available. The consequence of this is that we are unable to model outcomes for all ethnic groups and some compromised categories have to be adopted. Even with the superior geographical coding in the SAMs data it is still not possible to investigate neighbourhood patterns of ethnic minority migration in Scotland. As a result, we are not able to analyse the residential *choices* that individuals in different ethnic minority groups make with respect to the type of local neighbourhood they wish to live. Rather we make broader inferences about migration in Scotland in general. Secondly, the numbers of ethnic minorities in Scotland are very small, with the ensuing problem that when creating cross-tabulations of ethnic group against other characteristics small number problems quickly become apparent. Finally, the data used in this analysis are drawn from the 2001 Census, and the data are over 10 years old. The 2011 Census was conducted on the 27th March 2011 and, at the time of writing, the figures are yet to be released. When they are it will be possible to make a more up-to-date assessment of ethnic minority migration in Scotland.

Acknowledgements

The Census output is Crown copyright and is reproduced with the permission of the Controller of HMSO and the Queen's Printer for Scotland. The individual data used in the modelling is derived from the 2001 Census Samples of Anonymised Records, which is also Crown Copyright. We gratefully acknowledge the assistance of CCSR for providing and maintaining access to this data for the academic community. In addition, the delegates of the Minority Internal Migration in Europe conference, held at the University of Manchester, September 2011, are sincerely thanked for their comments and advice on this research. Thanks to Nissa Finney for her helpful comments on an earlier draft of this chapter.

Chapter 8

Internal Mobility of Immigrants and Ethnic Minorities in Germany

Sergi Vidal and Michael Windzio

Introduction

In North America and Britain, a standing tradition of analysis of residential mobility and internal migration of immigrants and ethnic minorities already exists. Research literature on this topic has found differences in patterns and determinants of internal migration among ethnic groups. In contrast, in Germany, which is also a country with a relevant multi-ethnic composition, the analysis of internal movement among ethnic groups has been partially neglected. The literature investigating the German context has been active in addressing questions related to initial settlement, spatial distribution and segregation at the local level (Gans 1987, Glebe 1997, Friedrichs 1998, Böltken, Gatzweiler and Meyer 2002, Schönwälder and Söhn 2009, Grüner 2010, Özüekren and Erzog-Karahan 2010), as well as the social and economic outcomes of minority spatial concentration (Clark and Drever 2000, Drever and Clark 2002, Drever 2004). Although internal mobility and migration may have an impact on the spatial distribution and residential segregation of minorities, as well as on their socio-economic achievement, very little research has addressed these issues (for examples see Schündeln 2007, Saka 2010, Rüger, Tarnowski and Erdmann 2011). Moreover, the picture presented in the scarce literature is incomplete, as it only accounts for a fraction of minorities. This is not surprising, given that data sources are rather limited for the analysis of both internal movement and ethnic groups.

The aim of this chapter is to add to previous research by presenting an overview of minority mobility in Germany. We focus on two sets of research questions aimed at shedding light on core concepts for understanding the internal mobility and migration of native and minority groups: 1) Are internal mobility rates of ethnic minorities similar to those of German natives? Do they differ between ethnic minority groups? 2) Which compositional factors, such as the socio-demographic and economic opportunity structure, are associated with deviances in internal mobility across ethnic groups? Do certain aspects of minority ethnic groups' immigration histories play a role in explaining actual internal mobility?

First, we compare mobility rates, overall and by distance, using German natives and ethnic minorities (i.e. a nation of origin other than Germany) as analytical categories. Second, we tackle compositional characteristics of ethnic populations in order to explore divergences in migration rates. In particular, we pay attention to the socio-demographic composition of ethnic groups, as age and household structure have been found in other national contexts to be the main reasons for ethnic group differences in migration propensities, as well as other socio-economic predictors of mobility (for example, for the U.S., Kritz and Gurak 2001, for Spain, Recaño 2003, for Britain, Finney and Simpson 2008, for the Netherlands, Zorlu and Mulder 2008). We also disaggregate ethnic minorities into several groups, in order to tackle heterogeneity within minorities or ethnic diversity. To solve the problem of the small sample size for single national origins, we create internally homogeneous groups by merging national origins with common immigration histories. We test to what extent immigration history conditions the evolution of diverse ethnic population structures in Germany and, thereby, has an impact on internal mobility heterogeneity across ethnic groups.

To answer these questions we present new evidence derived from two complementary data sources: the German Microcensus (cross-sectional, 2005) and the German Socio-Economic Panel Study (longitudinal, 1995–2008), whose suitability for the analysis of minority migration is addressed in the data section. The rest of the manuscript is structured as follows: first, we define the measurement of ethnic minorities in Germany and describe the ethnic composition of the German population. Subsequently, we raise some theoretical considerations on minority settlement and mobility in Germany. The analysis is presented after a description of the data sources. The chapter concludes with a discussion of the findings in addition to a future research agenda.

Ethnic Minorities in Germany

To study ethnicity, the German research tradition uses concepts of 'culture of origin', or migration background (*Migrationshintergrund*), linking one's cultural affiliation to the parents' national origin(s). Migration background is a comprehensive term encompassing immigrants as well as those born in Germany and German citizens who are descendants of immigrants to Germany after 1949. The German Federal Statistical Office has adopted this term for the production of new instruments to measure ethnic groups within the German population.[1] Particularly since the 2005 edition of the German Microcensus, the official survey for the observation of population dynamics (see data section for a detailed description), information on migration background of the population has been recorded. In 2005, it was estimated that about 18 per cent of the population

---

1   According to the Federal Statistical Office, individuals with a migration background are defined as those who moved to the German state after 1949, those born in Germany to non-German parents and those born in Germany with at least one parent who moved to Germany after 1949 or possesses a foreign passport/is a foreign national (Statistisches Bundesamt Deutschland 2009).

residing in Germany had a migration background. This definition includes very different groups, such as immigrants with a foreign nationality (36 per cent out of the population with migration background), offspring of immigrants with a foreign nationality (11 per cent), ethnic German repatriates – *Aussiedler* and *Spätaussiedler* – (12 per cent), German citizens with immigrant parents (20 per cent), and German citizens of whom at least one parent immigrated to Germany or is not a German national (21 per cent).

Following the German research tradition, ethnic minorities are defined as follows: (1) immigrants (after 1949) for whom German citizenship at birth was not an option,[2] as first generation migrants, and (2) those who, independent of their citizenship, were born in Germany to at least one immigrant/non-German citizen, as second (or higher order) generation migrants. We construct ethnic minority groups by merging national origins with similar immigration histories since the end of the Second World War – a time in which western Germany, in particular, witnessed a dramatic increase in the proportion of ethnic minorities. The aim of this approach is to create analytical categories which depict different immigration experiences and migrants' diversity to compare and contrast with the average minority behaviour. In Table 8.1 we describe the proposed ethnic groups' main characteristics. We also take an initial look at their compositional characteristics in the form of a statistical summary by immigration history characteristics (i.e. German nationality, birth in Germany, year of arrival in Germany), as extracted from the German 2005 Microcensus.

The largest group of national origins includes those who settled in the country after the Second World War and before the Oil Crisis in the mid-70s and their offspring. These early newcomers were mainly recruited in the 1960s via guest-worker programmes (*Gastarbeiterprogramme*) in order to palliate labour shortages. Guest-worker programmes were subscribed to by Southern European countries, as well as Morocco and Tunisia. Therefore, we label them the Mediterranean group. The programmes addressed labour supply for low-qualified jobs in the industrial sector. Most of the guest-workers were low-educated males recruited from rural areas. It was largely assumed that they would work for a limited period of one or two years in Germany and then return to their respective country of origin. While some workers returned home after earning and saving money, others did not and, instead, brought their families to Germany. Indeed, part of the enduring immigration from these countries after the end of the guest-worker programmes in 1973 and even nowadays is related to the immigration of spouses (and families) from the country of origin (Strassburger 2004). According to the

---

2 Nationality at birth is derived from the principle of *ius sanguinis* in Germany, meaning that children of non-German citizens born in Germany can acquire German nationality only through naturalisation. However, those born in German territory to foreign parents after 1st January 2000 can receive German nationality at birth if at least one of the parents has been a permanent resident for the last three years and has been residing in Germany for at least eight years.

statistical summaries, the Mediterranean group accounted for 6.2 per cent of the population residing in Germany in 2005. The largest nation of origin is Turkey, although considerable numbers of former Yugoslavians and Italians also reside in Germany. Although the highest share of the groups' current population immigrated to Germany in three evenly distributed arrival periods (i.e. until 1975, 1975–1989, after 1989), the Mediterranean ethnic group has the highest proportion of second generation individuals among ethnic minorities (around 36 per cent of the group-specific population), but relatively few possess German citizenship.

Table 8.1   Ethnic minority group definitions and summary statistics, Germany 2005

Ethnic minority groups	Most populous national origins (2005)	Main historical period of settlement (since 1949)	Main channels of entry (since 1949)	Summary statistics (2005)
Mediterranean	Turks Serbs Italians Croats Greeks	Guest-worker programmes (1955–1973)	Temporary work permits and family reunification	% of total pop.: **6.2** % females: **47.8** % German citizens: **18.6** % born in Germany: **36.3** % arrived 1949–1975: **18.8** % arrived 1975–1989: **15.8** % arrived after 1989: **20.9**
Eastern Europe (incl. Former USSR)	Russians Poles Ukrainians	After collapse of Soviet Union (since 1989)	Ethnic German repatriates: *Aussiedler / Spätaussiedler*	% of total pop.: **3.3** % females: **55.2** % German citizens: **64.0** % born in Germany: **9.0** % arrived 1949–1975: **4.9** % arrived 1975–1989: **22.3** % arrived after 1989: **55.0**
Western Europe	Austrians Dutch French British	Constant fluxes since World War II	Work permits and *freedom of movement of workers* within the EU	% of total pop.: **1.0** % females: **52.4** % German citizens: **26.1** % born in Germany: **33.9** % arrived 1949–1975: **18.2** % arrived 1975–1989: **13.9** % arrived after 1989: **21.6**
Rest of World	Vietnamese Iranians Afghans US-Americans Chinese	Mostly recent (since 1989)	Work permits, refugees / asylum seekers	% of total pop.: **2.0** % females: **50.8** % German citizens: **36.4** % born in Germany: **17.9** % arrived 1949–1975: **6.4** % arrived 1975–1989: **18.3** % arrived after 1989: **48.3**

Notes: Summary statistics calculated with Microcensus 2005.

The second largest group of ethnic minorities in Germany is composed of immigrants from eastern Europe and the former USSR. A large proportion of these individuals are ethnic Germans known as *Aussiedler/Spätaussiedler*. They are repatriates of German descent who emigrated from Germany or were residing in the former German territories in eastern Europe before the Second World War, and their descendants. After 1949, the German Constitution or Basic Law (*Grundgesetz*) conceded citizenship to ethnic Germans who returned from former German territories in eastern Europe to the current German territory. Up to 1989, the arrival of ethnic Germans (*Aussiedler*) and their integration into German society was quite straightforward due to their cultural proximity. However, after 1989, and coinciding with the breakdown of the Soviet Union, there was a massive increase in arrivals of younger ethnic Germans who have no such close cultural linkages (*Spätaussiedler*). This ethnic group accounted for 3.3 per cent of the population residing in Germany in 2005, and two other national origins accounted for more than 50 per cent of the population of the group: Russians and Poles. A remarkable characteristic of its composition is the very recent migration to Germany. Indeed, more than 50 per cent are immigrants who arrived after 1989, at the time of social and economic transition in the former USSR. The fact that this group has the lowest level of second generation migrants among minority groups (under 10 per cent of the group population) may be traced to their recent arrival. Still, more than 60 per cent of these populations are German citizens, suggesting a strict association of this group with ethnic German repatriates (*Aussiedler/Spätaussiedler*).

Groups of other national origins can be considered residual, not only because they are less populous, but also because they are less homogeneous in their immigration histories. We divided the residual national origins into two distinct groups: western Europe and Rest of World. Due to geographical proximity, there has been a constant flux of western European migrants to Germany, though it is relatively small compared to the above-mentioned groups. The most populous nationalities are the neighbouring Austrians and Dutch. The summary statistics indicate that those of western European origin are more likely to have settled in Germany relatively early, with a higher than average proportion of second generation migrants. Individuals of other world origins are very mixed, though mainly originating from Asia, including Vietnamese, Iranians, Afghans, Pakistanis and Iraqis. There also are Africans (mainly from former colonies) and north and south Americans. It is indeed very difficult to find homogeneity within this group, which encompasses very diverse individuals and histories. For instance, many Vietnamese, similar to those originating from other former communist countries, arrived under the auspices of an eastern German parallel guest-worker programme and are mainly concentrated in eastern German cities, such as Berlin. On the other hand, migrants originating from Afghanistan and Iran tend to have entered the country as asylum seekers due to persecution for political activism and are concentrated in western German cities. In general, the core of this ethnic group is likely to have arrived in Germany more recently: almost 50 per cent arrived after 1989, with a relatively high share of German citizenship and a lower share of second generation offspring.

Theoretical Considerations of Minority Settlement and Mobility in Germany

The portrait of the spatial distribution of foreign populations in Germany is rather uncommon among traditional immigration countries in Europe. Schönwälder and Söhn (2009) find that although populations with minority ethnic backgrounds tend to reside in urban areas (i.e. half of the foreign origin populations live in towns with at least 100,000 inhabitants), there is no strong pattern of spatial concentration of ethnic minorities across western Germany and/or in neighbourhoods within cities. First, the proportion of people with migration backgrounds in the *Regierungsbezirke*[3] of western Germany range from 13 per cent (the lowest) to 30 per cent (the highest) in 2009, with an average around 20 per cent (Statistisches Bundesamt Deutschland 2011). In contrast, the average proportion of foreign origin populations in eastern Germany is below 5 per cent, excluding the city of Berlin. These figures indicate that the distribution is relatively equal within the western regions of the country and that eastern Germany, where the income gap to western Germany is still prominent, is far from converging on the levels of foreign origin population. The main reason behind the even distribution of the foreign origin population across western Germany is state intervention on matters of immigrant settlement at the time of guest-worker programmes, as well as during the arrival of *Spätaussiedler* and refugees in the 90s, with apparent long lasting effects. Secondly, many studies (Glebe 1997, Friedrichs 1998, Schönwälder and Söhn 2009) find that there is no strong neighbourhood segregation by specific foreign origins and that people with a migration background tend to live in multi-ethnic neighbourhoods. The fact that affordable privately rented housing in Germany is widely available and evenly distributed within cities (Kirchner 2007) may have contributed to lower levels of ethnic concentration. Although all those with a Mediterranean origin live in areas of larger minority concentration (particularly in inner-city areas), this fact does not necessarily imply co-ethnic concentration.[4] Schönwälder and Söhn (2009) indicate that, differently to British or Dutch cities, concentration levels of co-ethnics in German cities do not surpass 10 per cent of the neighbourhood population and situations over 30 per cent of co-ethnic concentration are rare. However, low segregation does not prevent ethnic minorities from living in deprived neighbourhoods regarding social, economic and environmental aspects (Drever 2004, Statistisches Bundesamt Deutschland 2011).

---

3 *Regierungsbezirke* are government regions as administrative divisions only within the largest federal States – 39 units – EU standard NUTS II.

4 Particularly, guest-workers occupied inner-city areas in sub-standard flats, some of which were constructed in pre-war times. See Glebe (1997) for further insights on why the guest-workers presented relatively higher concentration levels within German cities.

Studies of minority mobility (beyond city borders) and their determinants in Germany are very scarce. Schündeln (2007) reports higher mobility rates of recent immigrants (i.e. arrival ten years before observation in his sample) than German natives, and he indicates that the younger age-structure among immigrants plays an important role. Moreover, he finds that eastern Germany is more likely to lose foreign minority ethnic populations moving to the western part of the country. Comparing ethnic minority populations by birth in Germany or by immigration, Saka (2010) identifies little difference in mobility levels between Turkish origins, former Yugoslavian origins and the other former guest-worker origins, while other ethnic groups show disparate mobility rates. She further reveals that a feature of all Mediterranean origins (not so important for other foreign origins) is the large gap in mobility rates between first and second generation migrants, where not only relative age difference, but also the difference in educational attainment, explain a great deal of variation in mobility rates within the group. This point is consistent with work with foreign national origins with similar migration histories,[5] where newly arrived groups of migrants, with a lower share of second generation migrants, depart from the settlement and mobility behaviour of older settlement groups.

Even if research on the topic has only presented a partial picture, it has indicated that migrant settlement is rather spatially homogeneous in Germany[6] (except eastern Germany) and that differences in mobility exist amongst ethnic groups. In particular, the mobility behaviour of ethnic minorities differs from that of German natives, and this seems to be determined by socio-demographic structure differences, also including different levels of education and occupational distributions, neighbourhood characteristics and difficulties in housing access for the ethnic minority groups.[7] In that vein, it is expected that compositional differences in averages of social and economic characteristics across ethnic groups, as well as difficulties to get equal access to job and housing opportunities across ethnic minorities, might be behind observed differences in mobility rates between the German natives and ethnic minorities.[8] However, they might not necessarily explain all differences across minority ethnic groups. Here is where historical immigration characteristics may play a more important role.

---

5   However, it should be pointed out that this strategy does not allow for analysis of hypothetical differences in preferences, practices, and specific origin discrimination among national origins averaged in an ethnic group.

6   See Schönwälder and Söhn (2009) for hypotheses on why settlement of immigrant population was rather spatially homogeneous in comparison with other countries with a similar proportion of foreign origin population.

7   See Horr (2008) for an assessment of the difficulties concerning housing access among Turkish origins.

8   For instance, it has been found that after controlling for educational credentials and other traditional predictors of wage and employment levels, economic achievement among native Germans and ethnic minorities remain marginal (Schmidt 1997).

It can be pointed out that time and other initial conditions of settlement can play a role in affecting mobility levels, as historical conditions of immigration of each ethnic group impact on the evolution of migrant population structures and, therefore, on the characteristics of the population known to be traditional determinants of mobility (e.g. age or duration of residence in a location). Early immigration means an older age structure of immigrants, but also a higher proportion of second generation migrants, who are more likely to resemble the native population in aspirations and behaviour. The condition of German citizenship can grant access to specific occupational opportunities, facilitate international mobility, or grant national identification. In addition to explaining the ethnic group composition, historical aspects can also directly impact on the migration behaviour of ethnic groups. For instance, those settling during the time of guest-worker programmes arrived from different parts of the origin countries and were homogeneously distributed across western Germany, reducing the chances of spatial concentration, due to lack of networking links among co-ethnics in different settlements (Schönwälder 2001). In this research, we are not able to go deeper into particular aspects for each ethnic group, as this would require considerable space, but instead we focus on whether there is an effect of immigration background aspects (period of arrival, migrant generation and German citizenship) on group-specific rates of mobility, along with other socio-demographic and economic characteristics. Evidence of significant effects of immigration background can bring some insight on the importance of conditions at settlement beyond group differences in average levels of traditional determinants of internal mobility.

Data Sources

Data for the analysis of internal migration in Germany are very scarce and generally derived from cross-sections that do not allow for the simultaneous analysis of both the determinants and consequences of migration. Farwick (2009) presented a detailed description of data sources for the analysis of internal migration in Germany. However, most of these data sources have limited information for identifying ethnic minorities, as only information on citizenship is available. Furthermore, none of the German data sources identify asylum seekers and undocumented immigrants. We will now briefly comment on the two data sources used for analysis, the German Microcensus[9] and the German Socio-Economic

---

9   http://www.destatis.de/jetspeed/portal/cms/Sites/destatis/Internet/EN/press/abisz/ Mikrozensus__e,templateId=renderPrint.psml (retrieved on 18 July 2011).

Panel Study.[10] In our opinion, these are the most reliable datasets for the analysis of minority migration within Germany.[11]

The German Microcensus is an official statistic of micro-data, conducted by the Federal Statistical Office, and is representative of the German population and labour market. Initiated in 1957 in western Germany and 1991 in eastern Germany, it covers around 1 per cent of the German population and has been conducted annually since 1995, surveying around 370,000 households and 800,000 individuals each time.[12] For research purposes, it is possible to obtain data of a representative sample (i.e. scientific-use-file) containing 70 per cent of the observations of the Microcensus. The respondents provide their current and previous place of residence (of the previous year).[13] Since its 2005 release, the German Microcensus has provided detailed information on the place of birth, nationality, form and year of acquisition of German citizenship and related issues for both parents of the respondents. The main drawback of the Microcensus is that we can only work on information for a cross-section, matching the information on internal migration to post-migration characteristics. It is only possible to identify mobility consequences, because little information is available on the situation before the move.

The German Socio-Economic Panel Study (GSOEP) is a longitudinal study that aims to analyse the socio-economic and health statuses of a representative sample of private households in Germany over time. The study was started in 1984 in western Germany with 5,921 households and 12,245 individuals (eastern Germany was included in 1990 with 2,179 households and 4,453 individuals), who are followed and surveyed on an annual basis. Refreshment samples, as well as calibrating weights, allow for a representative picture of German society

---

10   http://www.diw.de/en/diw_02.c.222508.en/soep_overview.html (retrieved on 18 July 2011).

11   Alternatively, the Institute of Labour Market and Occupations Research (IAB) provides the Regional Employment Sample, which is a database on the employment histories of about 2 per cent of workers and individuals affiliated with social security in Germany since 1975. However, information is available on workplaces, but not on places of residence, and minorities can only be identified by citizenship. To analyse spatial patterns, migration matrices of aggregated migration flows between administrative units are offered by the German statistical Office. In these matrices migration is measured as registrations and deregistration for few administrative levels and ethnic groups can be measured only by their citizenship.

12   The German Microcensus is a substitute for the 'traditional' population census, which has not been run in (western) Germany since 1987. The German overall population in 2005 was 82 million.

13   Since the lowest level of information on the geographical unit of residence is only publicly provided for a relatively large unit, the Federal States (*Bundesländer*, 16 units, with population per unit ranging from 0.6 to 18 million – European Union standard NUTS I), we will only consider changes of residence disregarding the distance of the move in our analysis with the Microcensus.

over time, with about 11,000 households and 20,000 individuals interviewed annually. The GSOEP has been widely used for the analysis of internal migration and residential mobility.[14] It provides yearly updated information on the date of residential relocation. With special permission, it is possible to work with sensitive geographical information at the county level (*Kreise* – 429 units – average population per unit: 435,000 – divided into urban and rural counties – EU standard NUTS III) and distance between counties, using remote access – *SOEPremote* – from the researcher's workplace or in a secure working environment on the DIW-Berlin premises.[15]

To enable an analysis of immigrants and ethnic minorities in German society, the GSOEP provides two minority-specific samples. In 1984, coinciding with the study's launch, a sample of immigrants from Mediterranean countries was added (N=1,392 households), leading to an overrepresentation of respondents with Turkish, Yugoslavian, Italian, Spanish and Greek origins. Moreover, in 1994/5, a sample of immigrants of all nationalities after 1984 was added (N=522 households), which ensures that new flows of migrants from all over the world are represented, especially those with intense flows of ethnic Germans (*Spätaussiedler*) who came from eastern Europe and former Soviet republics after 1989. A further refreshment sample was added in the year 2000, including additional new migrants in the panel study. We account for 6,500 individuals who either immigrated to Germany or whose ancestors did.[16]

Mobility Rates by Ethnic Group, Age-sex Composition and Settlement Size

In order to present an initial picture of minority migration in Germany, we calculated annual rates of mobility using the Microcensus 2005. Annual mobility rates are calculated as percentages of the population with a different permanent address of residence the year prior to the interview. In Table 8.2, we present crude rates by sex and ethnic group, as well as age-adjusted rates[17] in order to reduce the effect of different age-structures across ethnic groups. The average

---

14   See the English based website – *SOEPlit* – for updated and sorted publications derived from GSOEP data: http://panel.gsoep.de/cgi-bin/baseportal.pl?htx=/soeplit/soeplit (retrieved on 14 April 2011).

15   However, analysis of mobility origin and destinations is not possible due to data protection issues.

16   In order to assess the possible bias of the GSOEP sample, we contrasted the analysis with Microcensus data, the most reliable measurement tool. We found that internal migration rates calculated with GSOEP are downward biased because the recent immigrants are not accounted for.

17   Age-adjusted rates are computed using direct standardisation or a weighted average of age-specific rates using the overall population as a reference. Age-specific rates were calculated for five-year age groups except those aged 85 or more, who were coded as one age group.

mobility rate in Germany (around 7.4 per cent) is unevenly distributed across German natives (6.9 per cent) and ethnic minorities (11.2 per cent). Even after correcting for the younger age distribution of ethnic minorities, we still observe that minorities are more mobile (9.5 per cent) compared to the native group (7.1 per cent).

Mobility rates appear to be highly variable by minority ethnic groups. Among all minorities, Mediterraneans show below average crude and age-corrected rates. We suggest that relatively higher proportions of early arrivals in Germany, as compared with other minorities, may well offer an explanation of that result. The ethnic group formed by eastern Europeans and individuals from the former USSR shows slightly higher rates of mobility than the Mediterraneans. Although *Spätaussiedler* were under mobility restrictions during their first few years of residence in Germany, the relatively new immigration of eastern Europeans may explain the higher mobility of this group. As opposed to the averages for other ethnic groups before the age-adjustment of rates, females have higher rates of mobility than men. This may also be explained by the higher proportion of recent female immigration from eastern European countries. Mobility rates for

Table 8.2   Annual crude and age-adjusted mobility rates by ethnic group and sex, Germany 2005

Ethnic group	Overall		Male		Female	
	Crude mobility rate (%)	Age-adjusted mobility rate (%)	Crude mobility rate (%)	Age-adjusted mobility rate (%)	Crude mobility rate (%)	Age-adjusted mobility rate (%)
Total	7.4		7.6		7.2	
German native	6.9	7.1	6.9	6.9	6.8	7.2
Ethnic minorities	11.2	9.5	11.2	9.6	10.9	9.3
Mediterranean	10.4	9.0	10.7	9.8	9.9	8.2
Eastern Europe / Former USSR	11.3	10.2	10.9	9.9	11.4	10.3
Western Europe	8.2	8.0	8.0	7.4	7.9	8.3
Rest of World	15.2	12.3	15.0	12.0	14.2	12.4

Notes: Individuals with unknown residence in the prior year (5.52 per cent of the sample), those who migrated to Germany one year prior to the survey and individuals aged 1 or less are excluded. Age-adjusted rates calculated as percentages by direct standardisation with the overall population serve as a reference.

Source: Microcensus 2005.

those of western European origins are well behind the average migration rate of ethnic minorities, but still greater than the migration rates of those with German origin after rates are adjusted for sex and age. The results indicate that migrants of western European origin might behave similarly to native Germans, but that the effect of prior migration experiences may impact mobility rates positively, if only slightly. Lastly, people of other global origins show the highest migration rate before and after age adjustment. The results go hand in hand with the fact that half of these residents came to Germany recently and there is a relatively low incidence of second generation migrants in this group.

In Figure 8.1, mobility rates are shown over age intervals by sex and ethnic group, once again using the Microcensus 2005. One remarkable pattern across ethnic groups is that the natives are generally less mobile at all age intervals, although some exceptions are discernible. Mediterraneans, especially females, are less likely to move than German natives at young adult ages, meaning low incidence of student and labour market integration mobility, which in turn explains this group's low average rate of mobility. Western European individuals show lower mobility rates in young adulthood for males. In contrast, western European females show high rates of mobility in young adulthood, possibly the effect of young educated women spending time in Germany. The Rest of World and eastern European groups show higher rates of mobility at any age-interval than the other ethnic groups.

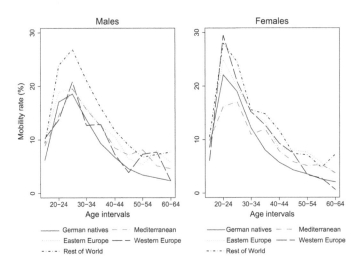

Figure 8.1   Age-specific annual mobility rates by ethnic group and sex, Germany 2005

Source: Microcensus 2005.

Mobility Rates by Ethnic Group and Distance of Move

Although most movements take place over short distances – which has implications for the analysis of segregation at local levels – long distance mobility is also influential in the re-distribution of the native and ethnic populations across urban and rural areas in the country. The Microcensus 2005 indicate that 23 per cent of minority populations are settled in large urban areas (i.e. over 500,000 inhabitants), while only 11 per cent of German natives are. In contrast, 45 per cent of the German population are settled in rural areas (i.e. settlements below 20,000 inhabitants), while only 24 per cent of ethnic minorities are. In general, internal moves are twice as likely to have a large urban destination compared to a rural one. However, the percentage of internal moves that originate in large urban areas is only slightly higher than the percentage of those that originate in rural areas. While the pattern commented on is true for German natives, ethnic minorities show a similar distribution of destinations among large cities, other urban settlements (i.e. between 20,000 and 500,000 inhabitants) and rural settlements. Differences in the size of origins and destinations of mobility among ethnic minorities are an initial indication that internal mobility serves to re-distribute the ethnic minority population more evenly. However, it has been reported that eastern Germany is more likely to lose than to gain foreign origin populations.

For the analysis of distances moved, regarding changes of administrative unit (i.e. county level or NUTS 3), we make use of responses to GSOEP interviews. In preliminary analysis we observed a decrease in mobility rates over time, but differences remain unchanged across ethnic groups for the period under analysis (i.e. between 1995 and 2008), and for that reason we pool these period-specific GSOEP samples for calculations. Overall mobility rates, inter-county mobility rates and the average distances of inter-county movements across ethnic groups are presented in Table 8.3. Moreover, we also present mobility rates for minority ethnic groups according to some minority characteristics.

Table 8.3 provides evidence that ethnic minorities do not make more long distance moves (i.e. between counties) than native Germans. Regarding inter-county mobility, rates of ethnic minorities and German natives are the same (2.2 per cent), but German natives move, on average, 17km further than minorities. If we break up migration rates by ethnic minority groups, we observe that the Mediterranean group has a rate of 1.9 per cent for inter-county moves, which is below the 2.2 per cent for the average rate for ethnic minority groups, while the other minority groups have rates surpassing those of Mediterraneans and German natives. Moreover, Mediterraneans move over shorter distances when they move between counties (average nearly 73km), while eastern and western Europeans show average distances of inter-county moves (139km and 143km) well above the German native ones (125km).

Table 8.3 also presents results breaking down minority backgrounds regarding characteristics such as migrant generation, period of last arrival (for those with migration experience) and German nationality. On average, second generation

Table 8.3   Annual mobility rates (percentages) by distance, ethnic group and migration background, Germany 1995–2008

	German natives	Ethnic minorities	Mediterranean	Eastern Eur. Form. USSR	Western Europe	Rest of World
Overall mobility (% moves)						
All	8.2	10.8	10.9	10.9	10.1	11.9
1st generation		10.3	9.8	11.1	9.8	11.8
2nd generation		13.0	13.9	8.0	11.6	12.4
Last arrival after 1989		13.2	13.5	12.9	15.6	15.9
German nationality		9.9	11.1	10.2	13.8	7.9
Inter-county moves (% moves)						
All	2.2	2.2	1.9	2.5	2.8	2.6
1st generation		2.1	1.7	2.6	2.9	2.4
2nd generation		2.4	2.4	1.5	2.5	5.1
Last arrival after 1989		2.9	2.7	2.8	5.0	3.5
German nationality		2.3	2.2	2.4	4.4	1.6
Average distance moved (km)	125.0	108.0	72.7	139.2	142.6	84.3

Notes: Average distance of move refers exclusively to inter-county moves.
Source: GSOEP 1995–2008.

migrants move more frequently than first generation migrants as far as internal moves are concerned, both overall and inter-county. The differences across generations could be explained by the different age-structures, because second generation migrants are on average younger than first generation migrants, but also by the higher educational attainment of second generation migrants. When we break down rates by ethnic group, we observe that the average minority pattern is reversed for the eastern European group, but the recent arrival of the large bulk of this group explains the low number of moves among second generation migrants, in this case individuals who are not yet adults and are, therefore, less mobile. Another exception can be found among western Europeans moving between counties; however, the differences in rates are smaller than for eastern Europeans. When we consider mobility rates for immigrants who arrived more recently in Germany, we find higher than average mobility rates for ethnic minorities in general, and for each minority ethnic group in particular. This result confirms the hypothesis that recent immigrants are more mobile people. This also confirms that behind the relatively low levels of inter-county moves among those of Mediterranean origin, there is a compositional effect of immigrants settling in Germany at a relatively earlier historical stage compared to other ethnic groups. However, there are still differences across ethnic minority groups in their mobility levels, where more recent migrants of western European and other global origins are comparatively more mobile than the other groups. Here, the mobility of students and professionals from western European and some Central Asian nationalities can offer an explanation. Lastly, German citizenship has no particular effect on mobility rates of minorities overall, but variation is to be found across minority groups. The most striking differences are evident for western Europeans, who move more frequently if they are German nationals, while the group of other global origins clearly moves less frequently if they have a German passport. This result requires more in-depth study, as many mechanisms may explain the result, like facilities for internal mobility and/or increased international mobility after acquiring German citizenship.

Regression Analysis of Internal Mobility in Germany

We now analyse, by means of regression analysis, whether the variation in mobility levels across ethnic groups can be further explained by several compositional characteristics. We use the longitudinal information of the GSOEP by pooling samples between 1995 and 2008 and by running hybrid panel models for binary outcomes – logistic model (Allison 2005). In a hybrid panel model[18] we can analyse the two components of variation for a covariate, whose values, such as German citizenship, change over time. On the one hand, a time-varying covariate includes information on the aspect of being in a given status at a given point of time (e.g. to be a German citizen) and allows for comparison among individuals who experience such a status at some point of time. This information is what we also capture with attributes or variables that do not change over time (like the place of birth) and is also known as *between-effects*. On the other hand, time-varying variables also have a component of change, which is the status change for a given individual, such as obtaining German citizenship, and which is separable from

---

18   The hybrid model takes the following form: $y_{jt} = \alpha + \beta X_j + \eta(x_{jt} - X_j) + v_j + \varepsilon_{jt}$, where y_{jt} is the dependent variable observed for individual $_j$ and year $_t$, where α signifies the intercept, β indicates the parameter estimate of between-individual differences (represented by individual means, X_j), η represents the effects of within-individual variation, and xjt represents the predictor for individual $_j$ at time $_t$. The individual-specific error term is represented by v_j, while the ε_{jt} denotes the model error term that contains the random variation for individuals over time.

being in a given status, also known as *within effect*. In separating both effects, we are able to study the impact that becoming a German citizen or getting married has (*within*-effects), as well as the average effect of the statuses of being a German citizen or married (*between*-effects). As our sample size does not allow for group-specific analysis, we pool all ethnic groups in the same model. Our modelling strategy is based on reporting changes in the odds of migration of each ethnic group with respect to the German native group (the reference category), when adding further covariates into the model.

Table 8.4 displays the outputs of the regression analysis. Models 1 to 4 in Table 8.4 present the odds of moving within Germany, independent of distance, and models 5 to 8 present the odds of inter-county moves. While the first two models for each type of move specify minorities as a single category (i.e. ethnic minorities), the other models for each type of move separate minorities into the four ethnic groups proposed above. On the one hand, we show models only with ethnic group categories and characteristics like migrant generation, period of arrival and German citizenship (*within*- and *between*-effects). These models (1, 3, 5 and 7 in Table 8.4) assess the impact of ethnic group characteristics on ethnic group-specific differences in the odds of moving. On the other hand, we specify models including socio-demographic and socio-economic covariates plus all covariates of the previous models. In particular, in models 2, 4, 6, and 8 of Table 8.4, the following covariates are added: age, age-squared, duration of residence in dwelling, house ownership (at t-1), educational attainment, residence in an urban/rural county (at t-1) and the proportion of the minority population in the county (in quartiles at t-1). In addition, the covariates used as *between*- and *within*-effects include: marital status, children, and employment status (at t-1). These models are intended to assess to what extent ethnic minority group effects are mediated by traditional predictors of mobility. Sex, calendar year and residence in eastern Germany are further covariates in all models.

Table 8.4 Determinants of internal mobility, Germany 1995–2008 (odds ratios (exp(β))

	All moves				Inter-county moves			
	Model 1	Model 2	Model 3	Model 4	Model 5	Model 6	Model 7	Model 8
German native (ref.)	*1.00*	*1.00*	*1.00*	*1.00*	*1.00*	*1.00*	*1.00*	*1.00*
Ethnic minorities	1.66***	1.10*			1.08	0.89		
Mediterranean			2.21***	1.24**			1.01	0.72*
Eastern Europe / Former USSR			1.71***	1.07			1.17*	0.91
Western Europe			1.70***	1.27*			1.63*	1.16
Rest of World			1.96***	1.20*			1.63*	1.04
Minority composition								
First generation (ref.)								
Period of arrival (ref.: after 1989)	*1.00*	*1.00*	*1.00*	*1.00*	*1.00*	*1.00*	*1.00*	*1.00*
1975–1989	0.80**	0.94	0.76***	0.92	0.75*	0.92	0.71*	0.91
1949–1974	0.46***	1.04	0.42***	1.00	0.44***	1.07	0.45***	1.12
Second generation	1.19*	0.89	1.08	0.85**	1.39**	0.94	1.44**	1.01
Foreign citizenship (ref.)	*1.00*	*1.00*	*1.00*	*1.00*	*1.00*	*1.00*	*1.00*	*1.00*
German citizenship (within)	0.94	0.84	0.96	0.85	1.37	1.01	1.33	0.98
German citizenship (between)	0.81**	1.09	0.98	1.19**	0.93	1.11	0.92	1.04
Socio-demographic and economic opportunity structure								
Age		0.98***		0.98***		0.97**		0.97*
Age²		0.99***		0.99***		0.99		0.99
Marital status (ref. Non-married)		*1.00*		*1.00*		*1.00*		*1.00*
Married (within)		0.92		0.92		0.85**		0.85**
Married (between)		0.67***		0.67***		0.66***		0.67***
Children in Household (ref. No children)		*1.00*		*1.00*		*1.00*		*1.00*

Table 8.4 Concluded

	Model 2	Model 4	Model 6	Model 8
Children in Household (within)	0.92*	0.92*	0.77***	0.77**
Children in Household (between)	1.06	1.06	0.71***	0.72***
Educational attainment (*ref: low*)	*1.00*	*1.00*	*1.00*	*1.00*
Medium	1.17***	1.17***	1.66***	1.64***
High	1.23***	1.23***	2.32***	2.29***
Employment Status (*ref: full-time employed*)	*1.00*	*1.00*	*1.00*	*1.00*
Self-employed (within)	1.09	1.09	0.81	0.81
Self-employed (between)	1.15*	1.14*	1.34*	1.33*
Non-employed (within)	0.87***	0.87***	0.81**	0.81**
Non-employed (between)	0.93*	0.92*	1.14	1.13
Atypical-employed (within)	0.92*	0.92	0.94	0.93
Atypical-employed (between)	0.90*	0.90*	0.82*	0.81*
House tenure (*ref: renter*)	*1.00*	*1.00*	*1.00*	*1.00*
Owner	0.26***	0.25***	0.34***	0.34***
County type (*ref: rural county*)	*1.00*	*1.00*	*1.00*	*1.00*
Urban county	1.03	1.02	1.35***	1.35***
Minority population (*ref: 1st quartile*)	*1.00*	*1.00*	*1.00*	*1.00*
2nd quartile county pop.	0.98	0.98	0.79*	0.79*
3rd quartile county pop.	0.96	0.96	0.72***	0.72***
4th quartile county pop.	0.88**	0.88**	0.65***	0.65***
N (respondents)	28.822			28.814
N (respondents-waves)	188.439			188.029

Notes: Hybrid panel logistic regression of ethnic group composition on internal mobility. *** probability <0.001, ** <0.01, * <0.05. Other controls in regression models 2, 4, 6, and 8 are calendar period, sex, residence in eastern Germany, duration of residence. A normally distributed person-specific random effect is included in all models.

Source: GSOEP, 1995–2008.

Models 1 and 5 of Table 8.4 indicate that after controlling for minority group characteristics, ethnic minorities as a group have a positive association with any kind of move and non-statistically significant differences with moving between counties when compared to German natives, as we observed in the previous section. We also observe the expected effects of minority group characteristics. Second generation migrants have higher odds of moving than those who immigrated to Germany. Among immigrants, the earlier they arrived in the country the more negative the association with moving, after duration of residence in the current location is controlled for. The acquisition of German citizenship (*within*-effect) has no significant effect on the odds of internal movements. However, being a German citizen (*between*-effect) has a negative association with moving in general, with the exception of inter-county moves. After the addition of the models of socio-demographic and socio-economic covariates (Models 2 and 6 in Table 8.4), we observe changes in the effects of minority group and minority characteristics. In particular, the odds of moving for minorities stay positive but are reduced from 1.7 to 1.1, while the odds of inter-county moves for minorities remain statistically insignificant, but the association turns negative. The results indicate that the covariates incorporated into the model capture part of the association between ethnic minority groups and internal mobility.

Comparing the average values of the socio-demographic and economic covariates included in the model by ethnic group and by distance of move (data not shown here), we find that younger age structures, lower levels of house tenure, lower educational attainment and residence in urban areas among minorities can explain why the positive associations with mobility are strongly reduced when compared to models not accounting for socio-demographic and economic covariates. This is true as traditional predictors of migration behave as expected in the models of Table 8.4. For instance, younger age and higher levels of education predict mobility positively, as more opportunities are open to the young and better educated, mainly regarding occupational changes. Also as expected, house ownership, being in a partnership and being parent (between-effect) are associated negatively with mobility. Having children or getting married (within-effect) diminishes the chances of mobility, but this is explained by the fact that people move in advance of these events. When compared to full-time employees, being non-employed increases the chances of long-distance mobility because of the need for new employment, but situations of no employment or atypical employment, as well as transitions to these statuses, correlate negatively with moves in general, probably explained by a lack of income to move. Unexpectedly, the status of self-employment correlates positively with mobility. This result can be explained by the fact that most of the self-employed in our sample are home-owners and after controlling for home-ownership we observe an income effect of being self-employed. Last, moves between counties normally take place from urban counties, and once controlling for that, the higher the foreign population in the county, the lower the probability of moving. The latter result can be basically explained by the fact that foreigners tend to live in the most prosperous counties, where they moved to take job opportunities. We do not disregard that some of the effects of socio-demographic and economic variables might have interaction effects with specific ethnic groups. However, sample size does not allow us to include multiplicative terms for ethnic minority groups.[19]

The addition of socio-demographic and economic covariates into the models collapses the effects of minority-specific characteristics as they become insignificant. The only exception is a weak negative effect (at <0.1 level of significance; not shown in Table 4) of acquiring German citizenship (*within*-effect) but a positive effect of being a German citizen (*between*-effect), as well as a weak negative effect of being a second generation migrant on overall moves. Generally speaking, characteristics of minority history do not directly impact on mobility rates, but they are associated with a minority group's particular socio-demographic and economic opportunity structure, which in turn affect mobility rates.

Models 3 and 7 of Table 8.4 present the overall moves and inter-county moves by dividing minorities into four ethnic groups. We observe that all minority groups have positive effects for both types of mobility. The only exception is Mediterraneans, who do not have different odds of moving between counties compared to German natives. After controlling for socio-demographic and economic covariates (Models 4 and 8 in Table 8.4), we observe a reduction in the internal mobility differences across ethnic groups. For all moves, effects are reduced but still remain significant and positive for all groups. For inter-county moves, eastern Europeans, western Europeans and other global origins change to have insignificant differences of moving when compared with German natives, while Mediterranean origin turns out to have significantly negative effects on mobility. Regarding minority characteristics, we find once again that after controlling for socio-demographic and economic characteristics, being a second generation migrant has a negative effect on overall moves, while obtaining German citizenship has an average positive effect. Although these effects seem to be residual, we believe that more in-depth analyses of these effects is necessary, as these are ethnic group-specific effects. Indeed, the remaining positive effect of being an ethnic minority on movement – in general – and the negative effect of being Mediterranean on inter-county movement, are due to more subtle societal mechanisms unaccounted for here, such as preferences, social norms, and unequal access to opportunities.

---

19   We did test for interaction effects of the ethnic minority groups with educational attainment and we found that minorities are more reactive to educational levels to predict residential mobility. Further research may want to extend on the differential access to opportunities of minorities not only by education but also by other variables associated with aspirations and opportunities.

Concluding Remarks

This chapter aimed to complement the literature on the geographic mobility of ethnic groups in Germany by putting forth an overview of the internal mobility levels of German natives and ethnic minorities, and by identifying some compositional characteristics of ethnic groups which may influence differences in mobility levels. Moreover, we also added to the identification of deviances from the gross minority average. Following the essence of the life-course tradition, we distinguished between minorities in several ethnic groups with regard to their immigration experiences in Germany since World War II. The logic behind this analytical strategy is to link the immigration histories of different national origins to the current experiences and evolution of the ethnic composition within the country. We identified four main groups: Mediterraneans (early settlement in the post-war period and larger component of second generation migrants), eastern Europeans/former USSR migrants (mainly recent immigrants and partly of German descent who were repatriated with rights to German citizenship), western Europeans (constant fluxes in immigration stemming mainly from neighbouring countries), and origins from the rest of the world (a mixed residual group of more recent settlement).

Results revealed that minorities are, on average, more mobile than German natives, even after controlling for differences on age structures. However, higher levels of mobility only apply to short distance moves and not to those that go beyond the county of residence. Short and long distance moves generally correspond to different logics: short distance moves may be a response to housing re-adjustments, and long distance moves may be a response to economic aspirations, in the form of searching for or starting a new job or educational track. According to this logic, our results indicate that ethnic minorities need housing re-adjustments because, as other research has found, they live in worse housing situations and in more deprived neighbourhoods. On the one hand, the fact that affordable private renting is widely available in Germany, where prices are almost comparable to social housing, and with high quality standards, may add to higher chances of residential mobility for non-high income groups, including a large bulk of ethnic minorities. On the other hand, German natives may also show relatively lower rates of inner city moves because they have slightly higher levels of home ownership than ethnic minorities. Owning a house is almost a one-time event in life in Germany, due to the relatively expensive cost of a mortgage and down payment, reducing considerably mobility levels after dwelling purchase for homeowners.

In comparison to other European countries, Germany presents scant levels of long distance internal mobility; this is partly explained by low levels of job or occupational mobility, due to national specific labour market dynamics. Indeed, long distance mobility reaches a higher peak before access to a full-time occupation, because of student mobility and/or trying out identity/occupational roles. However, we were not able to find differences between German natives and ethnic minorities at the descriptive level. It is very likely that higher proportions of higher education attendance and attainment among German natives, which increase aspirations and mobility levels over long distances, counterbalance lower attachment and unstable job situations of immigrant populations (i.e. first generation migrants), which lead to out-mobility from the county of residence. Here, we can identify the experiences of second generation migrants (i.e. those born in Germany to non-German parents) as being more similar to those of German natives, as we explain below.

Minority characteristics, such as recent settlement in the country (after 1989) or being part of the second generation (i.e. born in Germany to immigrant parents) were associated with higher mobility rates in our descriptive analysis. Recent settlement can occur under circumstances that require further housing re-adjustment, such as scant knowledge of the housing market or temporarily staying with relatives or friends. Those belonging to the second generation are comparatively more mobile than immigrants, not only due to an average younger age structure. The underlying explanation is that second generation migrants move more as students and/or to form a family. Moreover, second generation migrants achieve higher educational attainment than their predecessors that allows them to move to better-off housing, neighbourhoods, or places where they can meet occupational aspirations. In the regression analysis we tested whether after controlling for educational levels, age and other demographic and socio-economic characteristics, second generation migrants are less mobile than first generation migrants and even than German natives. The final result may indicate that even when second generation migrants are close to the German mobility levels, there are still structural differences between those born in Germany to German or foreign parents. We may speculate on income differences and levels of educational attainment, but one can also raise topics of preferences for living close to relatives as well as institutional discrimination concerning access to education and occupations. The regression analysis also confirmed that recent migration experience (i.e. immigration after 1989) and its consequences (e.g. initial housing experiences) might have an effect on further mobility, though not a very important one, due to low levels of statistical significance. We also find that holding German citizenship has a positive impact on mobility among minority groups after group composition is controlled for, but that acquiring German nationality alone has no effect. The result is indeed difficult to interpret and, as it only applies to short distance moves, can be associated with various positive dynamics of the housing market for German citizens. Overall, after the usual predictors of mobility are controlled for, there is still an average residual positive difference in internal mobility among minorities that might be attributed to ethnic minority effects and thus deserves further exploration.

The analysis of certain ethnic groups showed that, after age-adjustment, the mobility rates of Mediterraneans approximately correspond to those of German natives, and this is partly explained by low mobility rates at younger ages among the Mediterraneans. Other variations were identified, such as the higher mobility rates among eastern Europeans and immigrants of other global origins, and particularly of females from the former group, responding to more recent dynamics of immigration to Germany. Regarding long distance moves (i.e.

between counties), we find that Mediterraneans move less frequently than German natives, while the other minority groups clearly move more often than German natives. Results of the regression analysis confirm the Mediterranean difference, while the differences between the other ethnic minority groups and the German natives level off. Although no conclusive findings exist, it has been suggested that, even after controlling for traditional predictors, there are structural origin-specific differences regarding access to housing (i.e. discrimination) or preferences for living close to co-ethnics or relatives. It has generally been pointed out that the Turkish community (as well as migrants of North-African origin) are not preferred as house tenants, and they also have higher preferences for living close to relatives and friends. The so-called and unconfirmed Turkish specificity leads us to discuss the topic of the definition of certain ethnic groups in this research.

It may be debated whether grouping together different national origins with similar historical conditions of immigration, as we offered in this chapter, is an acceptable operational strategy to assess ethnic minority behaviour and outcomes. As commented above, the strategy allows analysis of a proxy of common features of the initial conditions of settlement with the trade-off of ignoring national origin uniqueness, such as socio-economic macro-indicators or average cultural practices of the sending countries. We think that our strategy here is not only justifiable because our small sample for analysis (in GSOEP) would limit our work to two Mediterranean countries: Turkey and all ex-Yugoslavian countries, where we may also find really different backgrounds and immigration conditions within these single national origins,[20] but also, because we aimed at studying direct and indirect associations of ethnic groups, objective immigration history conditions (i.e. arrival period, citizenship, second generation migrants) and observed internal mobility. In the end we found a scant direct effect of immigration history-related characteristics after controlling for traditional predictors of migration. However, the fact that socio-demographic and economic variables capture the effects of migration history characteristics on mobility levels suggests that the conditions of immigration are directly related to the current conditions of minority groups which in turn affect their differential internal mobility. Indeed, a more specific analysis of how immigration history led to the current composition of ethnic minorities is needed. For instance, the period of arrival can indicate not only how much exposure an ethnic group already has to the host society, but also how these individuals settle into the country, for example referring to the recruiting regulations of the German government at the time of the guest-worker programmes. Of course this includes examining specificities of single national origins and the interaction between national origin and historical conditions.

In addition to the need for further analysis of specific national groups to account for their idiosyncrasies, further research is necessary, though important

---

20 The single origin 'ex-Yugoslavian' is especially problematic, not only as this group channels different national origins within it, but also because it mixes former guest-workers and their offspring with refugees of the Balkan wars and more recent immigration.

limitations to analyses exist. It is critical to address the distinct geographical patterns of mobility of minorities and German natives. Do minorities converge with the native population over time regarding geographical patterns and mobility determinants? Are minorities more sensitive or less sensitive to the economic context and to the degree of ethnic concentration than natives? Moreover, it would be useful to analyse neighbourhoods as departure points or destinations and to analyse whether 'White flight' impacts on patterns of ethnic concentration on the national and local levels.

Acknowledgments

We are very thankful to Belit Saka (University of Duisburg-Essen) for her help with the GSOEP data preparation and to Gemma Catney, Nissa Finney, and participants of the Minority Migration in Europe Meeting 2011 in Manchester for valuable comments. Any mistakes are only attributable to the authors.

Chapter 9

Internal Mobility of the Foreign-born in Turkey

Ibrahim Sirkeci, Jeffrey H. Cohen and Neriman Can

Introduction

Migration scholarship is dominated by studies of international human mobility. The investigation of internal mobility of immigrants remains a low priority among students of migration (Ram and Shin 1999: 150). And while the theoretical divide between international and internal migration is known (Skeldon 2006), the volume of internal migration is unquestionably larger than that of the international movements (see, for example, Deshingkar (2006), Skeldon (2006) on China and India).

Despite some unique exceptions in the last decade or so, studies of Turkish migration mostly focus on Turkish emigration to Europe and ignore the importance of Turkey as a destination country and the internal migration of the foreign-born in Turkey (Akgunduz 1998, Kirisci 2000, 2007, Parla 2007, Sirkeci, Cohen and Yazgan 2012). In this descriptive study, we address this gap and analyse internal migration patterns of the foreign-born population in Turkey.

In their study of Mexican migration to the United States, Massey et al. (1993) argued that perpetuation of international migration can be explained by 'a process called cumulative causation' along with the growth of migration networks. Thus 'each act of migration alters the social context within which subsequent migration decisions are made' (Massey et al. 1993: 451). In particular, '[o]nce someone has migrated, he or she is very likely to migrate again, and the odds of taking an additional trip rise with the number of trips already taken' (Massey et al. 1993: 452). Regarding the internal moves of immigrants, several researchers pointed out that because immigrants are not attached to a specific location in their host country they tend to make frequent moves seeking better opportunities at least during the immediate period after their immigration (Bartel 1989, Ram and Shin 1999). Immigrants tend to settle in ethnic enclaves rather than in areas deemed to be economically more attractive (Massey 1986, Massey and Denton 1987, Zhou and Logan 1991, Ellis and Goodwin-White 2006). We expect international migrants (in our case, the foreign-born destined for, or settled in, Turkey) to make further moves within Turkey following their arrival and to settle in homogeneous enclaves. There are many reasons for internal migration following international moves. Previous migration experiences may facilitate further migration within a destination country. Also people, whether immigrant or native, often move around until they find a 'good fit' place in which to settle – what Newbold (1996) describes as 'fine tuning' of a destination choice after arrival. For example, foreign-born immigrants in Canada are found to have relatively higher interprovincial migration rates compared to the natives (Liaw and Ledent 1988, cited in Newbold 1996: 728). In a US study, Hempstead (2007) compared the internal migration of foreign-born from gateway states and non-gateway states to argue that gateway states were holding their foreign-born population. Kritz and Nogle (1994: 509) found that 'nativity concentration deters interstate migration but not migration within the same state' and argued that 'residing in a state where fellow nationals live is a more important determinant of internal migration than human capital, immigration status, or a state's unemployment rate' (see also Rebhun 2006). In other words, immigrants who arrive in areas where co-ethnics are concentrated are less likely to migrate away from them whereas those arrived in other areas are likely to move into those concentrated areas.

Alternatively an Israeli study found that internal migration of immigrants during their first year in the country is linked only to 'job-seeking' (Beenstock 1997). The level of literacy also has a positive effect on internal migration rates (Kearney and Miller 1984). Not surprisingly, some studies found a relationship between residential mobility and social mobility (Denton and Massey 1988, White, Biddlecom and Guo 1993), which can be indirectly measured by mobility propensities of different occupational classes. In a study on Canada, Newbold (1996: 728) admits that foreign-born have higher migration rates compared to the Canadian-born, but claims that 'selectivity with personal factors (i.e. education, age, sex, family type) is similar to the Canadian-born'.

Among the various factors influencing internal migration of the foreign-born (and others) we can include demographic background variables such as age and gender, marital status, home ownership, education and literacy (Kearney and Miller 1984), occupation and/or economic activity. Migration propensity decreases as people get older. Marriage is one important reason for migration but once married people are expected to settle. Migration of women is perhaps more linked to their partners' moves (e.g. Cohen, Rodriguez and Fox 2008). Similarly, some occupations require higher mobility while others are more settled, however until people find permanent posts they may wander. We examine the role played by citizenship of the host country which was identified as a less important factor in explaining the internal migration propensity of immigrants in Spain (Reher and Silvestre 2009). As Ram and Shin (1999) warn us, there are also barriers to migration: admission rules, dispersal policies, employment laws, citizenship laws as well as natives' perceptions of and attitudes to migrants may all influence the propensity of migration.

To test ideas about internal mobility we present characteristics of immigrants in Turkey. Second, we note popular destinations of the foreign-born within the country. Third, we describe the methods employed and the data used. Finally, we analyse the data on foreign-born population and their internal mobility.

Data and Methods

Our analysis is based on 2000 Turkish Census data. The Census data allow us to identify immigrants by their country of birth and their citizenship statuses. We note internal migrants by comparing their current place of residence and their place of residence five years prior to the Census date, as identified by two questions on the Census.

We define a sub-sample composed of the foreign-born population in Turkey, then we describe the total foreign-born population in Turkey. Then, for further analyses, we restrict our sample to those foreign-born who were resident in the country five years ago, while excluding foreign-born people aged 5 or younger as their moves are not identifiable in the Census data. Therefore, our subset used in the statistical analysis is composed of 935,088 out of 1,260,491 foreign-born people reported in the 2000 Turkish Census. We examine the internal mobility of these foreign-born between 81 administrative provinces in Turkey.[1] In our final model explaining determinants of internal mobility, the sample is restricted to omit those who were not employed on Census day.[2] This means eliminating everybody younger than 12, which is the legal age limit for employment in Turkey at the time of the Census. Our final working sample includes 902,532 individuals.

We define internal migration following the restrictions imposed by the data. Hence we compare the place of residence on the day of Census and five years prior to that date. We are aware that this leaves out all the moves which might have happened in the five year interval and prior to the Census date. For statistical analysis including the regression model in this chapter, we classify everybody who changed their place of residence from one village to another, one district to another, or from one province to another, as an internal migrant. However, we make it clear when we refer to movements between provinces only.

Potential factors that influence internal mobility, individual characteristics, household characteristics and province level factors are taken into account. The variables available in the Census and in the socio-economic development level reports we use are now discussed:

The dependent binary variable in our model is internal migration status, as described above. The Census questions enable us to identify the foreign-born population and the birth place of individuals. Although detailed birth place information (i.e. district and neighbourhood levels) is recorded for those born in Turkey, for those born outside Turkey, only the country and city/town of birth is available.

---

1 Provinces are relatively large administrative divisions in Turkey and the population sizes vary from about 70,000 (Bolu) to over 11,000,000 (Istanbul). In the 2000 Census, the average size of provinces was 670,000 when the three largest metropolitan provinces, Istanbul, Ankara, and Izmir are excluded.

2 We exclude the economically inactive populations as our model examines the role played by employment type, occupational class, and employment sectors.

Independent variables used are age (measured in single years), sex (women taken as the comparator), education (the last school completed ranging from no school completed to masters/doctorate where 'primary school graduate' is the comparator) and marital status (where those who are married, divorcees and widows are compared to never married singles).

We have also taken economic activity, which is reported in three Census questions, to indicate: a) economic activity status in the week prior to the Census (salaried employees is the comparator); b) occupational classes (the largest group, production and related workers, transport equipment operators and labourers, is the comparator for seven other categories according to ISCO 68 (ILO 1969); and c) industry of employment (where various categories are compared to non-agricultural manufacturing industry).

Other individual characteristics are also included in the model: citizenship (comparing non-Turkish with Turkish); household occupation rate (which is measured by dividing the number of members of the household by the number of rooms available to the household); and home ownership (where home owners is the comparison group against tenants and subsidised tenants). Type of place of residence comparing rural with urban is taken as a contextual variable. Similarly, the socio-economic development level of the province in which the individual lived five years prior to the Census date is included. It is a composite measure of development level differences in Turkey compiled by State Planning Organisation in 2003 (Dincer, Özaslan and ve Kavasoglu 2003).

Many of the foreign-born in the Census data are likely to be those who were born abroad to Turkish parents, for example the second and third generation Turks from Germany. Another group comprises those foreigners who are married to Turkish citizens. However, our selection criterion was simply place of birth of the individual as reported on the Census day.

We build a binary logistic regression model for migration propensity between provinces. We have also used an aggregated cross-tabulation (from the 2000 Turkish Census data) on internal migration motivations in Turkey. This allows us to see differences in motivations by gender for the foreign-born and those born in Turkey.

International Migration in Turkey

Turkey's population grew from 67,803,927 in 2000 to 73,722,988 in 2010 (TurkStat 2011). The population growth rate has declined for two decades, but Turkey remains one of the 20 most populous countries in the world and the third largest in Europe. Internal migration was strong through the second half of the twentieth century. Until the late 1950s, more than 70 per cent of Turkey's population was living in rural areas (Gedik 1997). However, by the end of 2010, more than 76 per cent were living in urban areas (TurkStat 2010). In this period, industrial urban centres such as Istanbul, Ankara, Izmir, Bursa, Adana and Antalya were the top destinations and accommodated about 40 per cent of the total population (TurkStat 2010). According to the 2000 Census, 27.8 per cent of the total population resided in a province other than their birth place, which indicates a very high internal mobility rate. The mobility (between provinces) rate between 1995 and 2000 was 7.9 per cent. The corresponding figure for the 1985 and 1990 period was 8.1 per cent (Gedik 1997).[3]

For example, although net migration to Istanbul was 407,448 individuals in the Census 2000, it was also the top origin province with more than 513,000 citizens moving to other provinces. Hence, internal migration in Turkey is no longer dominated by rural to urban moves (Gedik 1997). In fact, Gedik (1997) notes that the majority of these moves were intra-urban and to a lesser extent intra-rural migrations.

Internal migration is also described as 'stepwise' migration (Conway 1980, Zorhy 2005), or a move that facilitates an international migration at a later date. This is true for Turkey which has been a key emigration country over the last five decades, creating a strong Turkish diaspora in Western Europe. International immigration is an essential part of the Turkish mobility regime and dates as far back as the fifteenth century when Turkish speaking and Muslim populations moved to what is now modern Turkey following the collapse of the Ottoman Empire. This is a pattern that has continued to the present.[4] These immigrants have settled in Bursa, Istanbul, Izmir and Tekirdag, which are among the most popular destinations.

Mass labour migrations in the 1960s and 1970s and later political migrations have contributed to the creation of migration channels for Turkish movers with key destinations such as Germany. These links are likely to play a part in immigration to Turkey. A sizeable Turkish diaspora emerged in Europe and elsewhere. The total number of Turkish citizens and people of Turkish-origin (i.e. second and third generations) living outside Turkey proper is in the region of five to six million and includes those individuals who have acquired citizenship of their 'new' countries.

In the 1990s and 2000s, Turkey attracted a significant number of immigrants from Africa, the Middle East and Asia who often arrived as transit or clandestine migrants (Kirisci 2007) and included over 800,000 undocumented immigrants who

---

3   Once migrations within provinces are included these rates go up to 11.0 per cent (1995–2000) and 9.6 per cent (1985–1990).

4   Immigrants include the expelled Jewish populations arrived from Spain in 1492–93 (Diaz-Mas 1992, Masters 2004), as well as large volumes of inflows (mainly Muslims) from former territories into what was then left of the Empire (McCarthy 1996). Turkey saw large intakes of the Turks and Muslims through compulsory and voluntary exchanges from the Balkans (Pentzopoulos 1962, Sirkeci 2006). Nearly 700,000 immigrants arrived in large groups from Bulgaria from the 1950s to 1990 (Caglayan 2007). Similarly, mostly in the first half of the twentieth century, particularly non-Muslim populations left Turkey for the new countries such as Greece, Serbia, and Bulgaria (Karpat 1985, Akgunduz 1998).

were apprehended. Among all immigrants to Turkey, 'Turkish Origin Foreigners' (TOF) (Unal 2011) included numerous Azeris and Turkmen (Dedeoglu 2011). A prosperous country and close to some conflict zones (e.g. Iraq, Afghanistan), Turkey became an attractive destination for many migrants. A significant number of Europeans also settled in Turkey. Europeans of Turkish origin (those who were naturalised in destination countries as well as others who were born abroad to Turkish parents), along with Turkish return migrants, constitute a large portion of the total immigrant population in Turkey.

To contextualise our analysis of internal migration of the foreign-born in recent decades we examine the trends of immigration to Turkey. The data we present are from an array of summary statistics on undocumented migrant apprehension, asylum statistics and work permits statistics gathered from the Censuses as well as from other official sources.[5]

Immigration to modern Turkey began with the influx of Turkish and Muslim populations during the First World War, War of Independence, and through the population exchanges in the 1920s and 1930s (Table 9.1). Population growth in Turkey has been much faster than the growth of the immigrant population. Therefore, the share of the foreign-born has declined over the years. Table 9.2 shows migration flows to Turkey between censuses. The larger volume of arrivals in the 1980s is mainly due to Turkish speaking migrants fleeing Bulgaria (Vasileva 1992, Karpat 1995).

Table 9.3 breaks down foreign-born population in Turkey by country of citizenship according to the 2000 Census. Historic and current ties (e.g. political, trade, migration and ethnicity) with certain countries make Turkey a destination for citizens of those countries. Hence, EU citizens are the second largest group after Turkish citizens among the foreign-born. International migration ties are evident with these countries as they are also popular destinations for Turkish migrants (e.g. Germany). For example, German citizens top the list with more than 85,000 followed by Bulgaria and the United Kingdom (Table 9.3). Conflict (e.g. in Iraq), geographical proximity (e.g. from Ukraine), and cultural proximity (e.g. to Azerbaijan) are other likely reasons for choosing Turkey as a destination.

---

5   Despite the wealth of statistics compiled by several government departments, the lack of co-ordination and incomparability of the data hinders sophisticated analysis (Sirkeci 2009).

Table 9.1   Stock of foreign-born in Turkey, reported by censuses, 1935–2000

Year of Census	Total number of foreign-born residents in Turkey	Foreign-born as per cent of the total population in Turkey
1935	962,159	5.95
1945	832,616	4.43
1950	755,526	3.61
1955	845,962	3.52
1960	952,515	3.43
1965	903,074	2.88
1970	889,170	2.50
1975	134,746	0.33
1980	868,195	1.94
1985	934,990	1.85
1990	1,133,152	2.01
2000	1,260,530	1.86

Note: The last Census in Turkey was conducted in 2000 and no data on immigration have been released by Turkish Statistics Office since then. There is an irregularity in the data for which we could not have an official explanation; our educated guess is that for 1975, the Census data seems to have excluded those foreign-born with Turkish citizenship and reported only foreign born with foreign citizenship.

Source: Turkish Statistic Institute, Censuses.

Table 9.2   Foreign-born migration flows to Turkey, 1975–2000

Census interval	Foreign-born flows			Total population
	Turkish citizens*	Foreign citizens	Total	
1975–1980			254,171	44,736,957
1980–1985	398,801	11,419	410,232	50,664,458
1985–1990	287,986	99,855	388,994	56,473,035
1995–2000	182,000	50,244	234,111	67,803,927

Note: *Also includes those with dual citizenship. The last Census in Turkey was conducted in 2000 and no data on immigration have been released by Turkish Statistics Office since then.

Source: Turkish Statistic Institute, Censuses.

Table 9.3   Stock of foreign-born in Turkey by country of citizenship, 2000

EU countries	Total	Men	Women		Total	Men	Women
Germany	85,354	42,430	42,924	Belgium	3,033	1.530	1,503
Bulgaria	35,245	17,392	17,853	Poland	2,730	1.368	1,362
UK	11,139	5,454	5,685	Italy	1,785	1.040	745
Netherlands	8,950	4,644	4,306	Romania	1,761	875	886
Austria	6,059	3,192	2,867	Denmark	1,760	849	911
Greece	5,660	2,883	2,777	Finland	1,214	602	612
France	4,105	2,103	2,002	Other EU	2,985	1.530	1,455
Sweden	3,759	1,772	1,987	*Total EU*	*175,539*	*87,664*	*87,875*
Other Industrialised and Developed Countries (IDC)							
Russia	13,641	5,810	7,831	Switzerland	1,419	711	708
USA	7,120	4,296	2,824	Japan	1,238	595	643
Norway	2,851	1,397	1,454	Other IDCs	2,963	1,398	1,565
Israel	1,914	987	927	*Total IDCs*	*31,146*	*15,194*	*15,952*
Neighbouring Countries (NC)							
Iran	7,859	4,830	3,029	Ukraine	2,097	857	1,240
Azerbaijan	7,498	4,718	2,780	North Cyprus	1,363	719	644
Iraq	5,101	3,024	2,077	Syria	766	511	255
Georgia	2,941	1,392	1,549	Armenia	229	76	153
				Total NC	*27,854*	*16,127*	*11,727*
Others							
Saudi Arabia	6,334	5,137	1,197	Kazakhstan	2,638	1,310	1,328
Uzbekistan	3,180	1,595	1,585	Others	13,698	10,001	6,645
Afghanistan	2,886	1,924	962	*Total Others*	*28,736*	*18,341*	*10,395*
Total foreign citizens	*262,815*	*135,271*	*127,544*				
Turkish citizens	*997,.676*	*470,122*	*527,554*				
TOTAL	**1,260,491**	**737,599**	**779,911**				

Source: Turkish Statistic Institute 2000 Census.

Table 9.4 represents the top ten immigration destination provinces in Turkey and these provinces' rankings in the internal migration league (see Map 9.1, which indicates the provinces of Turkey). The rankings show that the majority of these provinces are equally popular destinations for internal migration. The first four provinces in the list are the largest and most industrialised provinces and traditionally have been the main source areas for emigration from Turkey. Konya can be added to this group as another area of origin for international migration. Bursa, Tekirdag, and Kocaeli historically attracted immigrants from neighbouring Balkan countries. On the other hand, Antalya is a major tourist destination, located on the South coast and with very attractive holiday resorts. International and internal migrants alike are often destined to the most advanced areas, cities and towns, which also explains the top ten list of receiving provinces in Turkey.

Table 9.4   Top 10 internal migration destinations (provinces) for foreign-born without Turkish citizenship in Turkey, 2000

Province	In-migration ranking by volume	Net migration ranking	Women	Men	Total
Istanbul	1	5	29,409	25,235	54,644
Izmir	3	7	9,440	8,585	18,025
Bursa	4	6	9,134	8,814	17,948
Ankara	2	10	10,055	7,605	17,660
Antalya	5	3	5,269	4,733	10,002
Tekirdag	10	1	2,855	2,718	5,573
Konya	8	22	3,104	2,196	5,300
Hatay	25	52	3,288	1,308	4,596
Kocaeli	6	23	2,457	1,822	4,279
Adana	9	46	2,442	1,694	4,136
All 81 Provinces			130,762	103,349	234,111

Source: Turkish Statistic Institute 2000 Census.

Map 9.1   Map of Turkey, provinces

Source: http://en.wikipedia.org/wiki/File:BlankMapTurkeyProvinces.png.

Although the Census is supposed to capture all residents in Turkey on the Census day including asylum seekers and undocumented migrants apprehended, the number of foreign-born is significantly smaller than what we might expect. The total number of non-Turkish citizen foreign-born in Turkey was 262,815 in 2000 while the total number of asylum applications and undocumented migrants (all non-Turkish citizens by definition) was 265,000 for the period of 1995–2000 (Sirkeci 2009). During the decade following the Census, nearly 700,000 more added to this total (Sirkeci 2009).[6] Therefore, we are cautious and aware that our subsequent analyses may not represent current foreign-born population movements within the country.

Determinants of Internal Migration of the Foreign-born

Studying the internal migration of immigrants is important because of the increasing volume of immigration to Turkey. The internal mobility rate in Turkey was 78.81 per thousand in 2000 (Table 9.5). This high rate of internal mobility has been observed in four previous Censuses (the corresponding migration rates were 70.35 in 1970, 65.47 in 1985 and 81.33 in 1990). For the foreign born, the rate of internal mobility (between provinces) was 94.74 in 2000.[7] This is significantly higher than the rate for Turkish born. It is in line with our expectation that people

---

6   See Sirkeci (2009) for details of potential undercount of immigrants in Turkish Censuses and registers.

7   In our estimation, we have only included those who were in the country five years prior to the Census.

with international migration experience are likely to have a higher propensity to migrate.

Not surprisingly, the migration rate for foreign born men is slightly higher than for foreign born women. Perhaps confirming what the migration literature would suggest, internal mobility rates for age groups younger than 35 are substantially higher than the overall rate. The highest rate of internal mobility (202.32 per thousand) is estimated for those aged 20–24 while the lowest rate (39.29 per thousand) is found among those aged 70–74. There are no large differences between men and women regarding age specific internal mobility rates. However, we have found that women constitute 51 per cent of total foreign born internal migrants whereas the corresponding figure for those born in Turkey is 46 per cent.

Table 9.5   **Internal migration rates, per thousand population, by age and gender for the foreign-born and Turkish-born, 2000**

Age groups	Foreign born			Turkish born		
	Total	Male	Female	Total	Male	Female
5–9	117.62	114.81	120.36	86.61	86.96	86.23
10–14	96.56	97.45	95.72	84.36	87.74	80.71
15–19	128.34	132.39	124.45	128.52	134.42	122.33
20–24	202.32	196.38	208.79	205.75	224.16	186.41
25–29	180.03	181.92	177.84	171.30	187.65	154.65
30–34	111.22	110.52	112.04	125.81	140.73	110.36
35–39	81.51	82.10	80.85	97.79	108.62	86.76
40–44	64.04	65.91	62.00	83.09	90.31	75.53
45–49	56.79	57.19	56.38	75.80	82.59	68.79
50–54	56.48	55.47	57.54	67.22	73.44	61.02
55–59	53.31	51.82	54.91	57.71	61.54	53.98
60–64	50.24	50.21	50.27	48.92	49.89	48.05
65–69	43.79	42.10	45.77	44.10	42.12	45.94
70–74	39.29	42.02	35.84	42.86	38.21	46.53
75 +	42.28	46.68	36.25	48.01	42.24	52.18
Total	94.74	94.46	95.06	110.20	118.74	101.48

Source: Turkish Statistic Institute 2000 Census.

The logistic regression model summarised in Table 9.6, using individual, household and province level indicators is an attempt to explain the variation in internal migration behaviour of the foreign-born (R square: 19.1 per cent. Modelling of the determinants of internal migration for the Turkish-born was not possible due to data availability). Understandably, some life-stage related variables appear to be significantly influencing the propensity for internal migration for the foreign-born. For example, married foreign-born people are less likely to migrate compared to others, while the highest log odds are recorded for widows. As individuals age, their likelihood of changing residence decreases. Foreign-born men are more likely to move than women. This difference is likely to be much wider for the Turkish-born, as men constitute more than 54 per cent among internal migrants.

Among all Turkish-born, the 20–24 and 25–29 age groups have the highest probability of migration internally. These are slightly lower than the corresponding probabilities among foreign-born. In the regression model, we have found that as the foreign-born individual gets older, the propensity to migrate decreases by over 2 per cent per year (Table 9.6). This complies with the literature, as young men aged 20 to 30 are expected to be most mobile (e.g. Finney and Simpson 2008, Stillwell, Hussain and Norman 2008, Dennett and Stillwell 2010a).

Marital status and gender (within a household) is believed to play a key role in migration decisions (Gubhaju and De Jong 2009). Married foreign-born individuals are found to be less likely to migrate than those who are not married. Similarly, the share of married Turkish-born persons among internal migrants is smaller than those never married, divorced, widowed or separated.

Recent studies provide evidence that education is positively correlated with migration propensity (e.g. Dennett and Stillwell 2010b, Morrison and Clark 2011). Our findings also suggest that the higher the level of education the higher the propensity for internal migration among foreign-born in Turkey. Internal mobility propensities are lower among those with no formal education and the graduates of the new combined primary school (years 1–8) compared to primary school graduates (years 1–5). Similarly, graduates of vocational secondary schools (years 6–8) are less likely to move between provinces.

On the other hand, individuals with upper secondary school and high school educations (years 9–11) are 1.7 times more likely to move to another province compared to their less educated peers. Foreign-born university graduates and those with higher degrees such as masters or doctorates are respectively 2.97 and 2.17 times more likely to migrate compared to their primary school graduate counterparts. One reason for this pattern is the higher education system in Turkey where a centralised examination system selects students almost randomly and thus most students move to another district and province (sometimes over very long distances) to enter university programmes. This often leads to further relocation into towns and cities where universities are, or to an industrialised or economically more developed province. Like the foreign-born, about 23 per cent of Turkish born university graduates have migrated internally. Nevertheless, the largest segments of Turkish-born internal migrants are primary and secondary school graduates (graduating at years 6 and 11) – 32 per cent and 30 per cent respectively.

Employment and occupational classes are also significant and influence the odds of internal migration. Salaried or waged employees and industrial workers are more likely to migrate compared to others. Foreign-born individuals employed in services are significantly more likely to move to another province compared to others working in manufacturing and industrial jobs. Those employed in the service sector constitute over 60 per cent of all Turkish-born internal migrants.

Home ownership, an indicator of wealth, significantly decreases migration odds. Migration is about four times more likely for tenants, while it is 6.8 times more likely for tenants on subsidised accommodation, compared to home owners. Those on subsidised accommodation are often public workers (teachers, military, doctors, etc.) who have to serve in different locations by rotational appointments, or those working on projects in other cities, or living in company owned accommodation.

At the household level, crowdedness of the household is also a significant factor. Interestingly, the more crowded the household the less likely the members are to migrate. This is likely linked to household formation. Couples with young children may live in crowded households but cannot or may not be willing to move because, for example, of schools or local care arrangements.

An interesting finding from this study is that there is a significantly lower migration propensity among the foreign-born without Turkish citizenship compared to the foreign-born with Turkish citizenship. The former is about 68 per cent less likely to migrate. Evidently, those who were born to Turkish parents abroad or naturalised as Turkish citizens are more likely to move within Turkey compared to foreign nationals.

Migrants tend to move from less developed and/or more deprived areas to relatively better off areas once they settle in destination countries. In the case of the foreign-born in Turkey, we found that at a macro level, the socio-economic development of the province where the individual was a resident five years prior to the Census has an impact on mobility. The higher the socio-economic development index score of the place of residence five years ago, the lower the likelihood of internal migration. Hence, those who reside in more developed provinces are less likely to move compared to those who live in less developed or deprived areas of the country. This is also consistent with the international migration trends in relation to socio-economic development levels (Sirkeci, Cohen and Yazgan 2012). It is possible to argue that the environment of socio-economic insecurity is one of the causes for internal, as well as international mobility (Sirkeci 2006, Cohen and Sirkeci 2011).

Table 9.6   Logistic regression model: factors affecting the internal migration of foreign-born in Turkey, 2000

	Exp(β)
Constant	0.359***
Comparator: Female	1.00
Male	0.974**
Age (12+)	0.977***
Comparator: Never married	1.00
Unknown/NA	0.044***
Married	0.980
Divorced	1.034
Widowed	1.407***
Comparator: Primary school	1.00
NA/illiterate	0.841***
No school completed	0.714***
Combined primary and lower secondary school	0.759***
Secondary school	1.298***
Secondary vocational school	0.970
Upper secondary school	1.764***
Upper secondary vocational school	1.702***
High School	1.869***
University	2.963***
Masters/Doctorate	2.177***
Unknown	0.407***
Comparator: Salaried employee	1.00
NA/Unknown	2.974***
Employer	0.819***
Self-employed	0.807***
Unpaid-family business	0.625***
Comparator: Production workers, transport equipment operators and labourers (a)	1.00
Professional, technical and related	0.996
Administrative and managerial	0.935
Clerical and related	0.775***
Sales people	0.905***
Service workers	1.087***

Table 9.6   *Concluded*

Agriculture, husbandry, forestry workers, fishermen and hunters	1.072
Unknown	0.790
Comparator: Manufacturing	1.00
Agriculture etc.	1.415***
Mining etc.	1.085
Utilities	1.057
Construction	1.084**
Sales and commerce	1.132***
Transport, Communication	1.030
Finance, Insurance, Estate	1.120***
Social Services	1.654***
Unknown	0.536***
Comparator: Home owner	1.00
Hotel, dormitory etc.	3.989***
Tenant	4.020***
Subsidised accommodation	6.826***
Tenant – unpaid	1.425***
Others	2.193***
Unknown	2.334***
Crowdedness of the household	0.998***
Type of place of residence rural *(Comparator: urban)*	0.250***
Citizenship foreign *(Comparator: Turkish)*	0.327***
Socio Economic Development level of the province lived 5 years ago	0.925***

Note: (a) ISCO68 classification (ILO, 1969). $*=p<0.05$; $**=p<0.01$; $***=p<0.001$.
Source: Turkish Statistic Institute 2000 Census.

Internal Migration Motivations

For the first time in Turkey the 2000 Census collected information about the reasons for internal migration. We have observed differences between foreign-born and Turkish-born populations regarding individuals' reasons for internal migration. Overall, about 25 per cent of the 4,788,193 Turkish born internal migrants moved from one province to another as a dependent. The corresponding figure was 17 per cent for men and 35 per cent for women. This confirms our argument that despite increasing female participation in international and internal migration, women typically move as dependents (Jolly and Reeves 2005, Cohen, Rodriguez and Fox 2008, DESA-DAW 2009). However, we found evidence that foreign-born women are more likely to migrate than foreign-born men (Table 9.6); more than a quarter of foreign-born internal migrant women reportedly moved as dependents (Table 9.7). Possibly, this is understandable as the composition of foreign-born migrants is skewed towards younger ages. Equally interesting is our finding that a significant portion of (Turkish-born and foreign-born) women migrated internally because of marriage, whilst the corresponding figure for men was low (1.2 per cent for Turkish-born men and 1.9 per cent for foreign-born men). Thus, marriage appears as the second most important motive for internal migration among women.

The single most important reason for Turkish-born men to move is a new job, job relocation, or job search, whereas for the foreign-born 'Other' reasons top the list, accounting for around a third of moves. Among foreign-born men, job related motives (new job and re-location) were around 26.5 per cent, while the corresponding figures for foreign-born women and Turkish-born women are 15.6 per cent and 16.4 per cent.

The conflict in the south east of Turkey is a concern and contributes to the 'environment of human insecurity', a likely cause for (internal and international) human mobility (Sirkeci 2006, Cohen and Sirkeci 2011). Nevertheless, only a small fraction of foreign-born internal migrants reported security as their reason for migration, according to the Census. The corresponding figures for Turkish-born movers are larger but still account for less than one per cent of the total. One possible reason for the small volume of internal migration among the foreign-born for security reasons is that the provinces where security risk is very high are receiving very few immigrants. For example, insecurity is explicit in south east provinces, but even the largest province (Diyarbakir) and the largest migrant sending provinces (Bingol and Elazig) in the region receive a very small share of total foreign-born arriving in Turkey. Only 0.1 per cent of immigrants preferred these three south eastern provinces, whereas 65 per cent immigrated to the five largest industrial provinces (Istanbul, Izmir, Ankara, Bursa, Antalya). Nevertheless, along with Istanbul, in eight eastern provinces (i.e. Diyarbakir, Mardin, Siirt, Bingol, Hakkari, Sirnak, Mus, and Tunceli), where most violent events occurred during the 1980s, 1990s and 2000s, the largest internal migration rates for security reasons were reported for the Turkish born. Large western provinces such as Istanbul, Icel, Izmir, and Ankara were at the receiving end of those internal migrants leaving their homes for security reasons.

Table 9.7   **Internal migration motivations of the foreign-born compared to the overall population in Turkey, aged 5 and over by gender, 2000**

Motivations	Gender	Born in Turkey		Foreign-born	
		Number of migrants	Per cent of migrants	Number of migrants	Per cent of migrants
New job/search	Men	955,323	26.7	7,165	17.0
	Women	263,519	8.8	3,608	7.8
Job re-location	Men	494,992	13.8	3,991	9.5
	Women	226,368	7.6	3,615	7.8
As dependent	Men	620,484	17.3	5,966	14.2
	Women	1,037,957	34.8	12,272	26.4
Education	Men	434,821	12.1	6,834	16.2
	Women	263,532	8.8	5,748	12.4
Marriage	Men	44,121	1.2	790	1.9
	Women	511,239	17.1	6,044	13.0
Earthquake	Men	102,850	2.9	1,816	4.3
	Women	102,148	3.4	2,115	4.5
Security	Men	33,369	0.9	230	0.5
	Women	22,991	0.8	191	0.4
Other	Men	801,257	22.4	14,499	34.4
	Women	484,376	16.2	12,056	25.9
Unknown	Men	92,957	2.6	836	2.0
	Women	71,713	2.4	869	1.9
	Men	3,580,174	100	42,127	100
Total	Women	2,983,843	100	46,518	100
	All	6,564,017		88,645	

Source: Turkish Statistic Institute 2000 Census.

Another interesting difference is the effect of the Marmara earthquake which happened near Istanbul in 1999. The Census was conducted just over a year after the earthquake. However, there are other cities with high earthquake risks in Turkey. A larger portion of the foreign-born (about 4.3 per cent) reacted to the risk and changed their place of residence, compared to a smaller proportion for their Turkish neighbours (3.1 per cent). One can perhaps speculate that this is because the foreign-born are less tied to the place of residence compared to the natives, or that the Turkish-born may be accustomed to the earthquake risk. While earthquakes were mentioned as a reason to move by only about 3 per cent of all internal migrants, it was much higher for individuals living in the Istanbul, Sakarya and Kocaeli provinces, which were the most affected by the 1999 earthquake: 30 per cent, 28 per cent and 15 per cent respectively. This is clearly strong supportive evidence for the effects of environmental changes and hazards on internal mobility.

Conclusion

In this chapter we have explored and revealed the characteristics of immigrants and their internal mobility in Turkey with the help of 2000 Census data. As an emerging destination country for immigrants, analysis of internal mobility of immigrants is warranted. As officially reported, 1.26 million immigrants constituted nearly two per cent of the total population of Turkey in 2000. A large volume of undocumented migrants apprehended and a relatively sizeable asylum seeker population suggest an even larger foreign-born population stock in Turkey.

Working with Census data does have shortcomings. Data are incomplete and the Census only reports the difference between the place of residence on the day of the census and five years prior to the Census day. Therefore, we deal with only a fraction of total moves in Turkey. Hence further surveys as well as qualitative studies would be useful in filling the gap in our understanding of immigration and internal migration in Turkey.

We have observed some similarities in internal mobility patterns as well as differences between the foreign-born and Turkish-born segments of the population in Turkey. Nevertheless, it was beyond the scope of our study to provide a comprehensive comparison of the two segments. Our regression model explains 19 per cent of the internal migration of foreign-born in Turkey, where education, home ownership, citizenship, gender, marital status, occupations, and socio-economic development level are all found to be significant factors in explaining internal migration behaviour.

Interestingly, we have found that foreign-born females are slightly more mobile than their male counterparts. This contrasts with the fact that men constitute a higher share of Turkish-born internal migrants. We do not have an explanation for this and further investigation is needed.

Education is surely one key determinant for internal mobility propensity of the foreign-born in Turkey. Those with higher education levels tend to be more mobile. This can be partly due to the structure of higher education in Turkey, as young people are required to sit in a national selection and placement exam and move around the country accordingly. This often favours large cities with many universities such as Istanbul, Ankara and Izmir.

We have found significant differences in internal migration propensities with regard to the citizenship status of the foreign-born in Turkey and this warrants further analyses (both quantitative and qualitative) to understand the differentiation process regarding citizenship. In this regard, family origins of the foreign-born should also be taken into account as a large proportion of the foreign-born appear to be Turkish citizens and likely to be the descendants of Turkish emigrants living abroad.

It is important to further investigate the number of moves within the destination country following the first arrival, for which surveys and qualitative interviews can be useful.

Also, we note that comparisons of migration motivations of foreign-born and Turkish-born by individual characteristics and geographies will contribute to our understanding of internal migration in Turkey. Further investigations are needed to understand the 'other reasons' reported in the Census data: 25 per cent of the foreign-born women and 34.4 per cent of the foreign-born men had other reasons to move to another province in Turkey.

Acknowledgement

We thank the Turkish Statistical Institute for granting access to Turkish Census data as well as some other data tables which were freely available to download from TurkStat (Turkish Statistical Institute) website at http://www.tuik.gov.tr [accessed May 2011].

Chapter 10

Here for Good: Immigrants' Residential Mobility and Social Integration in Athens During the Late 1990s

George Kandylis and Thomas Maloutas

Introduction: New Immigrants, Segregation and Inequalities in Athens

Post-war urbanisation in Athens is a story of high social mobility, usually involving comparatively low levels of residential mobility. Large segments of the urban tissue have been gradually renovated following the personal and intergenerational upward social mobility of their residents that had settled there just after their departure from rural areas (Maloutas 2004). Intergenerational resource allocation had a crucial role in the absence of public housing policies and in the broader context of a family-oriented reproduction model (Antonopoulou 1990, Maloutas 1990, Allen et al. 2004). As a consequence, urban social segregation remained relatively low and wide residential areas continue to be relatively mixed in terms of their social structure. On the other hand, the increasing suburbanisation since the mid 1970s gave birth to new suburban and peri-urban residential areas characterised by different and often increased levels of social homogeneity (Sayas 2006, Arapoglou and Sayas 2009).

The beginning of a massive immigration wave, mainly (but not exclusively) from the Balkans and Eastern European countries in the early 1990s, resulted in rapid inflows of new residents in Athens and the broader metropolitan area. In contrast with the dominant idea that international migration to Greece was an exceptional and temporary phenomenon (Marvakis 2004), the next two decades proved that immigrants were here to stay. Immigrants' labour force covered employment shortages in certain activities, mainly in the informal sector. Despite the policy of minimum regulation and the continuously ambiguous legal status that confine immigrants to quasi (il)legal conditions, quite a few individuals, households and migrant communities exploited opportunities in the labour market (especially in construction activities and domestic services) and in the housing market (especially in the private rented sector of the centre, but also in the low standard detached houses of the periphery). They thus followed successful paths of social mobility, involving family reunification, integration of children in the school system and waves of *ex post* legalisation. Moreover, shifting socio-economic hierarchies of ethnic groups emerged (Kandylis, Maloutas and Sayas (forthcoming); see Cavounidis (2002, 2004) for detailed data). Housing itineraries of the new population may be different from those of the Greek-born rural migrants in the first post-war decades, and may also lead to different relations between residential location and social mobility prospects.

The residential pattern of the immigrant population in Athens is one of low spatial segregation, at least if compared with the racial segregation level in the hypersegregated American metropolises (Arapoglou 2006, Maloutas 2007, Kandylis, Maloutas and Sayas 2012). Immigrants' settlement contributed to the further *reduction* of socio-economic segregation, as new immigrants, mostly belonging to the lowest strata of the socio-economic scale, reside in many and in socially diverse areas of the city. This is especially due to the dispersal of the Albanian population, by far the largest group within the total immigrant population in the metropolitan region (more than 50 per cent in 2001). Smaller groups are characterised by less even residential distribution patterns and tend to be concentrated in specific parts of the metropolitan area, especially in or close to the city centre and in certain municipalities of the distant peri-urban ring. Nevertheless, the residential concentration of these groups is not accompanied by isolation from the Greek population. Almost everywhere, even at the census tract level,[1] Greeks form the majority of the population, disproving widespread anxieties about the so-called 'ghettoisation' of specific places in the urban tissue (Arapoglou et al. 2009).

However, low segregation does not imply low levels of housing inequalities, as witnessed in other parts of Southern Europe as well (Arbaci 2008). Poor housing conditions were the rule for most newcomers in the early 1990s (Psimmenos 2000, Lazaridis and Psimmenos 2001, Lavrentiadou 2009). During the next two decades, these conditions have remarkably improved for many immigrants (Labrianidis and Lyberaki 2001). However, the members of most immigrant groups face housing conditions far below the Greek average in terms of available domestic space and housing age, either near the centre or in the periphery of the metropolitan area. Moreover, poor economic means and problematic legal status lead to proportionally limited access to home-ownership, in a context where the absence of public regulation of housing continues to render home-ownership a major component of social integration and mobility (Allen et al. 2004). The co-existence of immigrants and Greeks in the same places (even in the same buildings of the densely populated residential areas around the city centre), is consequently producing ethnically mixed but socially polarised residential areas (Maloutas 2007).

Many immigrant groups in the metropolitan region of Athens show residential relocation rates that are higher than those for the Greek population. Other groups appear much less mobile. The main question we deal with in this chapter is whether different rates of residential mobility are at the same time an indication of

---

1   The average census tract population in the metropolitan region of Attica is 750 residents.

unequal social status. Before that, we present the general picture of the spatiality of the relocations of different immigrant groups and the Greek-born population. In this way we propose some considerations about the complex relationship between residential relocation, segregation and social integration prospects.

Data and Definitions

The census data of 2001 include information about the permanent place of residence five years before the census. If the previous place of residence was in Greece, it was recorded at the Municipal level. If it was abroad, the recorded information is limited to the previous country of residence. Relocations within the same Municipality were not recorded. This leads to an underestimation of relocation rates for all ethnic groups, but it is not possible to estimate if this underestimation is more significant for some groups than for others.

The Municipalities (*Demoi* and *Koinotites*) constitute the first level of the Greek local administration system. In 2001 the country was divided into 1,034 Municipalities with an average population of 10,575 residents, but varying greatly between 756,652 residents in the biggest Municipality of Athens and only a few hundred in many rural municipalities.[2] The metropolitan region of Athens as defined for the purposes of this chapter consists of 115 Municipalities with an average population of 33,194 residents. This area is almost identical to that of the formal Attica region, leaving aside a few municipalities in remote areas.

Immigrants are defined on the basis of non-Greek nationality. We focus on certain immigrant groups that occupy different positions in the socio-economic hierarchy and exhibit various rates of residential mobility. We take into account two different types of residential mobility. First, the internal relocations of immigrants, either from rural areas (the initial settlement for many) or from other parts of the metropolitan region. Second, the recent arrivals from the countries of origin; although these new arrivals do not fall into the category of internal mobility, it is interesting to see how (if at all) newcomers follow the spatial pattern of those who settled earlier.

Residential Relocation, Segregation and the Importance of the Local Context

Under conditions of socio-ethnic inequality, high residential mobility rates can be considered to constitute the opposite of the spatial entrapment of immigrant groups. Segregated places of residence are often considered as loci where the concentration of deprivation, anomie, poor access to social services and low expectations limit the prospects for social integration and upward social mobility.

---

2   In 2010 the number of municipalities has been drastically reduced by a third, after an administrative restructuring decided by the central government.

Residential mobility is then the antidote to ghettoisation; physical and social mobility 'are often inseparable' (Houseman 1981: 12). Going back to a social ecology perspective, in which successive waves of minority residential mobility are thought to point out respective waves of social assimilation, this approach has led to the adoption of desegregation policy measures both in American and in European cities (Phillips 2010). However, as an important volume of research has shown, the relationship between segregation and integration is more complex (van Kempen and Ozuekren 1998, Musterd 2003, 2005, Jargowsky 2009, Bolt, Özüekren and Phillips 2010). Immigrant groups' residential congregation may also be attributed to practices aimed at fostering the bonds of their members, providing a social environment of solidarity, preserving cultural distinctiveness, promoting ethnically oriented economic activity, building local institutions and inventing forms of political participation in the host society. Thus, places of concentration may eventually enhance integration in the host society. Even if relocation leads to a less segregated residential pattern, this is not necessarily an indication of integration. On the contrary, immigration to Southern European cities provides evidence that residential dispersion may be associated with residential exclusion (Arbaci and Malheiros 2010).

Immigrants' residential relocations may be attributed to individual and household strategies. Immigrants will possibly move in search of better jobs or housing conditions, but they would also do so because of a new job or a change in the household structure. Moreover, we expect spatial mobility prospects to be mediated by ethnic community relations that reproduce or form common ties based on common origin. Informal and semi-formal networks have both motivational and practical consequences for immigrants' residential patterns. Networks provide information and support in looking for a house, while creating places of solidarity where collective appropriation of space becomes possible. They thus direct themselves to specific destinations that they already render attractive.

What the Athenian case has indicated in the past is that social mobility can be largely unrelated to residential relocation, at least with respect to the lower and lower-middle strata (Maloutas 2004). New immigrant settlers face a largely different context. The association between their residential mobility and their social status may indicate dissimilar paths of integration.

In the following sections of this chapter we seek to investigate the preconditions of immigrants' residential mobility in the metropolitan area of Athens by examining the spatial distribution, the socio-economic position and the housing conditions (housing space and tenure) of those who move vis-à-vis those who do not. As immigrants' relocations do not take place in a prearranged space that belongs solely to the established population, we consider the produced mobile geography in the light of tendencies for ethnic segregation/congregation. Comparing the relocation patterns of the Greek and the immigrant population, we obtain a more systematic picture of how immigrants' settlement contributes to the reformation of the metropolitan region.

Magnitude of Immigrants' Residential Relocation

14.6 per cent of the Greeks born before 1995 that resided in one of the 115 Municipalities in the mainland Attica region in 2001 had changed Municipality of residence between 1995 and 2001 (Table 10.1, column 3), i.e. they had come to their current location from any other Municipality of the same region or of the rest of the country. This compares to 20 per cent for the Albanian population and 19.3 per cent for the aggregate immigrant population, leaving aside those that came to Greece later than 1995. The residential mobility of other immigrant groups varies; some of them (e.g. Ukrainians and Bulgarians) exhibit comparatively high and others (e.g. Pakistani and Nigerians) comparatively low rates. Immigrants' residential mobility is of course far greater if relocations from outside Greece are taken into consideration (Table 10.1, columns 4 and 5). This kind of move is not literally part of the internal minority migration. However, excepting immigrants that settled in Greece after 1995, it is likely that we underestimate the internal residential mobility of the immigrant groups, as relocations are possibly more common in the early period right after the initial settlement (Bolt and van Kempen 2010).

As we suggest elsewhere (Kandylis, Maloutas and Sayas 2012), immigrant groups in Athens occupy unequal positions in the emerging socio-ethnic stratification. The relationship between the socio-economic hierarchy of immigrant groups and their internal residential mobility is quite ambiguous. Some of the most deprived (Pakistani, Bangladeshi, Indian) are also characterised by low relocation rates, but this is not true for people from Iraq who belong by and large to the same socio-economic stratum. Household structures – male boarding houses in the case of the former – combined with low education level and language skills may impede contact with the broader community and opportunities this may offer in terms of housing and employment. Some female dominated groups from Eastern Europe (Bulgarian, Ukrainian, Moldavian) appear highly mobile, while the opposite is true for female dominated groups from outside Europe, either settled in Athens much longer (Filipino) or rather recent ones (Ethiopian). Although these groups are employed mainly in domestic services, the former may be more mobile in terms of employment and housing (these often coincide) due to their higher education level that leads to seeking more effectively better opportunities. Immigrant groups from the Balkans and Eastern Europe that occupy similar hierarchical positions (i.e. Albanian, Polish, Bulgarians and Romanian) seem to differentiate on the basis of their time of settlement, as the Polish (the only Eastern European group that partly settled in Athens before the 1990s) presents lower mobility.

Table 10.1   Residential relocation rates by nationality in Athens, 1995–2001

National group	Population 2001	Internally relocated 1995–2001 %[*]	New arrivals after 1995 %[**]	Aggregate % of relocated[***]
Greeks	3,362,658	14.6		14.6
Albanians	206,978	20.0	54.0	63.2
Polish	11,509	15.9	44.1	52.9
Russians	11,324	17.3	45.9	55.3
Bulgarians	11,051	27.0	76.3	82.7
Romanians	10,222	23.9	73.0	79.5
Pakistani	10,130	13.5	70.0	74.1
Ukrainians	9,932	30.3	85.5	89.9
Egyptians	7,144	13.8	45.0	52.5
Filipinos	6,475	16.1	26.6	38.5
Bangladeshi	4,758	11.9	85.9	87.6
Syrians	4,699	18.4	60.0	67.4
Georgians	4,332	27.1	72.8	80.2
Moldavians	3,922	29,2	91.7	94.1
Indians	3,151	14.4	67.5	72.2
Armenians	2,856	21.9	58.0	67.2
Kazaks	2,132	17.4	42.3	52.3
Nigerians	1,823	7.6	61.1	64.1
Lebanese	1,402	18.8	30.5	43.6
Ethiopians	1,205	10.2	48.2	53.5
Total immigrant population	345,979	19.3	55.9	64.4

* Those that settled in Greece before 1995 (internal movers) and had changed Municipality of residence at least once by 2001, as a percentage of the respective pre-1995 population of 2001.

** Those that settled in Greece between 1995 and 2001 (newcomers), as a percentage of the respective population in 2001.

*** The aggregate percentage of internal movers and newcomers, as a percentage of the respective population in 2001.

Source: 2001 Population Census, EKKE-ESYE 2005.

Spatiality of Immigrants' Residential Location

Interestingly, the spatial structures of the Greek and the immigrant residential mobility are characteristically dissimilar. The Municipalities that received relatively significant numbers of Greek movers (i.e. the Municipalities with the highest values for the location quotient of the incoming movers) are generally located in the vast peri-urban area in the northern and eastern part of the Attica region (*Mesogheia*[3]), followed by some suburbs at the margins of the conurbation and by few traditional upper-middle class suburbs. Other middle class suburbs and the extended petit bourgeois residential areas around the city centre (i.e. those that received great numbers of new Greek settlers in previous rounds of suburbanisation since the 1970s, see Maloutas 2000) continue to receive new inflows but at a rate that is close enough to the city average. The central Municipality of Athens and most working class municipalities in the western part of the conurbation and in the western peri-urban part of the region received new settlers at a rate far below the average.

The case with Albanian movers was different. Leaving aside few high-status municipalities where new Albanian settlers represent an important part but only of a small local Albanian population, one can observe that Albanian relocation resulted in relatively important inflows mainly in middle and lower-middle strata municipalities, either on the east or on the west end of the central city. Most traditional working class areas attract Albanian relocations that are close and usually higher than the average rate, while peri-urban municipalities (with few exceptions) figure at the bottom of the list. With 16.2 per cent new Albanian settlers in its Albanian population, the Municipality of Athens lies slightly below the average.

In terms of distance from the centre, the spatiality of destinations for most other immigrant groups' inflows is not very different. High location quotients of incomers (on the basis of a local population of at least 50 people) are found in municipalities close to the urban core.

A comparable picture is produced when the net effect (i.e. internal inflow minus internal outflow) of relocation is taken into account. Map 10.1 presents the municipalities of the metropolitan area that benefited importantly and those that lost importantly from the relocation flows of the Greeks. The depopulation of the city centre and of some of its adjacent areas in the advantage of specific suburbs shape a centrifugal relocation pattern. As a matter of fact, between 1995 and 2001 the Municipality of Athens lost about 42,000 inhabitants or 6.5 per cent of its population in 1995. A few other neighbouring municipalities and fewer distant ones presented more moderate rates of population decrease.[4]

---

3   Residential relocations to this part of the peri-urban ring multiplied the size of already existing settlements. This second wave of suburbanisation can be represented as a case of urban sprawl in areas formerly specialised in traditional agricultural and/or in tourism activities, in the specific form of accommodating urban residents in private dwellings during their summer vacations (Sayas 2006).

4   The Greek population decreased in 18 out of the 114 municipalities of the metropolitan area. Apart from the quite impressive depopulation in the Municipality of

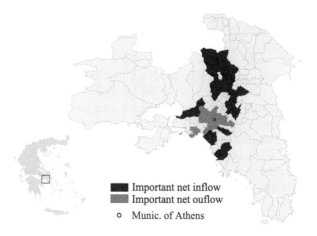

Map 10.1   Municipalities with important net effect of Greek relocation, metropolitan Athens, 1995–2001

Note: Municipalities of important net inflow are those where the increase rate of a group is at least twice the increase rate of the same group at the metropolitan level, while they represent at least 1 per cent of the total pre-1995 population of the same group in 2001. Municipalities of important net outflow are those where the population of a group decreased at a rate at least absolutely equal to the increase rate of the same group at the metropolitan level, while they represent at least 1 per cent of the total pre-1995 population of the same group in 2001.

Source: 2001 Population Census, EKKE-ESYE 2005.

Map 10.2 represents the net effects of the internal relocations of the aggregate immigrant population. There seems to be enough evidence that parts of the inner suburban ring of the urban core receive the main flow of internal immigrant movers. Interestingly, certain destinations of mobile immigrants adjacent to the centre are at the same time places where the Greek population shows tendencies to depart. As for the Greeks, the Municipality of Athens has again a negative relocation balance (although this is not so 'important' with the employed criteria). However, one should not fail to recognise that this map is composed by different relocation patterns for the different immigrant groups. In the case of the highly centralised Polish (as well as Nigerians) and of the moderately centralised Romanians (and Egyptians), a centrifugal pattern is again identified but leading mainly to less distant areas. The highly mobile Ukrainians tend to abandon not only the centre, but also the northern suburbs where many work as domestic personnel to Greek households. Albanian and Bulgarian immigrants that relocate move to nearby suburban destinations but they abandon not so much the centre of the city as other

Athens, this also affected some lower-middle class areas, some working-class areas and also some traditional upper-middle class suburbs.

suburban areas. Similarly, the most important relocation flows of the Pakistani group seem to occur between suburban municipalities in the western part of the city.

Map 10.3 represents the municipalities receiving the inflows of new immigrants, i.e. of those that settled in Greece after 1994. The importance of the city centre for these new settlers is clearly different; many newcomers (Polish, Bulgarian, Romanian, Pakistani and Ukrainian) have settled in the Municipality of Athens by 2001. Other newcomers of the same groups settle in municipalities where their co-ethnic predecessors relocated. Therefore, the urban core continued to receive newcomers, while not attracting massive internal resettlement.

On the other hand, there is no clear evidence that the ethnic communities in Athens tend to congregate systematically across their initial ethnic residential patterns. The correlation coefficients between (a.) the location quotient of the members of every immigrant group per Municipality in 1995 and (b.) the location quotient of the recently (later than 1994) arrived members of every immigrant group per Municipality are very close to zero or even negative. The newcomers do not exhibit any specific tendency to reside in areas that their co-ethnics already populated at the metropolitan scale. There is only a slight difference for the Pakistanis that came to Greece after 1995 who show a marginal tendency to concentrate in areas where the location quotient of their group was already high in 1995. No similar tendency is traced for the other groups.

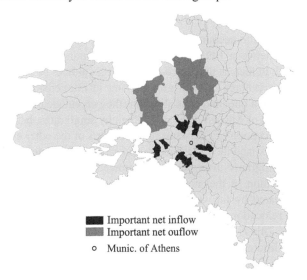

Important net inflow
Important net ouflow
o   Munic. of Athens

Map 10.2   Municipalities with important net effect of internal immigrants' relocation

Note: For the definition of the net effect of internal relocations see the Note on Map 10.1.

Source: 2001 Population Census, EKKE-ESYE 2005.

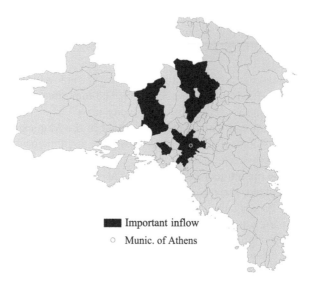

Important inflow
○   Munic. of Athens

Map 10.3   Municipalities with important inflows of new immigrants

Note: Municipalities of important inflow of newcomers are those where the location quotient of newcomers is bigger than one, while they represent at least 1 per cent of the new arrivals.

Source: 2001 Population Census, EKKE-ESYE 2005.

If we also take into account that the aggregate population of each immigrant group increased between 1995 and 2001, then a certain tendency for dispersion may be assumed. In fact, as illustrated in Table 10.2, indices of both concentration (the aggregate percentage of each group living in municipalities with location quotients higher than one) and centralisation (the percentage of each group living in the Municipality of Athens) decreased, albeit by small rates, between 1995 and 2001. With the exception of Pakistanis, the members of the other eight immigrant groups tended to expand their presence to wider parts of the metropolitan area.

However, immigrants' dispersal is not followed by further desegregation from the Greek population of the metropolitan area, i.e. to a more even residential pattern, as indicated by the values of the dissimilarity index (DI). Despite the relatively high relocation rates, these values did not change significantly between 1995 and 2001. This is partly attributed to the fact that the relocation of Greeks was directed to different parts of the metropolitan region. In this sense, immigrants' settlement in more municipalities does not lead to lower segregation vis-à-vis the Greeks, because of the above mentioned decrease of the Greek population in the same areas. The case is only slightly different if relocations from other countries are also considered, indicating a less even distribution of the newcomers.

Table 10.2 Indices of immigrant groups' residential segregation, 1995 and 2001

Group	% of group in Mun. with LQ>1 1995	% of group in Mun. with LQ>1 2001*	% in Central City (PCC) 1995	% in Central City (PCC) 2001*	Dissimilarity Index (DI) 1995	Dissimilarity Index (DI) 2001*
Albanians	56.7	55.1	35.1	35.0	0.239	0.234
Polish	70.8	69.6	65.7	64.2	0.489	0.485
Bulgarians	60.6	58.5	51.2	50.4	0.361	0.360
Romanians	64.4	62.8	46.6	44.3	0.323	0.312
Pakistani	68.2	87.2	20.0	20.0	0.421	0.431
Ukrainians	66.3	65.7	45.5	43.6	0.335	0.328
Egyptians	76.3	75.1	45.8	45.1	0.428	0.434
Filipino	82.9	82.2	68.5	68.3	0.597	0.604
Nigerians	91.6	90.6	89.1	88.5	0.701	0.710

* Values estimated on the basis of the internal relocations (i.e. excepting newcomers).

Note: $DI = \frac{1}{2}[\Sigma\, ABS[(x_i/X) - (y_i/Y)]]$

where, x_i the population of group X in Municipality i; y_i the population of group X in Municipality i; X the total population of group X ; Y the total population of group Y.

$LQ = (x_i/t_i)/(X/T)$

where, x_i the population of group X in Municipality i; t_i the total population of Municipality i; X the total population of group X; T the total population of the metropolitan area.

$PCC = Xc/X$

where, Xc the population of the group X in the city center (Municipality of Athens); X the total population of group X.

Source: 2001 Population Census, EKKE-ESYE 2005.

Albanian and Greek internal relocation patterns differ considerably with respect to the place of their departure. Almost 72 per cent of Greeks moved to their present place of residence from another Municipality of the Attica region and the rest from a Municipality of a different region. The attractiveness of the Greek capital has diminished for Greek residents since the early 1980s (Maloutas 2000) and this did not change in the 1990s. The respective proportions for the Albanians who moved during the same period (1995–2001) were 54.6 per cent and 45.4 per cent, providing evidence that they continued to resettle from rural to the urban areas in the late 1990s.

No other group presents a comparable proportion of people being attracted to the metropolitan region from the rest of the country – and in fact a two stage access to the metropolis – although some of these groups have a similar population distribution between Attica and the rest of the regions to that of Albanians (Table 10.3, column 4). Despite the moderate percentages (between 24.6 per cent and 28 per cent), attraction is low for Romanians, Ukrainians and especially Bulgarians, if compared with the proportions of people living in metropolitan Athens in the respective aggregate populations.

Albanians who relocated to Athens from other Greek regions are relatively evenly distributed between the latter: 16 per cent left from Peloponnese, 14.8 per cent from Western Greece, 12.5 per cent from Central Macedonia and 12.1 per cent from Thessaly. The opposite flow of Albanians from the metropolitan region of Athens to other parts of the country is much smaller. There seems to be enough evidence that in the long term international migration does not contribute to the reduction of the demographic imbalance between the capital and the rest of the country.

Table 10.3   Regional origin of relocated immigrants, Attica region, 1995–2001

| Group | Place of residence, 1995 | | % of population in Attica region (2001) |
	Other Municipality in Attica region %	Other region %	
Albanians	54.6	45.4	47.1
Polish	91.0	9.0	79.2
Bulgarians	72.0	28.0	28.5
Romanians	72.9	27.1	41.2
Pakistani	92.5	7.5	87.1
Ukrainians	75.4	24.6	58.0
Egyptians	86.1	13.9	82.1
Filipino	88.7	11.3	90.1
Nigerians	88.9	11.1	85.1

Source: 2001 Population Census, EKKE-ESYE 2005.

The exact destination of the internal movers shows some differentiation according to their former region of residence. Among the municipalities that received important net inflows of internal movers, some gained relatively more from intra-metropolitan inflows and others from inflows from other parts of the country. Importantly, a considerable proportion of internal movers from outside Attica settled in the Municipality of Athens, where the net inflow of immigrants from other parts of Attica was comparatively small. The centre of the city was more attractive for immigrants coming from other regions than for those relocating inside the metropolitan area.

Relocations and Socio-economic Position

Relocations of the Greek population (either intra-metropolitan or from other areas) seem to follow the expected pattern, in which those in higher socio-economic positions move proportionally more than the rest (Table 10.4). Against expectations, relocation of residence for Albanians is ambiguously connected to the socio-economic category of those who move. The distribution of the new Albanian residents into socio-economic classes is almost identical to that of those who did not move. Only small differences are observed in the two most important socio-economic categories of the lower technical and routine occupations, indicating a marginal tendency for the more mobile not to belong to the lowest groups of the classification. No such tendency holds true in the (significantly smaller) categories of those in lower service positions and the self-employed.

Socio-economic differentiations between those who moved and those who did not are relatively more important for the Polish, the Romanians and the Egyptians, indicating some tendency for movers to abandon the lowest employment categories. The strongest tendency of this type is found for Pakistanis and Filipinos with low mobility rates. Opposite tendencies are witnessed for Bulgarians and Ukrainians. For them the percentage of those employed in routine occupations is bigger among the mobile than the rest because of the significant proportions of female workers in domestic services that relocate between residences of Greek households, without moving to another occupation. Self-employed Nigerians (usually street vendors) participate proportionally more in the relocated than in the non-relocated population of their group, but in their case not belonging in the routine occupations category could hardly be considered as upward social mobility.

Similarly, Albanian residential relocations are not clearly related to the education level (Table 10.5). The proportion of secondary education graduates was only slightly higher among the mobile than among the rest. The opposite for the less educated and for higher education graduates indicates that there is no linear connection between education skills and residential mobility. However, one should bear in mind that immigrants are regularly heavily disqualified in the Greek labour market, and thus education skills that were acquired in Albania and other countries of origin were not evaluated as such in Greece (Lazaridis 1996,

Hatziprokopiou 2003). Evidence for smaller groups remains unclear. Some association between higher educational levels and higher residential mobility seems to exist for Ukrainians and Egyptians, but on the other hand low skilled Pakistanis are over-represented amongst those who relocated compared to those who did not. Once again, the association is clearly in the expected direction in the case of Greeks.

Table 10.4   Socio-economic classification of immigrants according to residential mobility, Attica region, 1995–2001

Group	Moved residence (1995–2001)	European Socio-economic Classes (ESeC) (%)									
		1	2	3	4	5	6	7	8	9	Total
Albanians	yes	1.6	1.9	1.7	5.4	0.6	0.6	7.2	31.8	49.3	100
	no	1.4	1.8	1.7	6.1	0.9	0.5	8.1	27.8	51.8	100
Polish	yes	2.7	5.1	5.3	9.1	0.4	0.5	5.8	25.9	45.2	100
	no	1.9	2.7	3.2	7.4	0.2	0.7	5.8	28.9	49.2	100
Bulgarians	yes	2.9	3.5	3.1	4.4	0.2	0.9	10.4	11.7	63.0	100
	no	6.0	3.9	3.0	4.8	0.0	0.7	10.9	11.2	59.5	100
Romanians	yes	11.7	12.5	5.5	3.6	0.0	1.0	9.1	22.9	33.8	100
	no	9.8	5.5	5.5	5.0	0.1	1.1	7.3	25.6	40.2	100
Pakistani	yes	11.7	12.5	5.5	3.6	0.0	1.0	9.1	22.9	33.8	100
	no	1.1	0.5	0.9	3.5	0.3	0.5	4.9	29.6	58.7	100
Ukrainians	yes	1.6	2.8	2.4	2.5	0.3	0.2	9.5	12.5	68.3	100
	no	5.9	4.6	3.1	4.9	0.0	0.2	9.3	13.4	58.6	100
Egyptians	yes	6.6	4.9	4.9	13.3	0.6	1.2	7.5	30.1	30.9	100
	no	5.0	3.3	5.0	12.6	0.7	0.6	5.4	32.0	35.4	100
Filipino	yes	1.2	6.9	2.5	0.9	0.2	0.3	5.0	24.8	58.3	100
	no	1.7	1.6	2.1	1.2	0.4	0.1	5.0	7.5	80.5	100
Nigerians	yes	3.1	9.4	0.0	43.8	0.0	0.0	3.1	3.1	37.5	100
	no	3.9	3.1	2.9	37.8	0.3	0.5	3.4	5.2	43.0	100
Greeks	yes	19.4	14.5	19.6	13.6	0.7	1.9	10.4	7.8	12.1	100
	no	14.0	12.0	19.0	16.9	1.0	1.8	10.6	9.6	15.1	100

Note: (1) Large employers, higher grade professional and administrative occupations, (2) Lower grade professional, administrative and managerial occupations, (3) Intermediate occupations, (4) Small employer and self employed occupations (exc. agriculture), (5) Self employed occupations (agriculture etc), (6) Lower supervisory and lower technician occupations, (7) Lower services, sales and clerical occupations, (8) Lower technical occupations, (9) Routine occupations.

Source: 2001 Population Census, EKKE-ESYE 2005.

Table 10.5   Education level of immigrants according to residential status, Attica region, 1995–2001

Group	Moved residence (1995–2001)	Less than primary	Primary	Secondary	Higher	Total
				Education level (%)		
Albanians	yes	3.6	21.5	63.5	11.4	100
	no	4.2	22.0	61.6	12.2	100
Polish	yes	1.0	8.6	67.9	22.6	100
	no	1.8	9.7	69.7	18.8	100
Bulgarians	yes	2.6	9.6	58.5	29.3	100
	no	2.7	12.6	57.3	27.4	100
Romanians	yes	2.4	7.8	60.8	29.0	100
	no	3.1	9.6	60.4	26.9	100
Pakistani	yes	24.0	22.0	47.3	6.8	100
	no	19.5	30.1	45.5	4.9	100
Ukrainians	yes	3.6	9.8	38.0	48.5	100
	no	3.5	13.8	46.6	36.1	100
Egyptians	yes	2.1	11.5	47.5	39.0	100
	no	6.5	13.9	47.9	31.7	100
Filipino	yes	4.3	10.4	59.9	25.5	100
	no	4.0	10.7	60.9	24.4	100
Nigerians	yes	0.0	5.6	58.3	36.1	100
	no	4.9	13.7	51.2	30.3	100
Greeks	yes	4.8	17.1	44.9	33.2	100
	no	7.5	26.4	44.1	22.1	100

Source: 2001 Population Census, EKKE-ESYE 2005.

Available domestic space and tenure differentiate the immigrant and the Greek-born population, as immigrants lived in significantly smaller houses and had limited access to home-ownership in 2001. Regarding housing conditions, relocated Albanians did not differ from those that did not relocate. Almost half of the Albanian population in Attica possessed less than 15 square meters per capita; this was true for both the mobile and the rest (Table 10.6). Most Albanians (84.9 per cent) lived in rented houses and once again this was not very different for the residentially mobile (Table 10.7). A small increase is recorded in the percentage of home-owners and a small decrease in the category of 'others' that includes those in gratis accommodation (i.e. those who are accommodated free of charge in dwellings owned or rented by others, usually family members for Greeks and employers for immigrants). Despite marginal improvements, residential mobility did not result in some increase in housing space nor in higher rates of home-ownership. In this sense, it does not seem to alter the position of Albanians in the housing market or to reduce inequalities between them and the Greeks.[5]

Table 10.6   Housing space per capita according to residential status, Attica region, 1995–2001

Group	Moved residence (1995–2001)	Domestic space per person (%)				
		Up to 15 sq.m.	15,1–25 sq.m.	25,1–40 sq.m.	More than 40,1 sq.m.	Total
Albanians	yes	46.2	39.3	11.0	3.5	100
	no	48.7	39.2	9.3	2.8	100
Polish	yes	19.8	41.8	25.6	12.8	100
	no	23.5	44.2	24.2	8.1	100
Bulgarians	yes	11.8	28.1	35.5	24.5	100
	no	15.3	34.8	30.4	19.6	100
Romanians	yes	21.9	35.0	27.5	15.7	100
	no	21.8	39.6	26.3	12.3	100
Pakistani	yes	48.0	35.2	10.7	6.0	100
	no	61.6	23.8	9.0	5.6	100
Ukrainians	yes	8.9	32.8	33.4	24.8	100
	no	18.0	34.1	29.0	18.9	100
Egyptians	yes	21.0	30.7	27.8	20.5	100
	no	24.1	35.1	24.8	16.0	100
Filipino	yes	23.0	22.4	19.4	35.1	100
	no	38.4	26.9	16.1	18.6	100
Nigerians	yes	34.2	28.9	23.7	13.2	100
	no	41.8	35.8	15.5	6.8	100
Greeks	yes	8.8	28.8	34.9	27.5	100
	no	8.1	34.4	34.7	22.8	100

Source: 2001 Population Census, EKKE-ESYE 2005.

---

5   Only 8.2 per cent of the Greeks lived in less than 15 square meters in 2001. Home-ownership among Greeks was almost 79 per cent.

Table 10.7 **Housing tenure according to residential status, Attica region, 1995–2001**

Group	Moved residence (1995–2001)	Renters	Owners	Total
			Tenure (%)	
Albanians	yes	83.8	16.2	100
	no	84.9	15.1	100
Polish	yes	82.4	17.6	100
	no	87.1	12.9	100
`Bulgarians	yes	59.3	40.7	100
	no	64.3	35.7	100
Romanians	yes	73.2	26.8	100
	no	69.6	30.4	100
Pakistani	yes	91.3	8.7	100
	no	89.9	10.1	100
Ukrainians	yes	59.2	40.8	100
	no	66.6	33.4	100
Egyptians	yes	73.7	26.3	100
	no	71.8	28.2	100
Filipino	yes	61.0	39.0	100
	no	79.5	20.5	100
Nigerians	yes	76.3	23.7	100
	no	91.8	8.2	100
Greeks	yes	40.7	59.3	100
	no	19.4	80.6	100

Source: 2001 Population Census, EKKE-ESYE 2005.

On the other hand relocated members of most other immigrant groups exhibit smaller proportions in the category of very limited domestic space than their non-relocated co-ethnics. The mobile participate proportionally more in the intermediate domestic space categories. Conversely, the proportion of the residentially mobile in the category of abundant domestic space is higher than the respective proportion of those who did not move for every single group. This finding gives a more clear indication that more affluent households exhibit higher relocation rates and that members of a middle-class component of every immigrant group relocate while achieving better living conditions (see Catney and Simspon (2010) for a detailed examination of the 'social gradient' regarding ethnic relocation in England and Wales).

Regarding housing tenure, relocation seems to be connected with increased access to home-ownership for some groups and not for others. Relocated Polish, Bulgarians and Egyptians enjoy slightly higher percentages of home-ownership than those that did not relocate. The variance is more significant for Ukrainians, Filipinos and Nigerians, although 'home-ownership' for the first two groups may refer to residences that are owned by their employers, due to high rates of domestic employment. On the contrary the relocations of the Pakistanis are not associated with access to homeownership but to some extent with the move from collective housing tenure (not included in the table) to private rented houses. For the Greek-born population the proportion of renters is much greater among movers, but in this case that must be an indication that home-ownership is the terminal of many housing itineraries.

In sum, residential mobility does not seem to reflect directly any substantial improvement in the socio-economic position, especially for the bigger and the most mobile groups. To some extent this might be because of the short period of the analysis that covers no more than ten years after the rapid increase of the migratory flow to Greece. However it may also be an indication that immigrants were able to benefit from housing opportunities in different parts of the Athenian housing market, while not having yet drastically improved their position in the socio-economic stratification of the city. In this way, they seem to invert the spatial stratification hypothesis (Bolt and van Kempen 2010), according to which ethnic groups find it difficult to match their improving social status with that of their residential location. Many immigrants in Athens resettled in a period when their socio-economic position did not change. Moreover, new immigration to Athens seems to follow a different relocation pattern from that of the immigration of the Greek-born population from the rural areas of Greece in the past decades, when upward social mobility was accompanied by low residential mobility (Maloutas 2004).

Discussion and Conclusions

In spite of different levels of upward social mobility and integration processes, by 2001 different positions in the socio-economic hierarchy and housing inequalities between the Greeks and most immigrant groups persisted in the metropolitan area of Athens. The varying rates of residential mobility of the immigrant groups are not in a linear relationship with their socio-economic status and do not imply similarly varying prospects for integration. While for some groups there seems to be a connection between residential mobility and some aspects of social status and/or residential integration, this is not the general rule, especially in the case of the single biggest Albanian group. Better socio-economic conditions seem to be more clearly linked with relocation in the case of the less mobile (and relatively deprived) groups of Pakistani and Filipino.

In the absence of positive regulation of their integration conditions, immigrants in Athens have to exploit the opportunities in the labour and housing markets, while being exposed to the restrictions created by them (Kandylis, Maloutas and Sayas 2012). The improvements that some who relocate achieve regarding housing space indicate that the private rented housing market provided opportunities that were not necessarily restricted by the conditions in the less favourable labour market. For some, residential mobility is even connected with access to home-ownership. On the other hand, for many others it is only a means to combat severe integration restrictions by exploiting opportunities in the private housing market.

The spatial patterns of immigrants' relocation show some degree of preference to the inner suburbs of the city. As Greeks relocate primarily to more peripheral suburbs (reaching the peri-urban ring), this does not provide a more even ethno-spatial distribution (notably of an already non-highly segregated immigrant population) at the metropolitan scale. It produces, however, an ethnic mix in the already socially mixed inner suburbs, expanding the ethnic diversity of these densely populated areas.

The residential areas of the city centre, an important place of residence for many immigrant groups, are not attracting a proportional part of internal relocations. They constitute, however, an important (albeit not the only) destination for newcomers of many groups. We should connect this preference with employment opportunities as well as affordable housing opportunities in the apartment buildings of the private rented sector. The central residential areas might continue to be a point of arrival for newcomers in the future. Recent migratory inflows from new countries of origin (such as Afghanistan and Somalia) and the increased inflows from countries such as Pakistan and Bangladesh seem to follow the same path, despite the absence of appropriate data.

Although massive immigration in Athens is quite recent, at first sight this pattern seems to imply a concentric model of residential mobility, quite similar to the Chicago School hypothesis that successive relocations from the inner city to the suburbs imply successive waves of integration. But the data we presented here help to identify three interrelated reservations. First, while residential mobility seems to be the spatial expression of social integration for some, it might be only a strategy to mitigate severe socio-spatial restrictions for others. People may move seeking better living conditions or simply following existing job and housing opportunities. The boundary between planned individual and household strategies and roaming in the city is of course far from clear-cut, but it may be useful to remember that relocations may be differently experienced, even by people 'similar' in socio-demographic perspectives. Second, ethnically mixed residential areas are not ipso facto spaces of less inequality between members of different groups. Low segregation of immigrant groups (indicated here by a quite even distribution together with non-isolation) may coexist with the emergence of intrinsically polarised residential areas. Third, the effect of residential mobility on the pattern of residential segregation between immigrants and Greeks is also conditioned by the residential preferences and choices of the latter. Immigrants' dispersion is not enough to reduce segregation, if it is accompanied by new waves of 'white flight'.

Based on relocations, there is not enough indication for community congregation in Athens at the municipal level, except for the Pakistani group. For a thorough examination of immigrants' tendency to congregate in ethnic enclaves, relocation data at a lower spatial scale should be available. 2011 census data will be useful for this purpose, as well as in order to cross-examine the expectedly significant impact of natural demographic change on segregation (Stillwell and Phillips 2006, Finney and Simpson 2009). However, the infamous self-segregation hypothesis (cf. Stillwell and Phillips 2006, Stillwell and McNulty, this volume) is not verified by existing data. The residential pattern of immigrants together with its transformations due to relocations at the scale of the municipalities show that the immigrant groups we examined, albeit unequal to Greeks, were in 2001 already quite far from the point of a temporary and isolated settlement of stagnated integration prospects. The study of residential mobility also reveals that despite certain similarities, different immigrant groups follow different paths of spatial integration, that being an aspect of increasing diversity of the immigrant population in Athens.

Chapter 11

Ethnic Differences in the Internal Migration of Higher Education Students in Britain

Nissa Finney

Introduction: Minority Internal Migration in Britain

Interest in the geographical location of people of different ethnicities in Britain has been evident in geographical and sociological scholarship for over half a century. The inclusion of an ethnic group question for the first time in the UK census of 1991 enabled researchers to quantitatively assess ethnic geographies (Peach 1996) and some researchers made use of these data to examine the residential movement of people of different ethnic groups within Britain (Robinson 1992, Champion 1996). It was not until the early 2000s, however, that debates about ethnic group population change entered the political fore.

The events that catapulted these issues up the political agenda were urban disturbances is several towns in northern England which were widely described as 'race riots'. The cause was identified as a lack of cohesion in the diverse communities involved and the problem summarised as such communities 'sleepwalking to segregation' (Cantle 2001, Phillips 2005). This shifted research interest to processes of neighbourhood population change (Simpson, Gavalas and Finney 2008, Stillwell, Hussain and Norman 2008) alongside work continuing to examine the behaviour of segregation indices (Simpson 2004, 2007, Johnston, Poulsen and Forrest 2005, 2010, Peach 2009, 2010).

The recent work on neighbourhood processes of ethnic group population change has provided a number of findings: first and foremost, ethnic differences in levels of residential mobility within Britain exist though there are similarities across ethnic groups in the individual characteristics associated with mobility (Finney and Simpson 2008). Ethnic similarities were also found in the geographical processes of migration and these were seen to hold at different scales and over time. In particular, counter-urbanisation and dispersal from areas of co-ethnic concentration are evident across ethnic groups (Simpson and Finney 2009, Simon 2010). Nevertheless, due to the uneven distribution of population by ethnicity in Britain, the absolute geographies of internal migration – the movement between particular origins and destinations – differs between ethnic groups, especially at small scales and in London (Stillwell and Hussain 2010). Overall, however, ethnic group population change is producing a picture of increased ethnic mixing (Finney and Simpson 2009, Rees et al. 2011).

The work summarised here gives a good understanding of ethnic differences and similarities in patterns of migration but our knowledge of exactly in what circumstances, for what population sub-groups and through what mechanisms ethnicity matters for migration – our understanding of why ethnicity matters – is limited.

It has been argued that a lifecourse approach that focuses on ethnic differences in norms of mobility and housing at different life stages and in association with certain life events is a fruitful way to further this field of research (Finney 2011a, 2011b, Wingens et al. 2011). This approach conceives the role of ethnicity (or ethnic group, or ethnic identity) as a cultural one in which the heritage, traditions, conventions and expectations of ethnic groups as communities of identity inform norms of where to live (and who with) at particular life stages.

This chapter is concerned with young adults, and particularly those who are students. There are two reasons for the focus on students. First, post-compulsory education in Britain is an important life juncture for young adults and, particularly for Higher Education (University) study, commonly represents leaving the family or parental home for the first time (Holdsworth 2006). Young adults are the most mobile population group and within that group students are highly mobile; thus, this paper examines ethnic differences in mobility amongst the most mobile section of the population. Second, it has been shown elsewhere (Finney 2011a, 2011b) that residential mobility is differentially associated with being a student across ethnic groups. In particular, students of some ethnic groups are more mobile than their non-student counterparts (Whites and Chinese) while students of other groups are less mobile (Pakistani and Black African). This raises questions about the persistence and causes of these patterns.

This chapter aims to further our understanding of the particularities of ethnicity's association with migration, with a focus on a specific sub-population and life-stage, by addressing the following questions: Are there ethnic differences in student mobility? Are there ethnic differences in the housing experiences of students? Are ethnic differences in student migration and housing experiences mediated by (a) whether an individual is an immigrant/overseas student; (b) religion; (c) gender?

The next section contextualises these questions in literature of migration, lifecourse, ethnicity and student choice, drawing out what might be expected given previous research. The data and measures used in this chapter are then outlined. The results are then presented in three sections: correlates of migration, correlates of housing experience, and the mediating effects of immigration, gender and religion. The final section discusses the implications of the results for minority internal migration research, for ethnic inequality and integration and for student choice in the changing environment of British Higher Education.

Ethnicity, Study and Migration

There are two broad literatures that inform the line of thinking in this chapter. First, lifecourse literature about mobility transitions in young adulthood, particularly that which focuses on differences between ethnic or immigrant groups. Second, Higher Education choice literature that examines different choice mechanisms across ethnic groups.

Transition to Adulthood, Migration and Ethnicity

Studies of internal migration in population geography and demography have in recent years been influenced by lifecourse theories (usefully reviewed by Bailey 2009) and re-theorisation of family. The aim has been to move beyond economic rationality explanations of migration and understand diversity of experiences. Despite recognition of the need to understand ethnic differences in migration experience (Bailey and Boyle 2004), little work has to date engaged with this issue (see Wingens et al. 2011).

A lifecourse approach to internal migration is concerned with how life events (or transitions) such as beginning or ending study or work, having a child, forming or dissolving a partnership, are associated with moving house (Rabe and Taylor 2009). This chapter is interested in how this association differs between ethnic groups for Higher Education students. Studies of cross-national heterogeneity in transitions to adulthood can provide theoretical direction for the study of sub-populations. For example, Fussell, Gauthier and Evans (2007) compared experiences of transition to adulthood in the USA, Canada and Australia. They found country differences which they attributed to 'a function of difference in values and marriage markets' but recognised that their explanations would benefit from 'analysis of sub-population within the United States since distinct ethnic groups exhibit very different family formation patterns' (Fussell, Gauthier and Evans 2007: 411, Gauthier 2007).

Of more direct relevance are studies of ethnic differences in migration in young adulthood which focus on family influence and cultures of home leaving. These studies are situated in family migration literatures which are concerned with the implications of changing family arrangements for residential mobility (Bailey and Boyle 2004). Mulder (2007) sets an agenda for this research, arguing that the family context and inter-generational transfers for migration decisions may be particularly important for non-western migrants who tend to be both more mobile and place greater importance on family than their western counterparts. Intergenerational transfer can have an effect on migration in terms of ethnic-specific preferences and behaviours (e.g. strength of family ties, traditions of home leaving) and in terms of status inheritance (socio-economic resources).

de Valk and Billari (2007) examine the effect of intergenerational transfer on home leaving for young adults of different ethnic groups in the Netherlands. They find few ethnic differences in factors associated with staying in the family home but greater ethnic difference in pathways out of the parental home. For example, 'being in a union was much less associated with leaving home for Moroccan, Antillean and especially Turkish young adults than was the case for the Surinamese and the Dutch' (de Valk and Billari 2007: 213). Ethnicity has also been found to affect the timings and pathways of home leaving in a North American context (Goldschneider and Goldschneider 1988, 1997, Mitchell et al. 2004). In the US, Hispanics have the highest rates of leaving home for marriage, followed by Whites, with Blacks having considerably lower rates. Whites have higher rates of leaving home for job-related reasons than Blacks or Hispanics. The probability of leaving home to attend university is consistently lower for Blacks than Whites over time in the US: 'leaving home for higher education is the family process most closely linked with the reproduction of socioeconomic differences...going away to school...remains difficult [for Blacks] in a deeply segregated society' (Goldschneider and Goldschneider 1997: 305).

Studies also point to the importance of examining gender: women can be expected to show greater family solidarity than men and are more likely to move long distances for reasons of marriage (Mulder 2007). In the Netherlands, girls of Turkish and Moroccan origin are likely to leave home at the point of marriage whereas Dutch girls tend to leave home before marriage to live independently (de Valk and Billari 2007).

Thus, we can expect to see ethnic differences in student mobility with minorities, particularly females, being less likely to live away from home. This may be expected to be particularly evident for ethnic groups whose culture (or dominant religion) fosters traditions of extended households particularly prior to marriage which is that case for Pakistani and Bangladeshi (Muslims) in Britain.

Educational Choice, Migration and Ethnicity

In the UK, entry to Higher Education is a common home leaving pathway (Faggian, McCann and Sheppard 2006). Given the findings of the studies discussed above, it is plausible to suggest ethnic differences in migration norms associated with being a student. Compulsory schooling ends after year 11, usually at the age of 16.[1] Young people then have the option to continue studying in schools or colleges via academic and vocational routes (further education). Further education qualifications are a pre-requisite for Higher Education (HE) study at University. It has been common, since the expansion of Britain's Higher Education system in the mid twentieth century, for university students to live away from home and to study at a university away from their home town. This student migration has resulted in 'studentification' of parts of British cities (Smith and Holt 2007). Indeed, living

---

1   There are some differences between the education systems of the four constituent countries of the UK (England, Wales, Scotland and Northern Ireland). In particular, the Scottish system differs in the breadth of subjects studies and qualifications gained pre-higher education and in the length and funding of degree courses.

away from home has become the social expectation, representing an important element of the student experience as a transition to independence: 'the experience of students taking different paths to university, which do not involve mobility, are excluded from popular images of going to university' (Holdsworth 2009: 1849).

Since the Labour governments of 1997 to 2010 there have been fundamental changes to the Higher Education system in the UK which have continued under the Conservative-Liberal Democrat administration elected in 2010. Two of these changes have implications for University choice and, concordantly, residential choice of young adults who are students: first, the programme to widen participation in Higher Education; second, the introduction of tuition fees for University courses.

The 'widening participation' agenda aims 'to promote and provide the opportunity of successful participation in HE to everyone who can benefit from it' including the objective '[t]o stimulate and sustain new sources of demand for HE among under-represented communities and to influence supply accordingly' (HEFCE 2009: 18). Widening participation is one of the strategic aims of the Higher Education Funding Council who state that it 'is vital for both social justice and economic competitiveness. We also believe in the benefits of learning in an environment with a diversity of students and staff' (HEFCE 2009: 19). In 2002 the Labour government under Tony Blair announced the target of at least 50 per cent of young people entering Higher Education by the end of the decade. The under-represented groups of concern were primarily lower socio-economic groups. The over-representation of ethnic minority populations in low socio-economic groups means that they are a particular target for the policy.

Alongside the widening participation agenda and, arguably, somewhat contrary to it, tuition fees for Higher Education were introduced and maintenance grants reduced. In 1998 university fees were introduced at £1,000 per year, later rising to £3,290. In 2010 it was announced that direct government funding for university teaching would be cut and universities would be able to charge up to £9,000 per year for undergraduate tuition fees. The extra cost of Higher Education study has implications for university choice, not just in terms of decisions based on the price of a course but on costs of living. Staying in the family home is one way to reduce the overall cost of Higher Education.

It can be argued that these changes have had (and will continue to have) differential effects across ethnic groups although '[a]mongst the preoccupation with gender and class within research into higher education in Britain, the operations of 'race' and ethnicity have been largely neglected' (Reay et al. 2001: 857). Participation of students from minority ethnic communities in HE is higher than for students from White communities with around a fifth of first-degree students being from minority ethnic backgrounds (Equality Challenge Unit 2009, HEFCE 2010). However, attainment is markedly lower for minority students even after controlling for factors which would be expected to have an impact on attainment (Broecke and Nicholls nd, Equality Challenge Unit 2009). There are also differences in institutions attended with minorities being over-represented in new universities with lower entry profiles (Reay et al. 2001, HEFCE 2010).

Different geographies of HE attendance are found across ethnic groups. In 2002 at least 20 per cent of students of minority ethnic groups came from and studied in London compared to 3 per cent of White students. More White students stayed in institution-maintained accommodation (halls of residence) in their first year of study than students of other ethnic groups (HEFCE 2010). Concurrently, it is common for Asian, particularly Pakistani and Bangladeshi young adults, especially women and those who are Muslim, to attend local universities because of parental preference for them to live at home while studying (Bagguley and Hussain 2007). Faggian, McCann and Sheppard (2006), in an analysis of migration for study and following graduation, concluded that Blacks and Asians were less likely than others to move away from their home region for study and to move again following graduation. This has implications for graduate employment opportunities which are seen to be enhanced by residential mobility.

These differences in participation and institution have implications for migration because they represent different choice processes and opportunity structures (Reay et al. 2001: 871): 'despite increasing numbers of working-class students, in particular those from minority ethnic backgrounds, applying to university, for the most part, their experiences of the choice process are qualitatively different to that of their more privileged middle-class counterparts... The combination and interplay of individual, familial and institutional factors produces very different "opportunity structures"'. Ball, Reay and David (2002) suggest that ethnicity acts in two ways on choice. First, in the consideration of the ethnic mix of the destination; second, in terms of socio-spatial perceptions. For some, '[s]patial horizons of action are limited, partly for reasons of cost and partly as a result of concerns about ethnic fit and ethnic mix and the possibility of confronting racism... Leaving London or leaving home is rarely an option for these students' (Ball, Reay and David 2002: 338). In addition, family influence on Higher Education decisions has been found to be greater for minorities than for Whites, particularly for Asians and particularly for females (Connor et al. 2004). Also, the relatively limited experience of older ethnic minorities of Higher Education study (in the UK) reduces the likelihood of intergenerational transfer of norms of university study for minority young adults (Brooks 2003).

Data and Measures

This chapter uses data from two sources to examine ethnic differences in student mobility: the 2001 Census Sample of Individual Anonymised Records (SAR) for Britain; and commissioned Higher Education Statistics Authority (HESA) data, 2009–2010 for the UK. The 2001 Census SAR is a 3 per cent sample of individuals for all countries of the UK with approximately 1.84 million records (Northern Ireland is excluded from this study because of very small non-White populations). The dataset includes demographic, health, socio-economic and household variables from the Census. The SAR is downloadable for registered users.[2]

Internal migrants are identified in the 2001 SAR through the Census question that asked for place of usual residence one year prior to enumeration day. If this address differed from the address at enumeration in 2001 the individual can be considered to be an internal migrant. The SAR provides three migration variables as standard: distance of move (in kilometers, banded), region of former residence and a migration indicator variable which details whether an individual moved within and between certain administrative boundaries. This chapter uses the migration indicator variable to construct a binary variable indicating whether or not an individual moved within Britain in the year prior to the 2001 Census.

The student population is extracted from the SAR by identifying those between the ages of 16 and 29 who are in full time education. Higher Education students would be better captured by selecting those aged 18 and over but this is not possible due to the age groups available in the SAR. Thus, the 16–19 age group will contain school pupils who will predominantly live in the parental/ guardian home as well as HE students. Student mobility for this age group is therefore likely to be underestimated. Through the selection specified here there were 54,817 students in the SAR.

Ethnicity was measured in the 2001 Census by a self-identification question with tick-box options and write-in spaces for other ethnicities. The SAR provides the full 16 ethnic group categories provided in the Census questionnaire for England and Wales and 14 for Scotland. These have been combined in the analyses for this chapter into a 13 category ethnic group variable for Great Britain. The ethnic groups are shown in Table 11.1 along with their population, age, socio-economic classification, religion and immigration characteristics and the main period of their arrival in Britain. In 2001 there were 4.6 million non-White people in Britain representing 8 per cent of the population (15 per cent were non-White British). The non-White groups have a young age structure with generally less than 5 per cent of their populations aged 65 and over compared to a fifth of the White population in this age group. This is a reflection of their origins as immigrant groups: immigrants arriving in the mid twentieth century will generally have been young and are only now entering old age. The proportion of the population that is immigrants (not born in the UK) is noticeably lower for the White British and Irish groups compared to other ethnic groups and it is as high as 80 per cent for the White Other and Other Asian groups. There is also considerable ethnic variation in the proportion of the population in managerial and professional occupations: for most groups this is around a fifth but the figure is lower for the Mixed, Black African, Black Other and particularly for the Pakistani and Bangladeshi groups (8 and 6 per cent respectively). The majority of the White and Black ethnic groups

---

2   The UK census SAR can be accessed by registered users, academic and non-academic, at www.ccsr.ac.uk/sars.

are Christians; the majority of Pakistanis and Bangladeshis are Muslims; Indians are predominantly Hindu or Sikh; and most Chinese do not affiliate with a one of the dominant religions in the UK (Christian, Muslim, Hindu and Sikh).

The three SARs variables of interest in addition to ethnic group – gender, immigrant status and religion – originate from self-completion Census questions. Immigrants are defined by whether or not they were born in Britain using data on country of birth.

The Higher Education Statistics Agency (HESA) data used in this chapter were commissioned for the purpose of this study. The data are derived from the HESA Student Record which collects information about all students registered at a reporting Higher Education institution who follow courses that lead to the award of a qualification(s) or institutional credit, excluding those registered as studying wholly overseas. The data used here are from the reporting period for the 2009/10 HESA Student Record (1 August 2009 to 31 July 2010). The data contain information on 2.53 million students.

A student internal migrant is identified in the HESA data as a student whose domicile postcode sector differs from their term time postcode sector. The domicile address is the student's permanent or home address prior to entry to the course. A postcode is a set of numbers and letters that identifies up to 80 addresses for postal mail purposes. The postcode sector detail in the HESA data indicate neighbourhood of residence. Thus the HESA migration information captures short distance (inter-neighbourhood) moves.

Students domiciled in the UK are required to report their ethnic origin to HESA. The HESA ethnic group classifications use Census 2001 ethnicity coding. The ethnic category groupings provided are White, Black, Asian (which includes Chinese) and Other. Religion is not recorded by HESA. Immigrants (or international students) are identified as those whose domicile is outside the UK.

The two datasets used in this chapter clearly provide a picture of different populations with slightly different definitions of students and immigrants. The SAR data allow analysis with a large range of covariates whilst the HESA data allow analysis of aspects of student identity and experience. Though results from the two sources are not directly comparable, together they build a fuller picture of the residential mobility of students in Britain.

Table 11.1 Demographic and socio-economic characteristics of ethnic groups in Britain (2001)

Ethnic Group	Population as per cent of 58.8 million total	Age (per cent)				Socio-economic classification: Managers and Professionals (per cent)	Religion (selected; per cent)				Immigration	
		0–15	16–29	29–64	65+		Christian	Hindu	Muslim	Sikh	Immigrant (born outside UK) (per cent)	Main Period of Arrival in Britain
White British	85.7	19.5	16.6	46.9	17.0	18.2	74.9	0.0	0.1	0.0	1.7	Pre 1900
White Irish	1.2	6.1	12.7	56.5	24.7	24.0	85.0	0.0	0.1	0.0	2.0	Pre 1900
White Other	5.3	13.6	27.5	48.5	10.4	28.3	62.0	0.1	8.5	0.0	79.2	–
Mixed	1.2	49.5	23.2	24.1	3.1	11.9	52.3	0.9	9.9	0.4	20.7	–
Indian	1.8	23.0	25.1	45.3	6.6	18.6	4.7	44.8	12.6	29.5	54.2	1965–1974
Pakistani	1.3	34.6	28.9	32.4	4.2	7.8	1.2	0.1	91.8	0.0	45.3	1965–1979
Bangladeshi	0.5	38.1	30.6	28.3	3.1	5.7	0.5	0.7	92.3	0.0	54.0	1980–1988
Other Asian	0.4	22.9	24.5	47.3	5.3	18.2	13.6	27.0	37.4	6.2	69.9	–
Black Caribbean	1.0	20.2	17.1	52.4	10.4	17.8	73.9	0.2	0.8	0.0	42.0	1955–1964
Black African	0.8	30.0	24.8	42.8	2.3	16.3	68.7	0.2	19.8	0.1	66.8	Since 1991
Black Other	0.2	38.0	23.8	35.3	2.9	12.6	66.7	0.2	6.0	0.1	21.2	–
Chinese	0.4	18.3	32.5	44.3	4.9	17.8	22.0	0.1	0.2	0.0	71.6	Since 1991
Other Asian	0.4	19.7	27.5	49.8	3.0	20.3	32.7	1.7	26.3	0.8	83.4	–

Source: 2001 Census; Simpson and Finney (2009) for period of arrival (not given for the diverse 'other' and 'mixed' groups).

Ethnic Differences in Student Migration and Housing Experiences

Correlates of Migration

In order to address the question of whether there are ethnic differences in student migration we can make use of HESA and SAR data. HESA data, in Table 11.2, show that nearly 40 per cent of UK domiciled Asian and Black students were internal migrants (had a term time postcode different from their postcode prior to study) compared to almost 50 per cent of White students. This is expected given the findings of previous studies discussed above. Only for White students is it the norm – the experience of the majority – to live away from the parental home while studying for Higher Education qualifications.

HESA data allow us to examine how mobility varies for students with different characteristics in terms of level of study, gender and age (Table 11.2). The percentage of postgraduate students who were internal migrants was lower – 36 per cent – than the figure for first degree undergraduate students – 52 per cent. Other undergraduates were particularly immobile: 21 per cent were internal migrants. A higher percentage of male students (51 per cent) were internal migrants compared with female students (44 per cent). Mobility decreases with age amongst students: 69 per cent of 16–19 year old students were internal migrants, 47 per cent of 20–24 year olds, 31 per cent of 25–29 year olds and 19 per cent of students aged 30 and over.

The pattern of mobility by level of study was consistent across ethnic groups: other undergraduates were least mobile, then postgraduate students, then first degree students. Also, the differences between ethnic groups were small amongst postgraduates (e.g. 36 per cent of Asians and 36 per cent of White postgraduates were internal migrants) and those studying for undergraduate qualifications that were not their first degree. Where ethnic differences could be seen was amongst students studying for their first degree. Amongst these students, Whites were more mobile (62 per cent were internal migrants) than Asians or Blacks (44 and 46 per cent respectively).

For all ethnic groups the proportion of students who were internal migrants was lowest for students aged 30 and older. Those aged 16–19 were most mobile amongst students of all ethnic groups but the proportion of internal migrants in this age group was lower for Black students (61 per cent) and particularly for Asian students (47 per cent) than for Whites (73 per cent). For all ethnic groups, student mobility decreased with increasing age. Ethnic differences in student mobility were least for ages 25–29 and 30 and over and greatest for 16–19 year olds.

Table 11.2 **Per cent of UK domiciled Higher Education students who are internal migrants by ethnicity, level of study, gender and age (2009–2010)**

	All Students	White	Asian	Black	Mixed
All Students		*48.6*	*39.8*	*39.3*	*49.8*
First Degree	*51.7*	61.5	44.2	46.4	58.2
Other Undergraduate	*21.0*	20.6	20.8	24.0	23.1
Postgraduate	*36.0*	35.6	35.8	32.3	37.6
Male	*51.5*	53.7	41.7	41.9	52.9
Female	*43.9*	45.2	38.0	37.6	47.5
16–19	*69.3*	72.7	47.4	61.0	67.0
20–24	*47.1*	49.1	36.1	43.4	43.1
25–29	*31.5*	30.9	32.1	33.5	33.4
30 and over	*19.2*	17.8	23.8	22.8	22.4

Source: Commissioned HESA data for 2009/2010. N=2.17 million.

These findings can be corroborated and additional correlates of student migration examined using the 2001 Census SAR. Whether or not a student (age 16–29) migrated in the year prior to the 2001 census was predicted with logistic regression by ethnic group, gender, tenure, highest qualification, immigrant status, whether the individual had dependent children, partnership status and religion. The results are displayed in Table 11.3.

The only non-significant variable in the model is dependent children. The R squared value for the model is 0.50, indicating that half of the variation in mobility between students is accounted for by the demographic and socio-economic variables included in the model. Higher likelihood of being an internal migrant amongst students is associated with renting, especially private renting, being female, being an immigrant (born outside UK), being single, having at least A level qualifications, living in Scotland and having a 'minority' religion or no religion. Lower likelihood of migrating is notably associated with being married and living in the North West, the East of England and London.

Ethnic group has a significant association with propensity to migrate. The results are significant for Pakistani, Caribbean and Chinese students who are all less likely to migrate than their White British counterparts after controlling for other demographic and socio-economic characteristics.

In summary, ethnic differences are seen in student mobility with Black and Asian students being less mobile than their White counterparts. The correlates of migration amongst students are the same across ethnic groups: young students and those studying for first degrees are particularly mobile. Within some student sub-populations, such as young students and those studying for first degrees, ethnic differences in mobility are particularly large. Ethnic differences remain (between the Pakistani Caribbean and Chinese groups and their White counterparts) after controlling for a number of demographic and socio-economic characteristics.

Table 11.3 Characteristics of student internal migrants in Britain (2001): Odds of being a migrant (versus not being a migrant) associated with demographic and socio-economic indicators

	Odds Ratio (exp(β))
Ethnic Group (Ref: White British)	
White British	1.000
White Irish	1.217
White Other	0.891
Mixed	0.873
Indian	0.849
Pakistani	0.674*
Bangladeshi	0.715
Other Asian	0.830
Black Caribbean	0.689*
Black African	0.900
Black Other	0.878
Chinese	0.730*
Other	0.853
Gender (Ref: Male)	
Male	1.000
Female	1.064*
Tenure (Ref: Own outright/with mortgage)	
Own outright/with mortgage	1.000
Part rent, part mortgage	1.793***
Social renter	1.221***
Private renter	7.083***
Highest qualification (Ref: No qualifications to GCSEs)	
No qualifications to GCSEs	1.000
A levels to degree level	3.455***
Other or unknown	1.483*

Table 11.3   *Concluded*

Immigrant status (Ref: Not an immigrant/born in UK)	
Not an immigrant/born in UK	1.000
Immigrant/not born in UK	1.235***
Dependent Children (Ref: no dependent children)	
No dependent children	1.000
With dependent children	0.858
Partnership (Ref: Single)	
Single	1.000
Married	0.258***
Cohabiting	0.876**
Region of residence (Ref: North East)	
North East	1.000
North West	0.845*
Yorkshire and the Humber	1.093
East Midlands	1.034
West Midlands	0.950
East of England	0.834*
South East	0.869
South West	0.908
Inner London	0.536***
Outer London	0.605***
Scotland	1.190*
Wales	0.904
Religion (Ref: Christian) r With Dependent C	
Christian	1.000
Hindu	0.896
Jewish	0.985
Muslim	0.881
Sikh	0.690
Other	1.680***
None/Unstated	1.309***
Constant	*0.126*

Note: Asterisks indicate coefficients are statistically significant at the following levels: *<0.05, **<0.01, ***<0.001. R squared = 0.50.

Source: 2001 Census SAR, GB, population aged 16–29 in full time education (N=54,817).

Correlates of Housing Experience

In addition to examining whether students are internal migrants, HESA data allow us to explore their accommodation situation. Table 11.4 presents the proportion of students living in the parental/guardian home crosstabulated with ethnicity and a number of other individual characteristics. Table 11.4 shows that the proportion of students living in the parental/guardian home varied between ethnic groups: 13 per cent of White students compared with 15 per cent of Black students and 32 per cent of Asian students. The highest proportion of those living in the parental home were studying for their first degree and aged 16–19 with the proportion decreasing with increasing age. There is no difference in the proportion of male and female students living in the parental home. 8 per cent of international students lived in a parental/guardian home compared with 17 per cent of UK domiciled students. These differences are persistent across ethnic groups.

Most students of all ethnic groups do not live in the parental/guardian home (results not shown in the tables). For Asians, in addition to the third of students living the parental home, 10 per cent lived in each of halls (institution maintained property), private sector halls and their own residence and a further 12 per cent lived in other rented accommodation. Of Black students, 19 per cent lived in their own residence, higher than the figure (15 per cent) living in the parental home. 14 per cent of Black students lived in other rented accommodation. For White students, living in private rented accommodation was the most common housing experience (17 per cent of White students); 12 per cent lived in halls; and 11 per cent in their own residence.

In summary, there are ethnic differences in the proportion of students living in the parental home with the proportion being particularly high for Asian students. Living at home is associated with level of study and age in the same way across ethnic groups and ethnic differences persist across level of study and age. Living at home is most prevalent for younger students studying for first degrees. However, most students of all ethnic groups do not live in the parental/guardian home: although this is the most common experience for Asian students it accounts for the experience of only a third. For Black students the most prevalent accommodation type is living in their own residence and for Whites it is living in private rented accommodation.

Table 11.4   Per cent of UK students who live in the parental/guardian home by ethnicity, level of study, gender and age (2009–2010)

	All Students	White	Asian	Black	Mixed
All Students		*12.8*	*31.7*	*15.3*	*18.4*
First Degree	*22.5*	16.2	39.4	20.0	22.2
Other Undergraduate	*9.2*	8.0	15.3	8.1	11.2
Postgraduate	*7.2*	6.3	12.6	6.0	8.6
Male	*17.3*	12.9	31.4	15.9	18.2
Female	*17.0*	12.8	32.0	14.8	18.6
16–19	*26.6*	19.2	42.0	28.5	25.1
20–24	*24.5*	16.8	36.9	25.8	24.3
25–29	*7.4*	6.3	10.8	7.8	8.4
30 and over	*1.7*	1.3	2.3	2.0	2.1
International Student	*8.1*	3.8	28.6	0.0	*
UK domiciled student	*17.1*	12.8	31.7	15.2	18.4

Note: *The category of 'Mixed' ethnicity was not recorded for international students.

Source: Commissioned HESA data for 2009/2010. N=2.53 million.

Mediating Effects of Gender, Religion and Immigration

The theorisation that ethnic groups operate as a mechanism of transmission of norms associated with migration and housing suggests that certain aspects of ethnic group identity can be identified as being associated with migration experience. Amongst these are gender, immigration status and religion. The literature discussed above has suggested that the traditions of remaining in the family home may apply particularly to females (prior to marriage) and to young adults of religions (particularly Muslim) that encourage close family networks.

From the HESA results above we have learnt that a higher proportion of male students are internal migrants than female students; but using the SAR shows that when demographic and socio-economic characteristics are accounted for, females are more mobile. There is no difference in the proportion of male and female students who live in the family home. In terms of immigration, immigrants (or international students) are more mobile than students who are not immigrants. This is expected in the HESA results as international students are defined as immigrants but SARs analysis confirms that, after controlling for other characteristics, students who are immigrants have a higher likelihood of being internal migrants than those who are not immigrants (Table 11.3). Regression models indicate that belonging to a minority religion or having no religion is associated with higher likelihood of being a student internal migrant compared with belonging to a major religion (Christian, Hindu, Jewish, Muslim, Sikh) (Table 11.3).

The question of interest here is whether the relationship between migration and each of the three mediating variables of interest – immigration, religion and gender – varies between ethnic groups. This can be discerned from the three-way crosstabulations of ethnic group, migration and the characteristics of interest in Tables 11.2 and 11.4 using HESA data. Considering gender first, for all ethnic groups a higher proportion of male than female students were internal migrants and the difference is greatest for White students (Table 11.2). There is very little difference in the proportion of males and females living in the parental home for any ethnic group, though the differences between ethnic group hold for males and females (Table 11.4).

The proportion of international students who are student internal migrants varies between ethnic group: 100 per cent for Asian, 33 per cent for Black, 77 per cent for White (Table 11.2). The proportion of international students living in a parental/guardian home also varied between ethnic groups. Most notably, this was the case for 4 per cent of White international students compared with 29 per cent of Asian international students (Table 11.4).

It is possible to further test for the mediating effects of immigration, religion and gender by including interactions between ethnic group and these variables in the model predicting migration. The results are shown in Table 11.5.

If an interaction between ethnic group and immigrant status is added to the model (Table 11.5a) these are the notable changes in results compared with the model shown in Table 11.3: gender is no longer significant (not shown); being Sikh becomes significant, with a lower the likelihood of migrating compared with Christians; being Black African becomes significant, with a lower likelihood of migrating compared with White Britons. Overall the interaction between ethnic group and immigrant status is significant. Individual coefficients are significant for Black African and Black Other students: being an immigrant and in these ethnic groups has the additional effect of increasing propensity to migrate.

Accounting for the relationship between religion and ethnic group through an interaction between these variables (Table 11.5b) emphasises the 'ethnic group effect' on migration: ethnicity is now significantly associated with students' propensity to migrate for the White Irish, White Other, Indian, Pakistani, Bangladeshi, Black Caribbean and Black African ethnic groups. With the exception of the White Irish, these groups are less mobile that their White counterparts, having accounted for demographic and socio-economic characteristics and the interaction between ethnic group and religion. Being Black African and White Other with no religion has an additional effect of increasing propensity to migrate. Being White Irish and Chinese with a religious belief but not of a major religion has an additional effect of decreasing propensity to migrate.

The final part of Table 11.5, part c, shows the results of the model predicting students' propensity to migrate with an interaction between gender and ethnic group. The interaction is not significant overall or for any individual ethnic group.

The gender main effect is also now insignificant (at the p>=0.05 level). In other words, after accounting for demographic and socio-economic characteristics and the relationship between gender and ethnic group, there is no statistically significant difference in the propensity of male and female students to be internal migrants and this is the case for all ethnic groups.

Table 11.5   The mediating role of gender, immigration and religion on the relationship between student migration and ethnicity: Odds of being an internal migrant for interactions of gender, immigration status and religion with ethnic group (Britain, 2001)

(a) Immigration

	Odds Ratio (exp(β))
Ethnic Group (Ref: White British)	
White British	1.000
White Irish	1.237
White Other	1.072
Mixed	0.930
Indian	0.940
Pakistani	0.694**
Bangladeshi	0.833
Other Asian	1.098
Black Caribbean	0.586**
Black African	0.518**
Black Other	0.600
Chinese	0.511**
Other	0.827
Immigrant status (Ref: Not an immigrant)	
Not an immigrant/born in UK	1.000
Immigrant/not born in UK	1.256*
Ethnic Group*Immigrant Status	
Black African	1.925*
Black Other	4.307*
Constant	*0.126*

Table 11.5 *Continued*

(b) Religion

	Odds Ratio (exp(β))
Ethnic Group (Ref: White British)	
White British	1.000
White Irish	1.430*
White Other	0.815*
Mixed	0.861
Indian	0.690***
Pakistani	0.593***
Bangladeshi	0.636**
Other Asian	0.721
Black Caribbean	0.651*
Black African	0.808*
Black Other	0.933
Chinese	1.008
Other	0.770
Religion (Ref: Major religion)	
Major religion	1.000
Non-Major religion	1.878***
No religion	1.276***
Ethnic Group*Religion	
White Irish*Non-Major religion	0.267*
White Other*No religion	1.359*
Black African*No religion	1.892*
Black Other*Non-major religion	1.000***
Chinese*Non-major religion	0.365**
Constant	*0.126*

Table 11.5 *Concluded*

(c) Gender

	Odds Ratio (exp(β))
Ethnic Group (Ref: White British)	
White British	1.000
White Irish	1.042
White Other	0.817*
Mixed	0.928
Indian	0.813
Pakistani	0.700*
Bangladeshi	0.684
Other Asian	0.882
Black Caribbean	0.575*
Black African	0.876
Black Other	0.790
Chinese	0.661**
Other	0.955
Gender (Ref: Male)	
Male	1.000
Female	1.037
Ethnic Group*Female	
Constant	*.127*

Notes: Asterisks indicate coefficients are statistically significant at the following levels: *<0.05, **<0.01, ***<0.001. Only significant interaction effects are shown in the table. Control variables included are gender, tenure, highest qualification, immigrant status, partnership, region of residence, religion. These variables act in the same way as the results in Table 11.3 indicate.

Source: 2001 Census SAR, GB, population aged 16–29 in full time education (N=54,817).

To summarise, descriptive results indicate that male students are more mobile than female students across ethnic groups but after accounting for demographic and socio-economic characteristics there are no gender differences in propensity to migrate for students of any ethnic group. Nor are there gender differences in the proportion of students who live in the parental/guardian home. Immigrants/ international students are more mobile than UK domiciled students and this holds after accounting for demographic and socio-economic characteristics. Being an immigrant has the additional effect of increasing propensity to migrate for Black African and Black Other students. When the association between religion and ethnic group is accounted for the 'ethnic effect' on migration is enhanced, becoming statistically significant for most ethnic groups.

Discussion and Conclusion

This chapter aimed to add to minority internal migration research by furthering our understanding of what it is about ethnicity that is associated with migration. To do this, 2011 Census microdata and 2009/10 Higher Education Statistics Agency data have been analysed to investigate whether there are ethnic differences in student mobility; whether there are ethnic differences in the housing experiences of students; and whether ethnic differences in student migration and housing experiences are mediated by (a) whether an individual is an immigrant/overseas student, (b) religion, (c) gender. The focus on students builds on previous research including by the author (Finney 2011a, 2011b) and allows examination of ethnic differences for the most mobile of the most mobile section of the British population.

The chapter has shown that, as for the population more generally, there are ethnic differences in the mobility of students in Britain. In particular, Black and Asian students are less mobile than their White counterparts. Young students and those studying for first degrees are particularly mobile, and it is amongst these groups that ethnic differences are greatest. Ethnic differences in mobility remain (between the Pakistani, Caribbean and Chinese groups and their White counterparts) after controlling for a number of demographic and socio-economic characteristics.

There are also ethnic differences in the proportion of students living in the parental home with the proportion being particularly high for Asian students. Living at home is most prevalent for younger students studying for first degrees, for all ethnic groups. However, most students of all ethnic groups do not live in the parental/guardian home: although this is the most common experience for Asian students it accounts for the experience of only a third. For Black students the most prevalent accommodation type is living in their own residence and for Whites it is living in private rented accommodation.

Understanding these ethnic differences is important for informing how we think about internal migration, ethnic integration and student experiences. It is clear that there are diverse experiences of mobility and housing across ethnic groups; there is not a common 'student experience'. The 'norm' of students as mobile (Holdsworth 2009) does not apply across ethnic groups. Indeed, this is a very White-British perspective on the student experience. Young White undergraduates are very mobile, but they also have high levels of living in the parental home indicating a rather dichotomised experience. Amongst White students who are not in the youngest age brackets and not studying for their first degrees, under 50 per cent are internal migrants. For Black and Asian students, mobility – in the sense of having moved house in the past year or having a term time address different from the address prior to study – is the experience of the minority.

It is likely that there will be greater diversity in how students manage their location and accommodation choices within ethnic groups as changes to Higher Education funding, particularly the rise in tuition fees, are implemented. For example, it is easy to envisage more White students choosing to remain in the family home while studying bringing their experience closer to that of minority students and, potentially, reducing ethnic differences in student mobility.

Ethnic differences in mobility are particularly marked for young students (aged 16–19 and 20–24) studying for first degrees. This is unsurprising given the potential for Higher Education study to be a transition from the family home (Holdsworth 2006, Faggian et al. 2006). The ethnic differences observed for young, undergraduate students supports literature about differences in home leaving and particularly the tradition amongst Asian communities for young adults to remain in the family home while studying (Bagguley and Hussain 2007). This can be interpreted in relation to the mechanisms of how ethnicity acts on migration: parental influence, intergenerational transfer of expectations and spatial horizons, as well as economic concerns, may all have an influence.

Whilst this chapter has not investigated these mechanisms directly, and there is thus still much to investigate about the role of ethnicity in migration decisions, it has aimed to dig a little deeper into what it is about ethnic group that is related to migration by investigating the roles of immigration, religion and gender. The premise here is that the immigration history that is particular to each ethnic group, the religions and religious practices that characterise ethnic groups, and perceptions about gender and gender roles that may vary between ethnic groups are part of the cultural impact that ethnic group has on norms of migration in young adulthood, including in association with Higher Education study.

The findings on the impact of immigration/international student status are as expected: immigrants are more mobile than UK domiciled students and this holds after accounting for demographic and socio-economic characteristics. This is the finding for the population as a whole and can be expected to hold for students because fewer will have an option of living in a parental/guardian home and they may change their residence as they adapt to student housing markets. It is notable that for most ethnic groups a small proportion of international students live in a parental/guardian home but for Asian students this is the experience of almost 30 per cent. This implies that the family networks that operate to stabilise mobility rates of Asian students generally, apply internationally in providing housing for relations who wish to study whose usual residence is outside the UK.

When the association between religion and ethnic group is accounted for the 'ethnic effect' on migration is enhanced, becoming statistically significant for most ethnic groups. In other words, the ethnic group association with migration is something more than the religious character of ethnic groups. This analysis did not find any evidence to support the hypothesis that belonging to a religion that promotes strong family households (particularly Muslim religions) is associated with lower residential mobility.

Despite previous findings for the population as a whole in Britain and elsewhere that there are significant gender differences in mobility (de Valk and Billari 2007, Mulder 2007, Finney and Simpson 2008), with differences across ethnic groups, this chapter found that, after accounting for demographic and socio-economic characteristics, there are no gender differences in propensity to migrate for students of any ethnic group. Nor are there gender differences in the proportion of students who live in the parental/guardian home. This challenges theories that for some ethnic groups – Pakistanis and Bangladeshis – female young adults in Higher Education are encouraged to remain in the family home more than their male counterparts.

An additional finding worthy of note is that region of residence was significantly associated with student mobility: the highest likelihood of being a student internal migrant was for students in Scotland; the lowest likelihood of being a student internal migrant was for students in Inner and Outer London. This could be interpreted as a reflection of the regional variation in Higher Education offer. With fewer Higher Education institutions in Scotland compared to London, student location or institution choice without migrating is relatively restricted. Given that a high proportion of ethnic minorities in Britain live in Greater London, their relative student immobility may be partly explained by choice being met by their local offer which does not necessitate a move away from the family home (Reay et al. 2001, Ball, Reay and David 2002). Sub-national variation in student mobility may well alter as funding changes (particularly the increase in tuition fees) take effect in 2012. In particular, no tuition fees for Scottish students attending Scottish Universities may be an incentive for them to remain in Scotland for Higher Education.

The lifecourse perspective taken in this chapter by focusing on minority internal migration at a particular life stage allows us to begin to tease out what it actually is about ethnicity and how it matters for internal migration. This chapter has argued that theoretical development about the selectivity of internal migration is best advanced through focused study of specific sub populations. The findings of this chapter allow us to confirm our expectations: that there are ethnic differences in student mobility; that minorities are less mobile than Whites; and that immigrants/ international students are particularly mobile. They also encourage us to challenge our expectations: that a mobile student experience is the norm; and that female students, particularly of Asian ethnic groups, are less mobile than male students.

The key issue is whether it is choice or constraint that is driving the ethnic differences in student mobility; whether inequalities represent differential opportunities. The literature suggests that there are different norms of home leaving which likely account for these patterns. This chapter has been limited in its ability to address these theories: the migration measures do not allow home leaving to be captured; there is no longitudinal dimension to the study; the covariates available are limited; and, more generally, the meaning of ethnic group categories are debateable especially in the HESA data where they are aggregated. It is possible for these deficiencies to be addressed with use of the forthcoming *Understanding Society* panel survey and 2011 Census data publication. These sources, particularly in combination with qualitative methods in a mixed-methods framework, hold great promise for understanding minority internal migration in Britain.

Acknowledgements

The 2001 UK Census Samples of Anonymised Records (SARs) are provided through the Cathie Marsh Centre for Census and Survey Research (The University of Manchester), with the support of the ESRC and JISC. All tables containing Census data, and the results of analysis, are reproduced with the permission of the Controller of Her Majesty's Stationery Office and the Queen's Printer for Scotland. Higher Education Student Records data are provided by HESA Services Ltd. All interpretations of these data in this chapter are the responsibility of the author. I am grateful to Gemma Catney and Phil Rees for helpful comments on this chapter, and more generally to Gemma for a happy collaboration on this book project.

Chapter 12

The Internal Migration of Foreign-born Population in Southern Europe: Demographic Patterns and Individual Determinants

Joaquín Recaño-Valverde and Verónica de Miguel-Luken

Introduction

The massive arrival of foreign immigrants since the 1990s constitutes a transcendental geo-demographic and social phenomenon in Italy, Portugal and Spain. These countries, with a common emigration history, have experienced a fast transition from the 1980s that has turned them into some of the most important immigration destinations in the European Union. These Southern European countries hosted more than 10.4 million foreigners in January 2009, a dramatic increase from the number of less than 2.9 million in 2000. Immigration has therefore seen rapid and accelerated growth, with an inflow of 618,300 foreigners in 2000 to a maximum of 1,205,500 in 2007 (OECD 2011). In this intense process, Italy, Portugal and Spain share a series of common characteristics: intensification and acceleration of the flows, diversification in the demographic structure by age, sex and geographical origin and a rising quantitative importance of the irregular flows (Recaño-Valverde and Domingo 2006, Domingo and Gil-Alonso 2007).

Three groups of factors that help to explain immigration in Southern Europe can be identified (King and Zontini 2000). The first one relates to geography; the countries in this area are located on the routes of access to other destinations for the people crossing the South Mediterranean border and the people crossing the Eastern European border (Italy). During the 1960s and 1970s, Spain and Italy played the role of transitional countries for emigrants moving from the North of Africa whose main final destinations were France, Switzerland, Germany, Belgium and The Netherlands. Besides this fact, Italy, Portugal and Spain have also maintained historical links with Latin-America due to their cultural and linguistic relationships (Spain and Portugal), together with a past of strong migratory exchanges that includes Italy. The second factor corresponds to economic motivations, the modernisation of the economy and the growing relevance of some specific productive sectors. Since the 1990s Spain, Portugal (Corkill 2001) and Italy have begun to experience a phenomenon that already occurred in the Northern Europe three or four decades before: a rising prosperity, which was associated in Spain and Portugal with their entry to the European Union in 1986 and an accelerated process of population ageing that implied the opening of a professional sector around personal services. Besides, the shortage of workers in certain low paid services in sectors such as tourism, the hotel industry, agriculture and construction has activated a great global demand for non-qualified workers.

The third group of reasons corresponds to the socio-demographic factors that King and Zontini (2000) called the 'border demographic gradient'. That is to say, the situation of progressive ageing in these reception countries which have very low fertility as compared to the full cohorts from the South of the Mediterranean and other Latin-American origins. The issues associated with international migration in Southern Europe since the 1980s are further discussed in Arango (2000), King (2000), King and Zontini (2000), Corkill (2001), Domingo (2002), Ribas-Mateos (2004), Solé (2004), and Aja and Arango (2006) (see also Chapters 4 and 14).

The impact of these numerous arrivals has been acutely noticed in all spheres of these countries. The modification of the internal migration patterns is one of the many consequences that stem from this phenomenon. However, this topic has generated limited interest in the new destinations of this immigration, in contrast with the situation in Western countries with longer traditions of immigration, such as the United States, Canada and Great Britain. In these latter countries the research on the internal migration patterns of the foreign or foreign-born populations has given rise to abundant literature, mainly since the late 1980s.

In this chapter we present results of our research which has focused on answering the following questions: Are the demographic patterns of internal migration of foreigners similar to those of natives by age and sex? Do these migration patterns differ by country of origin? Are the observed demographic patterns by specific national groups always the same or do they vary according to the country of destination? And lastly, what are the effects of the individual characteristics on the internal migration of foreigners as we compare by country of residence?

In brief, the objective is to study which demographic characteristics and individual factors play a part in explaining the internal mobility of immigrants, when we consider the behaviour of the native-born population as the comparative element.

Up to now, the studies carried out in Canada, the United States, Germany, Belgium and Great Britain have arrived at the following conclusions: immigrants[1] tend to be more mobile than natives because of their demographic and social characteristics, such as their age and their life cycle stage when they entry the destination country, the duration of residence, the situation of the labour market

---

1   Italy, Portugal and Spain provide information regarding country of birth and country of citizenship. Portugal and Spain offer detailed data about both categories (52 and 120 countries, respectively), but in the Italian data categories are grouped as regions such as Eastern Africa, Northern Africa and so on (15 in total), which limits considerably the applications of these variables. Note also that year of arrival is only asked in Italy to those who are not Italian. For the Italians, even if born abroad, date of arrival is not recorded.

and their academic attainment (Bartel 1989, Bartel and Koch 1991, Nogle 1994). On the other hand, several authors have pointed out that foreign-born people show lower elasticity than the native-born population to adapt to the factors of the regional market[2] that have a strong influence on the medium and long distance changes of residence, such as unemployment levels, salary differentials and employment growth differentials (Kritz and Nogle 1994, Nogle 1994, Liaw and Frey 1998).

An important finding of previous research is that social networks have an intense influence on mobility: the presence and territorial location of already existing communities of the same immigrant origin lessen the costs associated with the migration process. These communities represent the immigrants' main source of information about the potential internal destinations (Frey 1995, Gurak and Kritz 2000). The concentration of the natives of a particular community in a specific region also constitutes an element of attraction for those of the same geographical origin. By integrating the effect of contextual economic factors and the action of the social networks, Gurak and Kritz (1998) show that immigrants move less frequently from regions with high economic growth rates, with high proportions of workers in the manufacturing sector, and with high concentrations of immigrants from the same national origin. Attending to these arguments, the concentration of nationals from the same country in a region acts, thus, to reduce the internal migration of these immigrant groups. Newbold (1996) has stressed, in his work about Canada, the capacity of some regions to attract and keep foreign immigrants from other Canadian regions, and this result is confirmed by recent research by Krahn and Derwing (2005).

In this chapter we study Spain, Italy and Portugal for a number of reasons. First of all, we find the absence of comparative studies about the mobility of the foreign or foreign-born population in the academic literature: the existant works focus on individual countries rather than international comparison. The second reason is the structural comparability of the three countries. They have a common international migration dynamic, with an intense emigratory past that has turned, at present, to a situation of intense immigration. They receive flows which are very diverse in terms of immigrant origins, and they are also countries with moderate or low internal mobility (Rees and Kupiszewski 1999, Módenes 2002), in which the incorporation of the foreign population has meant the increase of this internal migration (Recaño-Valverde and Roig 2006, Mocetti and Porello 2010). Finally, they have very similar demographic and labour structures. These facts make these Southern European countries interesting laboratories to assess the effects of foreign born population internal migration in geographical contexts of low mobility.

---

2   These results, however, have been obtained in countries with high mobility, where the native-born population shows an intense migration response to economic incentives, both at the individual and regional levels. We advance that the situation in the Mediterranean countries (Spain and Italy) is not the same.

In Spain and Italy, the existing research shows some similarities with the results highlighted by previous international literature about other destinations (Recaño-Valverde 2003, Recaño-Valverde and Roig 2006, Mocetti and Porello 2010). However, these are countries with low internal migration intensity, where the differences in mobility between foreign-born and native-born populations are more noticeable than in countries with higher internal mobility, such as the United States and Canada. Altogether, international researchers have collected a series of socio-demographic and economic variables that have a decisive impact on the foreign or foreign-born population mobility. For this work, we tackle some of these aspects from a more comparative perspective. To achieve this objective, we assess the demographic structure, the migratory intensity and the individual factors that have an influence on the mobility of the different foreign-born groups in Southern European countries.

Data and Methods

At present, the available data for the study of the internal migration of the foreign population (or non-native born population) differ considerably for the different countries that have been included in this work. In this regard, Spain and Italy count on population registers, the *Padrón Continuo* for Spain and the *Anagrafe dei Comuni Italiani* for Italy,[3] which provide data on migratory flows up to a municipality level. In Portugal, internal migration data are limited to those provided by the decennial Censuses.[4] In the Spanish case, the information about migration is derived from the flows that the Statistics on Residential Variations (*Estadísticas de Variaciones Residenciales – EVR*) establish according to the data received from the population register (*Padrón Continuo*). Registration in a municipality implies an automatic removal of the same person from the register of the previous municipality of residence. In the Italian case, the information about origin and destination of the migration movement is obtained through the *iscrizioni* (registration) and *cancellazioni* (cancellation) because of the *trasferimento di residenza* (change of residence), in a very similar way to that described for Spain.

---

3   A detailed description of the characteristics of the Italian data on internal migration can be found at: http://demo.istat.it/bil2006/index03.html; with regards to the Spanish data, the migratory information is elaborated in the Statistics of Residential Variations (EVR) that comes from the population register (Padrón Continuo) (see http://www.ine.es/daco/ daco42/migracion/notaevr.htm). The Italian data about mobility provide information about the academic attainment, the marital status or occupation, which cannot be found at the Spanish EVR.

4   The demographic profile of the foreign population at a local level that asked for resident status can be checked at http://www.ine.pt/xportal/xmain?xpid=INE&xpgid=ine_ base_dados. Similar information about population with a legal residential status is provided by the Italian municipalities for the database of the ISTAT. However, those data are not available in Spain for the municipal level, although they are for the provincial level.

However, there exist some essential differences between both sources despite being population registers. The Italian data just refer to the population with a legal status of residence, who are the only ones allowed to be registered. In contrast, the Spanish Continuous Register (*Padrón Continuo*) includes both immigrants with legal status of residence and immigrants in an irregular situation (with no residence or work permits). The scope of the Spanish population register is, thus, greater than the Italian one with regard to foreign born population, providing information by country of birth and country of citizenship, which is restricted to the latter in the Italian register. As we have already pointed out, the variations in the characteristics of the Spanish and the Italian information, and the absence of some of it in the Portuguese case, take us to reject data on flows and just consider some homogeneous information that is available for the three countries. This is the reason why we use the 2001 Census information.

Nonetheless, it is difficult to compare census data for different countries (Courgeau, 1973a and 1973b, Long and Boertlein 1990, Bell et al. 2002, Bell and Rees 2006, Bell and Muhidin 2009). Realities of each context, geographical divisions, priorities of the specific administrations and years of collection change, thus research questions and hypotheses have to be adapted to these disparities.[5] However, our effort to homogenise the datasets has been facilitated to a great extent by the *Integrated Public Use of International Microdata Series* (IPUMS) (Minnesota Population Centre 2009), which has provided us with the harmonised data files for the countries we have included in the analysis for this chapter (Table 12.1).

The microdata of the IPUMS Census database allow us to obtain two samples: one with data about the population born abroad and one with the population by citizenship. We have chosen to analyse the sample by country/place of birth. There are two motivations for this. First, the characteristic of place of birth remains stable across time in contrast with the numerous acquisitions of citizenship by the population of Latin-American origin that are registered in the three considered countries. Second, indirectly linked to the previous motivation, it gives rise to a larger sample for foreign-born (Table 12.1). The major inconvenience of this decision is the fact that a large proportion of the foreign-born population, corresponds to the children of Portuguese, Spanish or Italian parents born abroad during the intense emigration processes of these countries in the 1960s and 1970s. We argue that this factor does not alter the sense of our results.

---

5   Apart from the differences in the socio-economic and demographic contexts, definitions of migration are much affected by the particularities of the spatial administrative division and the time intervals used in the Census to obtain the category of migrants.

Table 12.1   Characteristics of the IPUMS data files

Country	Sample fraction (%)	Sample size	Foreign-born population subsample	Foreign population subsample	Census date (d-m-yr)	Major administrative unit	Minor administrative unit
Italy	5	2,990,739	117,890 (3.9%)	70,462 (2.4%)	21/10/2001	Region (20)	Municipality(8101)
Portugal	5	517,026	32,136 (6.2%)	11,440 (2.2%)	12/03/2001	Subregion (22)	Municipality(308)
Spain	5	2,039,274	107,394 (5.3%)	77,631 (3.8%)	01/11/2001	Province (52)	Municipality(8111)

Source: Own elaboration based on the Integrated Public Use of International Microdata Series: version 5.0. Minneapolis: University of Minnesota, 2009.

Regarding our specific research objectives, we also have to mention the approaches followed in the different countries with regards to the questions on mobility. In Italy and Portugal the Census inquired about the place of residence one year ago.[6] For Spain, we have information about the last place of residence and the year of change of residence so, even if conceptually it is not exactly the same, because census data on internal migrants by year of arrival in 2001 are close to the migration in the Spanish Population Register for the same year, we can still build up a proxy for the dependent variable that can be understood as the situation one year ago (Figure 12.1).

It was necessary to adjust the explanatory variables to the degree of detail supplied by each Census, while maintaining the possibilities of cross-national comparisons. This led us to a greater simplicity in the categorisation of the covariates than we would have used for country specific models. Since educational attainment was not coded in the same way in the three Censuses, we have re-coded it in such a way that it allows comparison (for the re-coding we studied the country differences in the distribution of responses for the dependent variables). The most difficult explanatory variable to harmonise was place of birth. First of all, not all countries include detailed information on this.[7] Secondly, those that do provide some sort of detail about geographical origin emphasise the places of birth of their own interest, which are not necessarily coincident across countries. So, even if our main research question focuses on the similarity or dissimilarity of the internal migration patterns by region of birth, we have to limit the number and types of categories to those available for all countries of study.

The problem with some of the items is not related to the selected categories for the responses in each country but to the specific population that has been asked about them. For instance, employment status and academic attainment have been treated differently in the various Censuses, depending on the age of the interviewee and his/her situation as an active/non-active citizen. In order to avoid the bias associated with missing data we have constrained our initial database to people aged 25 and over. In Table 12.2 we present the figures for the sample sizes and percentages of migrants by main individual characteristics and place of birth (native born and non-native born populations).

───────────────

6   In Portugal information was also collected about the place of residence five years ago. We have kept the year interval as that of our interest because when the question on migration is about the place of residence one year before, the number of migrants registered by the census and the number of migration movements for the same time interval are numerically close, according to Courgeau's inequality (Corgeau 1973a).

7   For instance, the 2001 Italian Census microdata only distinguish 15 places/ regions of birth, compared to the 52 for Portugal and 120 for Spain, which has forced the aggregation of information according to the limitations of the Italian information in order to make it fully comparable.

Italy (Census date 21/10/2001)

4.7 Indicate whether one year ago (21 October 2000) the person had a steady residence

[Question 4.7 was asked for persons over 1 year of age.]
[] 1 In the
dwelling
[] 2 In this municipality, but in other dwelling or institutional
household

[] 3 In another Italian municipality

____ Indicate which municipality

____ Indicate the abbreviation for the province

[] 4 Abroad

____ Indicate which foreign country

Source: https://international.ipums.org/international-action/source_documents/view/enum_form_it2001_tag.xml

---

Portugal (Census date 12/03/2001)

9. Where was your place of usual residence on December 31, 1999?

[] 11 Not yet born - End of completing

[] 12 In the parish where you live

[] 13 In other parish of the municipality where you live

[] In other municipality, please specify: ____

[] 15 Timor

[] 16 Macau

[] 17 Angola

[] 18 Mozambique

[] 19 Cape Verde

[] 20 Germany

[] 21 France

[] 22 Brazil

[] 23 Venezuela

[] Other country, specify: ____

Source: https://international.ipums.org/international-action/source_documents/view/enum_form_pt2001_tag.xml

---

Spain (Census date 01/11/2001)

5. From what year resides (although is from birth) in:

[] Spain

[] This Region

[] This municipality

[] In other country/municipality, please specify: ____

Source: https://international.ipums.org/international/resources/enum_materials_pdf/enum_form_es2001.pdf

---

Figure 12.1   Questions about internal migration in the Italian, Portuguese and Spanish Censuses of 2001

Source: Own elaboration based on the Integrated Public Use of International Microdata Series: version 5.0. Minneapolis: University of Minnesota, 2009.

Table 12.2 Internal migrants by place of birth and the main individual characteristics, people aged 25 and over, Italy, Portugal and Spain, 2001

Explanatory variables	Italy Native-born	Italy Foreign-born	Italy % Foreign-born	Portugal Native-born	Portugal Foreign-born	Portugal % Foreign-born	Spain Native-born	Spain Foreign-born	Spain % Foreign-born
Sex									
Male	1,015,057	39,991	3.8	157,092	10,186	6.1	661,965	36,395	5.2%
Female	1,121,794	48,172	4.1	177,562	1,171	0.7	717,240	36,614	4.9%
Age-group									
25–29	206,952	13,578	6.2	34,166	5,361	13.6	159,677	13,319	7.9%
30–44	647,773	44,513	6.4	101,421	9,602	8.6	445,054	35,545	7.4%
45–59	554,954	16,130	2.8	90,169	3,696	3.9	347,338	13,725	3.8%
60–74	482,588	8,985	1.8	74,584	1,914	2.5	280,842	7,503	2.6%
75+	244,584	4,957	2.0	34,314	684	2.0	146,294	2,917	2.0%
Marital status									
Single	420,249	20,597	4.7	40,822	5,227	11.4	292,446	21,704	6.9%
Married/in union	1,397,811	57,373	3.9	248,742	14,075	5.4	898,490	42,003	4.5%
Separated/Divorced	88,013	4,945	5.3	12,162	1,101	8.3	51,395	5,242	9.3%
Widowed	230,778	5,248	2.2	32,928	854	2.5	128,722	3,528	2.7%
Educational attainment									
Less than primary	170,941	5,654	3.2	195,977	4,448	2.2	246,982	8,423	3.3%

Table 12.2 *Concluded*

Explanatory variables	Italy			Portugal			Spain		
	Native-born	Foreign-born	% Foreign-born	Native-born	Foreign-born	% Foreign-born	Native-born	Foreign-born	% Foreign-born
Secondary completed	575,072	32,427	5.3	35,332	5,352	13.2	330,203	25,150	7.1%
University completed	170,426	9,051	5.0	25,803	4,574	15.1	103,059	8,000	7.2%
Housing tenure									
Owned	1,588,215	41,748	2.6	256,277	14,996	5.5	1,174,372	39,863	3.3%
Not owned	528,909	44,407	7.7	72,420	5,632	7.2	196,681	32,614	14.2%
Employment status									
employed	957,344	49,274	4.9	181,526	15,447	7.8	620,650	40,775	6.2%
unemployed	140,313	8,091	5.5	11,092	1,072	8.8	87,088	6,960	7.4%
inactive	1,039,194	30,798	2.9	142,036	4,738	3.2	663,315	24,742	3.6%

Source: Own elaboration based on the Integrated Public Use of International Microdata Series: version 5.0. Minneapolis: University of Minnesota, 2009.

We use different approaches to answer our research questions, according to the available data. First, we explore the data at an aggregated level and we calculate the Gross Migraproduction Rate (GMR). We also build up migration profiles by age and sex. Then, we move to the micro perspective through some logistic models.

The Gross Migraproduction Rate is analogous to the total fertility rate in that it is the sum of age specific migration intensities and it is interpreted as the mobility a person would experience in his life if he or she followed the pattern observed at a specific time point (by sex, age and whatever variables are considered to compute the rates). The GMR measures the intensity of migration between two regions at a particular point in time (Rogers and Willekens 1986). In its simplest form it is defined as:

$$GRM = \sum_{x=0}^{z} m_{x,x+n},$$

where $m_{x,x+n}$ are the age-specific migration rates or transition probabilities.[8] This is a way of standardising age and gender structure that is sensitive to the starting and ending ages of summation (Bell et al. 2002).

Finally, we centre our attention on the individual characteristics that have an effect on the probability of having changed residence since one year prior to the Census (Portugal, Italy and Spain). In this case we are not measuring migration intensity, but focusing on the personal circumstances that may act as push effects for migrating. In particular, we are especially interested in grasping the differences in behaviour according to the geographical origin (place of birth) of the migrants and whether their patterns are similar (or not) across countries.

For this purpose, we apply two sets of logistic models depending on the territorial unit under consideration. First, medium and long distance movements, defined by IPUMS International as changes between 'major administrative units' and, then, short-distance movements, defined as changes between 'minor administrative units'.[9] Information provided in the former case is available for all

---

8   In order to improve the robustness of the GMR estimates we have used ten-year groups from 0 to 80 and over. We have used different weights for males and females for the '80 and over' group according to the differences in the life expectancy by sex. Weights for 0 to 79 (by ten-year groups) = 10; for 80+ weight = 5:

$$GRM = 10 * \sum_{x=0}^{70} m_{x,x+10} + 5 * m_{80+}$$

The data for Spain have been recoded in order to make data from the three countries comparable. Obviously, the GMR for the diverse groups are just comparable within each country.

9   The model with place of residence a year ago (different major administrative unit) corresponds to the interprovincial migrations with a mean migration distance over 100km.

the countries analysed. We are aware that these minor and major administrative units differ, even if not substantially with regards to their size and population density, but since in this step we are studying individual propensities to move, instead of migration intensities, the territorial differences should not disturb our results too much.

The variables included in the logistic regression models are: sex, age group, place of birth, marital status, academic attainment, housing tenure and occupational status. Thus, our dependent variables are:

> Model 1: Migration status -1 year ago. Same major administrative unit, value 0. Different major administrative unit, value 1. Obviously, people who lived abroad at the time point of reference (1 year ago) are excluded from the data file.
> Model 2: Migration status -1 year ago. Different minor administrative unit within the same major administrative unit, value 1; value 0, otherwise. Obviously, people who lived abroad at the time point of reference (1 year ago) are excluded from the data file.

Migration Intensity and Continent of Birth

Demographers have observed important regularities in the migratory profiles by age in a wide set of developed countries (Rogers and Willekens 1986).[10] This migratory profile is characterised by the higher mobility of young adults, between 20 and 39 years old, linked to work, marriage and house searching, and the relevant mobility of children and teenagers (0–16 years old), that reflect their parents' mobility. Migration rates are higher at the youngest ages because they are often children of young parents that belong to the age segment with the highest mobility. Low mobility is seen after age 40, when job searching and household formation are considerably reduced. Finally, there is likely to be appearance of a second mobility maximum, of minor intensity, around those ages in which people retire. One of our research questions is: to what extent does this general pattern remain as we consider the migration rates by places of origin (continents of birth) and destination countries?

In order to compare the distribution of the migration rates by age, we have to avoid the scale factor by obtaining the weight of each age group over the GMR total. The difficulties of comparing measures of migration in countries

---

We consider them as medium-long distance migration. On the contrary, the model for different minor administrative unit within the same major administrative unit corresponds essentially to migratory movements associated with changes of residence.

10 Demographers have associated these regularities with the influence of different events and life cycle stages: job search, getting married and family formation, migration at dependent ages and low labour mobility from certain ages.

with different geographical coverage (Courgeau, 1973a and Bell and Muhidin, 2009) force the standardization of the rates in order to make them comparable, $m_{x,x+n}$ (standarized) = $m_{x,x+n}$/GMR. The addition of these rates is 1. What we really compare is the schedule of rates.

As we can observe in Figure 12.2, the migration schedule presents relevant differences between native and non-native born populations for all the analysed countries (Italy, Portugal and Spain). Even so, profiles by age are very similar among non-natives in the three countries (Figure 12.2).

The most important differences are found in the schedule of the native-born population in the different countries. Also, the peak for young adults is higher among the native-born population, spreading to a wider range of ages for the foreign-born population. This suggests less life-cycle stage dependency of migration for those born abroad. In other words, socio-economic factors affecting mobility are active for more years for the foreign-born population. Finally, we do not find important differences by sex in the age-migration schedule of native and foreign-born populations, although these results will be further explained through Figure 12.3.

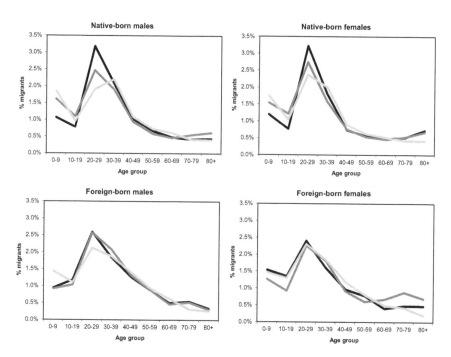

Figure 12.2   Age Standardised rates of internal migration of Southern Europe by sex, age and place of birth

Source: Own elaboration based on the Integrated Public Use of the Microdata International Series: version 5.0. Minneapolis: University of Minnesota, 2009.

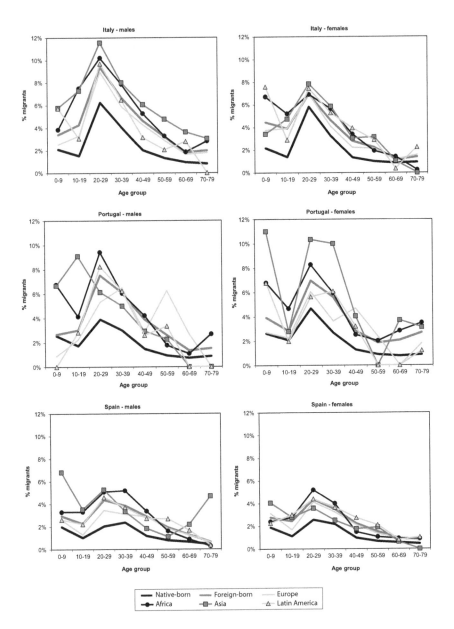

Figure 12.3   Internal migration rates by sex, age and continent of birth

Source: Own elaboration based on the Integrated Public Use of the Microdata International Series: version 5.0. Minneapolis: University of Minnesota, 2009.

In Figure 12.3, rates by sex, age and continent of birth are shown for the selected countries. For the reasons already stated in the methodological section, the results by group of immigrant origin are comparable only within each country. As it can be observed, the groups analysed present different age-migration profiles, both in intensity and shape. In terms of place of birth, we can argue that the Europeans have a very similar age-migration profile to that of the native-born population in most of the countries, and low differences by sex are observed. The people of African origin manifest a pattern which is predominantly masculine at all ages, especially in Spain and Italy, where there are remarkable differences in the intensity by gender. The important mobility experienced by the Africans, aged between 20 and 49, means the existence of a hyper-mobility pattern that contrasts with the migration profile that can be found in the native-born population (Recaño-Valverde 2003). The Latin-American pattern is characterised, on the contrary, by the more important protagonist role of the females and their more outstanding trend towards family migration. Finally, the Asians concentrate a great deal of their migratory intensity around the young people, with males more mobile than females.

There is variation in the internal migration intensity of native and non-native born populations (Table 12.3). Generally, the changes of residence by the foreign-born population are appreciably more numerous than those by native-born individuals. Differences in mobility between native and foreign-born populations are greater in the medium and long distance migration for males and females in each country. Another important distinction is found in the extreme variation in GMR as we consider the continents of origin. In sum, the internal short-distance residential mobility of the population born is 70–90 per cent higher than that of the native-born population in Italy, Spain and Portugal. On the other hand, the previous differences in Italy and Portugal remain and are even more highlighted now in countries like Spain, where long distance mobility of males born in another country is 2.12 times higher than that of the Spanish-born population.

People born in Africa, Latin-America and Asia have the highest mobility levels, much higher than those of the Europeans. North Americans represent a special case, since they seem to transfer the high mobility in their countries of origin to the countries to which they have emigrated.

Table 12.3 Demographic indicators by type of migration and continent of birth (2000–2001)

		Gross migraproduction rate (GMR)						
		Same major, different minor administrative unit (Short -distance)						
Country	Gender	Native-born	Foreign-born	Europe	North America and Oceania	Africa	Asia	Latin America
Italy	Males	1.20	2.36	1.95	1.84	3.17	3.15	2.61
	Females	1.23	1.95	1.74	0.87	2.55	2.01	2.52
Portugal	Males	0.86	1.57	1.31	0.85	2.08	2.72	1.45
	Females	0.90	1.60	1.40	0.57	2.00	2.32	1.63
Spain	Males	0.71	1.23	1.16	1.17	1.21	1.47	1.31
	Females	0.70	1.16	1.11	1.18	1.17	0.76	1.24

		Different major administrative unit (medium–long distance)						
Country	Gender	Native-born	Foreign-born	Europe	North America and Oceania	Africa	Asia	Latin America
Italy	Males	0.76	1.26	1.25	0.72	1.14	2.45	0.89
	Females	0.58	0.93	0.95	0.61	0.81	0.90	1.08
Portugal	Males	0.72	1.36	1.36	0.52	1.53	1.05	0.88
	Females	0.81	1.49	0.96	0.43	1.85	2.33	1.39
Spain	Males	0.37	0.82	0.55	1.43	1.14	1.40	0.82
	Females	0.39	0.72	0.63	0.59	0.71	0.96	0.81

		Sex ratio of GMR						
		Same major, different minor administrative unit by continent of birth						
Country	Gender	Native-born	Foreign-born	Europe	North America and Oceania	Africa	Asia	Latin America
Italy	Sex ratio	0.98	1.21	1.12	2.11	1.24	1.57	1.03
Portugal	Sex ratio	0.97	0.98	0.93	1.48	1.04	1.17	0.89
Spain	Sex ratio	1.02	1.06	1.05	0.99	1.03	1.93	1.05

		Different major administrative unit by continent of birth (2000–2001)						
Country	Gender	Native-born	Foreign-born	Europe	North America and Oceania	Africa	Asia	Latin America
Italy	Sex ratio	1.31	1.35	1.32	1.19	1.40	2.70	0.83
Portugal	Sex ratio	0.89	0.91	1.42	1.20	0.83	0.45	0.63
Spain	Sex ratio	0.95	1.14	0.87	2.41	1.61	1.46	1.01

Source: Own elaboration based on the Integrated Public Use of the Microdata International Series: version 5.0. Minneapolis: University of Minnesota, 2009.

A second factor to bear in mind is the existence of important gender differences in the migration rates of the foreign-born populations. Whilst among Europeans the mobility intensity is similar for males and females, the migrations of Africans and Asians show much higher intensities among men (Table 12.3). On the contrary, Latin-American women change residence with higher intensity than their male counterparts. In brief, immigrants from Asia and Africa present an internal mobility pattern primarily masculine, a fact that is reversed as we consider the Latin-American and European populations. Nonetheless, the composition by nationalities of the selected groups in the countries of destination has an important effect on these results. In Italy, where the Albanians are the most represented European origin, the gender differences are more obvious and do favour males, especially in the medium-long distance migration.

Two reasons help to explain the interactions between gender, place of birth and country of residence. With regards to the differences by country of origin, preliminary works about internal migration in the developing countries (United Nations 1993, Hugo 1993, Bilsborrow 1993), and specifically the data by sex and age estimated by Singelmann (1993) for 47 countries, show a strong association between the gender differences by age, the change in the status of women during the life cycle and the high differences in internal migration intensity, in favour of males in the African and Asian countries (mainly in the Arabic countries of the Asian continent). This relationship is modified by the influence of some cultural determinants that have to do with the female condition, not easy to quantify for the moment, that seem to explain the great spatial variability of gender relationships in the countries of origin. These cultural differences may have been transferred to the countries we study and may be fostered by other socio-economic factors, especially linked to the role that non-native males and females have in the labour markets of their countries of residence. For instance, in the case of Spain, the seasonal low qualified work (Recaño-Valverde 2003, De Miguel, Solana and Pascual 2007) that African and Asian male labourers assume in sectors such as agriculture and construction mean continuous longer or shorter migrations to the places where this labour demand is generated. The higher differences in the migration intensity by sex found for the medium and long distance mobility among the Africans and Asians in Italy and Spain lend weight to his interpretation.

The Individual Determinants of Internal Migration

The results obtained from the micro perspective confirm those previously discussed for the aggregated data for sex, age and place of birth. The general pattern of most of the explanatory variables is similar across countries when we study medium-long and short-distance migration (Table 12.4), although the magnitude of the coefficients varies. The odds ratios of having experienced medium-long distance migration in the previous year are always lower for females than males, although Italian women move much less than those in the other countries. In terms of short-

Table 12.4   Odds ratios (exp(β)) for migrating, by type of migration, in Italy, Portugal and Spain (2001)

Explanatory variables	Short distance			Medium/long distance		
	Italy	Portugal	Spain	Italy	Portugal	Spain
Sex						
male	1.000	1.000	1.000	1.000	1.000	1.000
female	0.919*	0.904*	0.911*	0.687*	0.861*	0.893*
age group						
25–29	1.000	1.000	1.000	1.000	1.000	1.000
30–44	0.575*	0.487*	0.512*	0.518*	0.555*	0.692*
45–59	0.254*	0.225*	0.232*	0.233*	0.265*	0.376*
60–74	0.195*	0.136*	0.179*	0.170*	0.280*	0.335*
75+	0.259*	0.228*	0.164*	0.164*	0.275*	0.344*
place of birth						
native-born	1.000	1.000	1.000	1.000	1.000	1.000
non-native born						
Africa	2.153*	1.678*	1.958*	1.315*	1.789*	2.283*
Latin-America	1.696*	1.612*	1.796*	1.271*	1.558*	1.869*
North America and Oceania	0.851	.614	1.08	1.149	0.691	1.378
Asia	1.934*	1,485	1.124	2.107*	1.971*	1.769*
Europe	1.391*	1.194	1.690*	1.551*	1.540*	1.253*
marital status						
single/never married	1.000	1.000	1.000	1.000	1.000	1.000
married/in union	.920*	1.994*	1.511*	.500*	1.092	1.000
separated/divorce	2.431*	3.563*	2.765*	.925*	2.140*	1.693*
widowed	1.420*	2.698*	1.874*	.706*	1.593*	1.123
educational attainment						
less than primary completed	1.000	1.000	1.000	1.000	1.000	1.000
primary completed	1.166*	1.498*	1.269*	1.091	1.510*	1.217*
secondary completed	1.434*	2.285*	1.936*	1.905*	2.384*	1.792*
university completed	1.771*	2.989*	2.175*	3.944*	4.091*	2.683*
housing tenure						
owned	1.000	1.000	1.000	1.000	1.000	1.000
not owned	1.235*	1.175*	1.339*	1.529*	1.689*	3.372*
employment status						
employed	1.000	1.000	1.000	1.000	1.000	1.000
unemployed	0.632*	1.011	0.944	0.936*	1.682*	1.614*
inactive	0.788*	0.772*	0.854*	1.305*	1.300*	1.123*
constant	0.029*	0.010*	.010*	.017*	.008*	.004*

Note: *p<0.05; ** p<0.01; *** p<0.001.

Source: Own elaboration based on the Integrated Public Use of the Microdata International Series: version 5.0. Minneapolis: University of Minnesota, 2009.

distance mobility, women still move less, although in general the estimators are now closer to one, pointing out that the gap with regards to men has reduced. This result was expected since this kind of migration is mainly associated with residential mobility and not so much with labour market adjustments. Change of municipality is often a response to changing house requirements which push the family unit (or just some members) to move to a new dwelling.

As we showed above with the aggregated data, the younger group (25–29) is more likely to move for all of the time intervals considered and the probability of having migrated in the previous year/s decreases with age. The gap between the baseline category and the next one (30–44) is lower in Spain for medium-long distance migrants (regarding migration in the last year). Also in Spain, the estimates for people over 74 are higher in this type of migration, explained by a high incidence of migration strategies associated with entry into widowhood and the search for geographical proximity (if not cohabitation) to children who migrated in medium-long distance in the past. Return movements of former inter-regional emigrants could have some effect on the results for this group, but this partial effect should explain more about propensity to move of people aged 60–74, at least in the countries where the time point reference is one year ago. Something similar is observed for the influence of age groups in short-distance migration. They follow the general trend already discussed for inter major administrative unit migration, but we also find slight differences for older groups in Italy and Portugal that reveal the increase in the probability of having changed municipality of residence during the previous year for those aged 75 and over, in relation to the preceding category. That is, maybe a situation of more dependency explains this discrete increase in their mobility. Residential strategies linked to a deterioration of health conditions may be one of the main reasons for this finding: people moving to one of their children's home or getting a place to live that it is closer to them.

In medium-long distance migration, considering the effect of place of birth, we observe that geographical immigrant origin does not have exactly the same effect in the three countries of residence. Generally, nonetheless, the pattern observed with the aggregated data, of a higher mobility of the non-native born people, persists after controlling for other socio-demographic variables, except for those born in North America and Oceania. Asians' propensity to migrate in the last year is higher than that of the other immigrant origins in Italy and Portugal, and it is also quite high in Spain. The history of immigration in each destination helps to clarify the differences. Immigration flows of Asians are recent in the majority of Southern-European countries, for instance. The longer the time spent in the country, the lower the likelihood of changing region of residence. Mobility of people born in Africa is double that of people born in Spain, and almost double that of people born in Portugal (despite the fact that major collectives in this category are, for both destinations, originally from different African countries). We cannot confirm, thus, that groups sharing this continent of birth have the same internal migratory patterns in the countries where they live. Europeans tend to migrate more than the native-born population. We have to take into account that, due to the variability in data sources, we have not been able to disaggregate further the categories of the place of birth. Europe, as for the other continental origins (except maybe North America and Oceania) groups a heterogeneous profile of immigrants from very diverse origins. In Spain, for instance, where Europeans' mobility is higher than that of natives, the presence of foreign-born people from Western European countries that change residence for reasons frequently associated with the improvement of their quality of life (climate, etc) share the category with the so-called labour immigrants from Eastern European countries of birth.

After Africans, Latin-Americans' mobility is particularly high in Spain, after controlling for explanatory variables, and almost double the mobility of the Spanish-born population. This immigrant group also shows a high probability to have experienced a recent move in Portugal, with odds close to 1.6.

With regards to place of birth, the position of Latin-Americans has been modified and, for short-distance migration, Africans and Asians have lower estimators of probability to migrate. Latin-Americans move, controlling for covariates and compared to the other continents of birth, more at short than long distances. The role of Africans is especially interesting since this group has the highest probabilities of having moved in Italy, Spain and Portugal. Also in the three countries (Portugal, Spain and Italy), Latin-Americans have high odds ratios of moving. Europeans, on the other hand, are more likely to change municipality in Spain, whilst Asians are more mobile in Italy. In general, however, these three Western-Mediterranean countries (Portugal, Spain and Italy) show similar patterns of mobility for immigrants according to their continent of birth.

The effect of marital status differs across countries. Relationships within the family do vary depending on the cultural norms prevalent in the different contexts. Those categorised as single are more willing to move in Italy, but separated or divorced individuals have a higher odds ratio in almost all countries of having experienced a medium-long distance migration in the last year, maybe often as a consequence of their entry into this status. After them, widows are most mobile, probably for the same reason: maybe they have fewer commitments that link them to the place of residence or maybe it is the change in their marital condition which implies the new mobility.

As we saw for medium-long distance migration, marital status does not have the same influence across countries for short-distance migration. In fact, the divorced and separated are those with highest coefficients in all countries. As we suggested before, mobility in these cases could be partially understood as a consequence of a change in marital status.

In general, the higher the academic attainment, the higher the odds ratio of having migrated in the period considered. People with a university degree move four times more (all other variables set to zero) than people with no completed studies in Portugal and Italy and around three times more in Spain (which shows the smallest differences between the extremes). It is interesting to highlight this effect of education since inter-regional migration in certain countries, such as Italy and Spain, was in the recent past associated with labour mobility, following to some extent the same patterns that international immigrants would eco years after. In 2000–2001, controlling for explanatory variables, medium and long distance migration was more frequently experienced by those who were best prepared in terms of formal education.

There is no doubt (despite the differences in the magnitudes across countries) about the influence of academic attainment on internal migration. For long distance mobility, and also for short-distance, those who are more likely to migrate are those who are educated to the highest levels. The differences are more noticeable in Portugal and less relevant in Italy, but the results are consistent for all datasets and territorial perspectives of analysis. The higher the formal education received, the higher the chance of migrating, regardless of place of birth, sex and age. This breaks a pattern that characterised some of these countries in recent periods in the past, when labour migration would affect persons with low qualifications.

Not owning a dwelling has a relatively important positive effect across the selected countries. Ownership of a house is the most relevant explanatory element in Spain, a country where the incidence of home ownership is particularly high. Having a property prevents people from migrating to another major administrative unit.

The ownership condition (even if the property is not totally paid for) prevents migration, since this circumstance normally roots the person (or the family unit) to the place of residence. However, coefficients are in general lower for non-owners in the case of medium-longer distance than in the case of short-distance migration, indicating that those who do not own a dwelling have higher odds of living in a different major administrative unit a year ago than to be living in a different minor administrative unit, other variables kept constant. Obviously, part of the explanation relies on the fact that part of the inter-municipal mobility is an effect of the acquisition of a house.

People who are unemployed or inactive at the time of the Census are, in general, more likely to have migrated (medium-long distance) in the previous year than employed people. It is reasonable to state that persons who have a stable employment situation would be more reluctant to change province/region of residence (unless it is a job requirement) than a person who is jobless or does not have a tie to a place (students, retired people, etc.). Something similar happens in Italy for the unemployed, although the coefficient is quite proximate to unity, indicating that differences are modest. Finally, the behaviour observed according to employment status is the same for all countries for short-distance migration. People who are unemployed by the time of the Census data collection have estimators very close to one in Portugal and Spain (showing no relevant differences with regards to the employed population) and even lower than one in Italy. Short-distance migration, as we have mentioned before, is not so much related to the labour market demand as is medium-long distance, so it is somehow predictable that change of residence within the major administrative unit in the countries where these units do not imply much distance correspond more often to people who are employed and can afford a new house. In sum, results for short-distance mobility do not differ much from those for medium and long distance.

Conclusion

The research questions we have proposed at the beginning of this chapter have been partially answered. The response to whether or not the demographic patterns of internal migration of foreign-born are similar to those of natives by age and sex is negative. The pattern by age of native and non-native populations differs significantly. Also, the intensity of internal migration of the foreign-born population is notably higher than that of the native-born population. The answer to the second question, whether the migration patterns differ by immigrant origin, is positive. The population born in Africa, Asia and Latin America extends their internal mobility to all active age groups, in contrast to the migration pattern of the native-born; the profile of age migration rates of the Europeans and native-born population are quite similar. However, important differences in intensity are found for the foreign population. In particular, people born in the Asian and African continents show the highest intensities of internal migration. In addition, male mobility predominates amongst immigrants of African and Asian geographical origin, but this trend is reversed for the Latin-American population. The European countries do not display important differences by gender.

Are the observed demographic patterns by specific national groups always the same or do they differ by the country of destination? Our findings with regard to this question are ambiguous. The sex and age structures of internal migrants by continent of origin and country of residence are very similar (and possibly affected by the limitations of the sample), but the intensities vary according to the country of residence and the region of birth. We presume that the national composition of each continental group, which we have not been able to analyse in detail, may explain this result. We have to bear in mind that the nationalities that constitute the European, Asian and Latin-American immigrant groups are not the same in the three countries of residence we have compared.

What are the effects of the individual characteristics on the internal migration of foreigners in Spain, Italy and Portugal? In short, our findings confirm the results offered by the international academic literature for other contexts. Females tend to move less than men and gender differences are higher for medium and long-distance migration than they are for short-distance mobility. In Italy, the gap between females and males for this sort of mobility is the most noteworthy. The effect of educational attainment is regular across countries of residence: the likelihood of having experienced a change of residence increases with level of academic attainment, for any distance of move. Being a home owner diminishes the likelihood of having migrated in all selected countries and this influence is relatively higher for long-distance migration. Marital status and employment status have a less homogeneous relationship with internal mobility as we compare across countries of settlement. Single people are more likely than those in relationships to move long-distance in Italy, but the high mobility of separated and divorced people is outstanding in Spain and Portugal. For example, this is the group most prone to migrate short-distance. In general, the unemployed migrated more than employed people over long distances, but this relationship differs by country of residence for short-distance migration. We have to take into account short-distance mobility being more associated with the housing market, whilst long-distance moves are more often a response to other motivations such as job searching.

The new 2010–2011 Census data will allow us new possibilities for the study of minority internal migration due to the augmentation of the Census samples for the immigrant population. This will enable a more detailed geographical disaggregation of the data by origins and allow us to consider other substantive questions such as: how have the processes of geographical assimilation evolved after more than a decade of permanence of the immigration? Finally, we will be able to assess the impact of the present economic crisis on the internal migration patterns of native and non-native populations in the South of Europe.

Acknowledgements

This paper has been carried out in the framework of two research projects: La movilidad geográfica de la población extranjera en España: factores sociodemográficos y territoriales (SEJ2007-61662/GEOG) and Inflexión del ciclo económico y transformaciones de las migraciones en España (CSO2010-19177), both funded by the Ministry of Education and Science, National R+D+I Plan 2004–2007 and 2007–2010. We would like to acknowledge the help of the Editors (Nissa Finney and Gemma Catney) who provided insightful comments and suggestions for improvement.

Chapter 13

Understanding Ethnic Minorities' Settlement and Geographical Mobility Patterns in Sweden Using Longitudinal Data

Roger Andersson

Introduction

Existing Swedish research shows that crude migration rates (i.e. not standardised by age) are clearly higher for most categories of immigrants compared to the native Swedes (Andersson 2000, Fischer et al. 2000). Early studies proposed that the (compulsory) refugee dispersal policy introduced in 1984 (Hammar 1993, Robinson, Andersson, and Musterd et al. 2003) could have triggered unnecessarily big volumes of secondary migration. There are however some indications that ethnic minorities' migration rates might have been equally high already in the years preceding the introduction of this policy and that they have remained high also after an important liberalisation of the placement policy in 1994. It is therefore unclear whether the policy as such really made the anticipated difference (Andersson 1998). It is one important aim with this chapter to bring Swedish research on the ethnic dimension of internal migration a step further. By employing multivariate statistical methods and using rich individual longitudinal data it should be possible to explore in more detail and explain different aspects of the ethnic variation in geographical mobility.

But does it really matter whether ethnic minorities move more often and in other directions than natives? Yes it does, and not least for policy reasons. First, migration is a key demographic factor deciding which regions, cities and neighbourhoods that will face population growth or decline. Hence, having good knowledge about migration is a fundamental point of departure for making reasonably good population forecasts for the short- and long-term future (Finney and Simpson 2009). The uneven national and local distribution of immigrants makes this even more important. Secondly, different types of settlement steering of new immigrants and in particular of refugees are discussed and implemented in many countries. Knowing whether 'secondary migration' (migration after placement) supports or counteracts such policy initiatives is therefore highly relevant. Thirdly, internal migration could be important for integration processes. So far we do not know whether the internal migration of minorities is related to upward social mobility or to social marginalisation, discrimination and segregation.

Behavioural differences within countries, such as group variations in the likelihood of migrating, could have 'ethnic' explanations. At least it could not be ruled out that cultural factors have an impact on the experiences, life situations, preferences and strategies of people, and that such factors produce different social outcomes, including variation in geographical mobility. However, it is the experience of this author, that what often is referred to as ethnic differences have nothing to do with ethnicity or culture but rather can be explained by demographic and social class features that co-vary with, for instance, the country origin of people (i.e. compositional differences between ethnic groups, see also Finney and Simpson 2008). This chapter will focus on this issue and it will in particular address three questions: do internal migration frequencies in Sweden differ between native Swedes and immigrants? If so, are such differences similar if we focus on inter-regional migration compared to intra-regional migration? Can differences be explained by the variation in demographic and socio-economic composition of natives and immigrant categories or can factors attributed to the specific experiences of immigrants or their background characteristics explain part of such variations?

The chapter is organised into five sections. Following the introductory section, section 2 provides some basic data on the regional and local settlement patterns of immigrants in Sweden and also data on how this pattern has changed over time. The subsequent section provides theoretical arguments for why we could expect migration to vary between natives and immigrants. The fourth section reports empirical results from a study using comprehensive longitudinal Swedish population data. This section provides a descriptive overview but also a multivariate analysis of the likelihood of moving (a) between labour market regions, (b) from one neighbourhood to another within the same labour market region. Some attention will also be paid to intra-neighbourhood migration. The fifth section concludes the chapter and sums up the main findings.

The Geography of Immigration to Sweden

The Swedish population currently grows by around one percent per year and reached 9.4 million people in 2010. It has been estimated that the entire growth since 1970 (+ 2 million) is due to immigration (Ekberg 2000). Statistics Sweden (2004) has calculated that more than three quarters of the population increase since 1945 is due to immigration. It is also a striking fact that Swedish regions with net immigration are the only regions experiencing population growth over the last half a century. Because these growing regions are often more urban than others, one might think that it is urbanisation and not immigration that is the basic process reshaping population balances across the country. Another way of putting this is to say that urbanisation is driven not by internal migration of the native Swedes but to a large extent by people immigrating to the country. Larger cities have a higher share of immigrants than the national average. The over representation of immigrants in the capital region and in other major cities is by no means unique to Sweden. According to the OECD, the density index of immigrant population shows that in many countries the density of the immigrant population is at least 1.5 higher in capital regions than in any other region (OECD 2011).

In terms of absolute numbers in 2008, Stockholm municipality had 170,000 foreign-born residents, Gothenburg 106,000, Malmö 82,000, Uppsala 30,000 and Botkyrka in the southern part of the Stockholm region had 28,000. These are followed by Södertälje, Lund, Västerås, Huddinge and Örebro, all having 18–25,000 foreign-born. Of these ten, Uppsala has the lowest share of first generation immigrants (14.8 per cent) while Botkyrka has the highest (35.7 per cent). Of the three biggest municipalities, Malmö has 29 per cent, followed by Gothenburg and Stockholm (both 21 per cent). The relative growth of foreign-born in these municipalities 2000 to 2008 ranges between 22 percent in Botkyrka and 36 per cent in Malmö, to be compared with the overall net national growth of 27 per cent (+269,000). This means that although these ten municipalities (having the largest absolute number of immigrants) have increased their number of foreign-born from 436,000 to 531,000 in eight years, their share of all foreign-born in Sweden in fact declined from 43.5 per cent to 41.7 per cent. In other words, the relative growth of Sweden's foreign-born population was faster outside of the largest concentrations.

Figure 13.1 shows the geographical distribution of immigrants in 2008 across municipalities (left), and the relative change 2000–2008 (right). First of all, 35 municipalities have a proportion foreign-born exceeding 16 percent (national average was 13.8 per cent in 2008). Beside some border municipalities (along the Norwegian and Finnish borders) we find concentrations also in regions that experienced labour migration in the 1950s and 1960s. However, most of these 35 are found in the three metropolitan regions, Stockholm, Gothenburg and Malmö, which all have been primary destinations for refugee migrants arriving since the 1970s. Secondly, many more municipalities, 125, have a proportion foreign-born clearly below the national average. These are predominantly small, rural municipalities in the northern part of Sweden.

The map (Figure 13.1, right) showing relative change 2000–2008 tells an interesting story. It seems like many of these small, rural municipalities have had a rather fast expansion of immigrants. This message is corroborated by Table 13.1, showing detailed population change according to country origin and type of municipality. People born in Sweden is the only category decreasing in sparsely populated municipalities, while those born in Eastern Europe (many with Bosnian and Russian origin) and countries outside of Europe more than double their numbers during this rather short period of time. The increase is from low levels and it can only compensate for about half the loss of native Swedes. The most remarkable compositional change is the fast expansion (+51 per cent) of people born in Asia, Africa and Latin America, and their presence increase substantially in all types of municipalities.

The vast majority of new immigrants persistently end up in the three metropolitan regions and in the larger cities. Immigrants comprise about 75 per cent of the total population increase in these two municipality types. Suburban municipalities mostly expand due to increase (natural growth) of native Swedes (see also Kulu, Boyle and Andersson 2009).

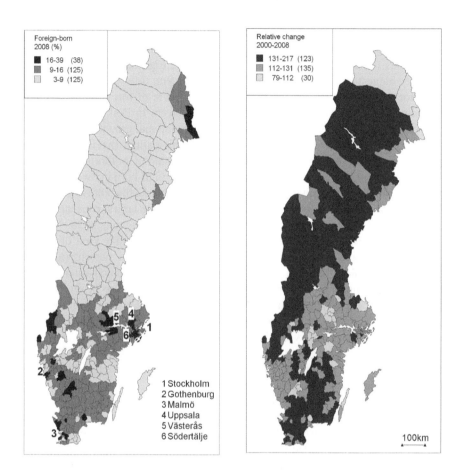

Figure 13.1   The percentage foreign-born in Sweden´s municipalities, 2008, and relative change 2000–2008 (Index, 2000=100)

Source: Geosweden. Source: Data from Geosweden, Institute for Housing and Urban Research, Uppsala University.

Table 13.1 Population numbers and population change according to country of birth and municipality types 2000 to 2008

Country of Birth	Population	Metro-politan munici-palities	Suburban munici-palities	Large cities	Commuter munici-palities	Sparsely populated munici-palities	Manufactu-ring munici-palities	Other munici-palities, more than 25,000 inh.	Other municipalities, 12,500–25,000 inh.	Other municipalities, less than 12,500 inh.	Total
Born in Sweden	Stock in 2000	1,189,625	1,148,888	2,175,620	512,568	302,201	538,287	1,144,644	600,756	253,487	7,866,076
	Stock in 2008	1,208,870	1,226,555	2,204,796	524,878	288,082	532,443	1,134,985	586,456	244,845	7,951,910
	Net change	19,245	77,667	29,176	12,310	-14,119	-5,844	-9,659	-14,300	-8,642	85,834
	Relative change***	1.6	6.8	1.3	2.4	-4.7	-1.1	-0.8	-2.4	-3.4	1.1
Born in Western countries	Stock in 2000	85,542	82,003	92,909	22,204	10,041	24,707	41,966	25,203	15,599	400,174
	Stock in 2008	93,688	81,712	97,459	24,134	11,668	26,958	44,224	27,282	17,501	424,626
	Net change	8,146	-291	4,550	1,930	1,627	2,251	2,258	2,079	1,902	24,452
	Relative change	9.5	-0.4	4.9	8.7	16.2	9.1	5.4	8.2	12.2	6.1
Born in Eastern European countries	Stock in 2000	72,633	33,966	62,673	10,404	1,375	19,346	26,189	9,871	3,109	239,566
	Stock in 2008	85,933	46,421	74,830	14,235	2,821	24,414	32,007	13,440	4,188	298,289
	Net change	13,300	12,455	12,157	3,831	1,446	5,068	5,818	3,569	1,079	58,723
	Relative change	18.3	36.7	19.4	36.8	105.2	26.2	22.2	36.2	34.7	24.5
Born in Asia with Turkey, Africa, Latin America	Stock in 2000	129,103	72,849	101,093	10,456	2,741	10,593	24,758	8,864	2,828	363,285
	Stock in 2008	178,101	99,405	162,899	17,385	5,986	18,441	44,349	17,205	5,574	549,345
	Net change	48,998	26,556	61,806	6,929	3,245	7,848	19,591	8,341	2,746	186,060
	Relative change	38.0	36.5	61.1	66.3	118.4	74.1	79.1	94.1	97.1	51.2
Total population	Stock in 2000	1,476,903	1,337,706	2,432,295	555,632	316,358	592,933	1,237,557	644,694	275,023	8,869,101
	Stock in 2008	1,566,592	1,454,093	2,539,984	580,632	308,557	602,256	1,255,565	644,383	272,108	9,224,170
	Net change	89,689	116,387	107,689	25,000	-7,801	9,323	18,008	-311	-2,915	355,069
	Relative change	6.1	8.7	4.4	4.5	-2.5	1.6	1.5	0.0	-1.1	4.0

Source: Geosweden database. Institute for Housing and Urban Research, Uppsala University.

Local Settlement Patterns

Some politicians argue that Sweden is ethnically more segregated than other comparable developed countries. Such proposals lack empirical justification and they do so primarily because of the problems of measuring residential segregation across countries (and often also between cities within countries). The problems concern both the definition of population (ethnic) groups, the identification of such groups in official statistics, and not least the spatial definitions needed to divide urban space into a comparable neighbourhood typology (see Musterd 2005).

Like most of Europe, Sweden does not have mono-ethnic clusters of immigrants (i.e. one single ethnic group dominating a neighbourhood or district). It is indeed difficult to find neighbourhoods with more than ten percent of the population originating in a specific foreign country. What is a fact, however, is that residents from some parts of the world – typically from Muslim countries – are highly concentrated into neighbourhoods which have high overall immigrant density. Such areas are often located in the periphery of towns and cities, and many were constructed as part of the Million Homes Programme initiated by the Social Democratic government in the mid-1960s. This programme was a national political project that aimed at modernising the housing stock and reducing scarcity of housing following urbanisation and the demographic shift towards smaller households. Rental housing dominates in these areas but cooperative housing (a market form whereby the right to dispose of a particular dwelling is tradable while the property is owned and managed collectively by all the residents in the cooperative) is not unusual and sometimes amounts to 20 to 30 per cent of the dwellings.

These immigrant-dense housing estates have been characterised as transit areas, or ports of entry, for newcomers to a city and it is a fact that turnover rates are high and that typically half the population of a particular area will have moved out within a few years (Andersson and Bråmå 2004, Andersson 2008). The reproduction of the areas' characteristics of being income-poor and immigrant-dense is due to a constant influx of newly arrived refugees (and subsequent immigration due to family ties) and selective migration (see Andersson and Bråmå 2004, Bråmå 2006), whereby those who succeed relatively well tend to quickly leave the areas. In this sense, the spatial assimilation thesis seems to hold true also for the Swedish case (for literature on the spatial assimilation thesis, see Massey (985, South, Crowder and Pais 2008, and for a Scandinavian study, see Skifter Andersen 2008).

The concentration of different minorities in these areas differs quite substantially and the sorting results in something that could be called an ethnic hierarchy. Such a hierarchy is discernable in Table 13.2, showing index of dissimilarity values for a range of population groups in Gothenburg. Residents originating in Somalia, Iraq and some other primarily Muslim countries live more at a distance from native Swedes than do for instance European migrants, an important part of the reason being their much lower presence in the labour market, and consequently their low work income.

Table 13.2   Illustration of the ethnic hierarchy in Gothenburgh city: average work incomes, employment, and residential segregation by country of birth 1995 and 2006

Country of origin	Average work income*, 100 SEK		Perc. Employed age 20 to 64		Segregation (*D* Index)	
	1995	2006	1995	2006	1995	2006
Sweden	1449	2273	72	79	Ref.	Ref.
Norway	1136	2008	59	69	25.7	24.6
Germany	1334	2052	65	67	23.2	26.5
Denmark	1227	2292	65	75	28.6	27.9
Finland	1137	1807	61	69	40.4	34.9
Poland	921	1472	52	61	41.4	38.6
Iran	337	1096	23	51	57.2	45.5
China & Taiwan	609	900	41	41	68.9	53.4
Yugoslavia	701	1318	40	56	55.6	54.4
Chile	726	1326	47	63	63.1	56.4
Turkey	452	960	35	51	75.0	68.4
Bosnia-Hercegovina	87	1366	6	61	80.6	68.6
Lebanon	289	820	19	44	74.6	73.1
Ethiopia	380	1142	25	57	74.7	74.1
Vietnam	708	1035	44	53	84.5	74.3
Iraq	211	637	13	34	77.6	77.2
Somalia	102	432	7	24	83.5	85.0
All w. foreign background	776	1304	44	55	41.2	43.3

Note: The country of origin groups are ranked in descending order of the Segregation Index in 2006. The D Index calculation is based on a SAMS neighbourhood division. The average population of a SAMS is 1000 inhabitants. The D Index compares the spatial distribution across neighbourhoods of each country of birth category with the distribution of native Swedes. The index runs from 0 (no difference) to 100 (maximum difference).

Source: Geosweden database. Institute for Housing and Urban Research, Uppsala University.

Subsequent sections will shed light on the issue of whether intra-urban migration is affected by the neighbourhood context. It is of course plausible that the ethnic composition can be either a pull or a push factor in the context of intra-urban migration. Immigrant-dense areas are stigmatised, which typically means that people with a real choice in the housing market in general tend to avoid living there. The presence of co-ethnics might at the same time be attractive, at least for a period after immigration. It is not the primary focus of this chapter to disentangle the intricate causality that might be at hand here, but the overall hypotheses would be that (a) living in high immigrant concentration will increase the probability of moving, and (b) living in high own group concentrations will reduce the propensity of moving. The subsequent analyses will contribute to testing the first of these hypotheses but the second one will be left for later studies.

Why Can We Expect Internal Migration to Differ Across Ethnic Categories?

There are a number of reasons for why we should expect geographical mobility to differ between natives and immigrants (see for example Robinson 1992; Champion 1996; Andersson 1998; Bailey and Livingstone 2005; Stillwell and Hussain 2010; Stillwell, Hussain and Norman 2008; Finney and Simpson 2008; Catney and Simpson 2010). These reasons could be grouped into three broad sets of factors: demographic, cultural, and socio-economic.

Demographic

It is a well-known fact that the propensity to migrate varies greatly over the life cycle and that age captures the characteristics of these life cycle stages very well (Rossi 1955, Clark and Dieleman 1996, Mulder 2006). The most mobile are in their 20s. Although immigrants get older over time, countries with a high proportion of more recently arrived immigrants most certainly will have quite big age differences between natives and immigrants, the latter composing disproportionately high numbers of young adults.

Other demographic characteristics, such as household composition and sex, also co-vary with propensity to migrate. Ravenstein (1885) found that males dominated international migration while females migrated more frequently within the countries (England and Wales). This is of course no 'law' and actual migration frequencies can vary by gender in different ways, being different for different periods of time and types of origin-destination contexts. Household composition has been found to be very important. Singles tend to be more geographically mobile than couples and adults living with children less mobile than those without. In countries with high divorce rates and a growing awareness of the importance of taking joint care of children after a divorce – like in Sweden – divorced people may face severe migration constraints; having something like a common settlement strategy after a divorce is probably more difficult than negotiating such things while being a family (Stjernström 2011). Like for age, and partly related to age, the household composition of a particular immigrant group may sometimes differ quite substantially from natives' overall household composition, causing more or less migration solely because of such compositional variation.

Cultural

The concept of culture is often contested and rightly so. It is often misused to imply something static and primordial and people are categorised into cultural groups on fuzzy indicators. There is however reason to believe that factors like religion, food habits, and language under certain circumstances (for instance a period after immigration) influence people's settlement and migration decisions (Mulder 1993, Abramsson, Borgegård, and Fransson et al. 2002). Even more importantly is probably the fact that majority residents and institutions of host societies may discriminate immigrants with reference to their 'Otherness' (Phillips 2010). Racism and discrimination can foster actions for defense (such as geographical clustering) but it can also block members of certain minority groups from entering specific types of housing or neighbourhoods (Molina 1997). Swedish research on housing discrimination is growing but is still limited in volume and few studies have been made accessible for non-Swedish readers. A paper by Ahmed and Hammarstedt (2008) presents strong evidence of discrimination in the rental housing market and their results confirm studies carried out by others, including the Ombudsman against discrimination (DO 2008).

Socio-economic

Households' economic resources have significant impact on migration behaviour, partly because economic resources increase the degree of freedom of choice someone has in the housing market. Class positions are strongly but not entirely related to education, but somebody with a higher level of education does not only earn more money and accumulate more wealth, (s)he could also be hypothesised to have better access to information and job opportunities which could be more geographically concentrated. If migration researchers would like to control for individuals' socio-economic positions when comparing mobility for different categories of residents, including both income measures and information on education are recommendable.

Also other class-related characteristics are of relevance when studying migration, such as housing tenure (Hamnett 1991, Clark and Dieleman, 1996). In Sweden, as probably elsewhere, home owners migrate more seldom than renters. The causality can sometimes be difficult to disentangle: someone who plans to, or needs to, move within a couple of years (for example a person moving into a city for study or temporary work) might not risk buying a house in a volatile housing market and therefore prefers rental accommodation even if (s)he could afford to buy a house. The selection of people into different tenure forms is therefore not only related to income but also to long-term settlement strategies and the preferences of households. In a country like Sweden, intra-urban migration frequencies are often twice as high for renters as for home owners (see also Clark and Dieleman 1996), and immigrants – especially from parts of Africa and from the Middle East – are disproportionately concentrated into the rental segment (Andersson 1998, Andersson, Magnusson Turner and Holmqvist 2010; for an international comparison, see van Kempen and Özüekren 2002).

To sum up: there are many compositional differences between the native population and most immigrant categories. Demographic, cultural, and socio-economic variations are often inter-related (recently arrived immigrants are younger, earn less money, have less secure jobs, live more often in rental housing, differ in household size, etc.) and how each one of these factors may impact upon migration might in principle be well known. However, when combined, it can be difficult to know from descriptive statistics whether there also exists an ethnic variation in migration frequencies that cannot be explained by compositional variations.

Before turning to the empirical sections of the chapter, we should consider one important but seldom recognised factor that often distorts analyses of ethnic variation in internal migration. A broad concept introduced to understand the high level of immobility found in many European countries compared to the United States is *insider advantages* (see Fischer et al. 2000). Insider advantages are not transferable to other places of work and residence and they are 'obtained within a specific learning process which requires time, information and temporary immobility. Mobility turns such investments into *lost* sunk costs, i.e. costs which are tied to a specific project or – in this case – a specific location and which are lost in case of out-migration.' (ibid: 10). The concept of insider advantages is similar to what DaVanza (1981) calls *location-specific capital*, which 'refer to such diverse things as: homeownership; job-related assets such as an existing clientele (of, say, a well-regarded doctor or plumber), seniority, specific training, or a nonvested pension; knowledge of an area; friendships; and indeed any factor that "ties" a person to a particular place. Such "assets" are costly (or impossible) to replace or transfer to another locality' (DaVanza 1981: 47).

It is an old empirical finding that a person who has made one move is more prone to migrate than someone who has not moved, everything else being equal (Goldstein 1958, DaVanzo 1981). By definition, (first generation) immigrants have all made a move; they even changed country of residence. The native population comprises a substantial share of people who have never left their home village, town, or city. In Sweden, around 30 percent of the population of a random municipality are born there and have never left the place of birth. This factor alone speaks in favour of expecting a higher – at least inter-regional – level of migration for the foreign-born compared to the native population. Fortunately, it is possible to partly control for these variations in migration history in the multivariate models introduced later on in the chapter.

Geographical Mobility of the Foreign-born Population

The Swedish Population Registers

Finney and Simpson (2008: 63) state that 'Relatively little is known about the internal migration of different ethnic groups'. This is true for most countries but to a lesser degree for Sweden. The existence of a population address register (Register over totalbefolkningen, RTB) and also of a parallel event-driven registration of all demographic events (births, deaths, migration, marriages, divorces), that can easily be combined with the RTB, provide excellent opportunities for both contemporary and historical geo-demographic dynamic and group-comparative studies. Detailed accounts of internal migration in Sweden have been provided since the 1950s but mostly published in Swedish; see for instance Hannerberg, Hägerstrand and Odeving (1957), Jakobsson (1969) and Bengtsson (1991).

The Swedish population registers have existed for centuries and are of a high quality. More reliable migration records have been in place since 1916. Until relatively recently it was not the lack of available data but the laborious effort to make use of handwritten records that posed a great deal of problems to Swedish demographic researchers. Nowadays, most materials are digitalised and are more easily accessible. It also helps that Swedish law provides generous rights for researchers to access individual records and complete individual databases.

These registers provide excellent opportunities but they do have their limits. Some of these limitations affect the possibilities of studying the ethnic dimension. Unlike the UK and some other countries, Swedish researchers cannot access information on peoples' self-reported identity. The only type of information available (in a more complete form since 1968) is an individual's country of birth and citizenship. Via a specific generational register it is also possible to track people having one or two parents born in a specific country. The quantitative study of 'ethnic groups' suffers from these shortcomings and quantitative Swedish research therefore has to base ethnic studies not on groups in a more sociological sense but on categories that sometimes are ethnically heterogeneous.

Descriptive Analyses

Data and Definitions

All analyses presented below are based on individuals having been registered as living in Sweden on December 31st 2005, as well as three years later. It goes without saying that only residents having a known address are included in the population register, and the authorities (County administrations and Statistics Sweden) cannot be certain that a person actually lives on the reported address; some might at least have temporary residence elsewhere. Uncertainties are bigger for young people, who leave their parents' home for temporary work or studies but fail to report their new address. Others may have an unsecure tenure (second hand renting) and move quite often for a period of their life. Due to the fact that many immigrants, especially in metropolitan regions, find it difficult to find appropriate housing upon arrival one might suspect that the figures reported below underestimate the real level of mobility for categories comprising large numbers of more recently immigrated individuals.

Migration frequencies are calculated as the percentage of people fulfilling the basic requirement of being in the registers in 2005 and 2008, and who have moved house at least once during this three year period (having moved either in 2006, 2007, or 2008). Using a three year period is in a sense arbitrary, but any period could be said to be arbitrary. It is however one way to avoid problems relating to the representativeness of the registers. It is far more likely that people who actually have moved will be picked up in the address register if one allows for more than one year between the measurement points. Applying a three year period also means that 'return movers' are disregarded and that those moving more than one time will be identified as having made only one move.

The definition of a move is that the location in 2008 should be different from the one in 2005. Three types of location change are studied in more detail: change of labour market (LM) region (100 LM regions in Sweden; normally clusters of several adjacent municipalities which are identified using commuting statistics), change of neighbourhood within labour market regions (based on the SAMS (Small Area Market Statistics) classification scheme, all in all around 9,200 units with an average population of around 1,000 people), and finally change of address within neighbourhoods (SAMS). The SAMS classification scheme is defined by Statistics Sweden and designed to identify relatively homogeneous areas by taking into account housing type, tenure and construction period. They vary somewhat in terms of population, even within metropolitan areas.

Due to space limits intra-neighbourhood moves will be given less attention. Only people aged 16 to 71 (in 2005) are included in the analyses. This reduces the population under study from the country total of 9,029,000 in 2005 to about 6,174,000 (see Table 13.3).

Overall Migration Rates

Table 13.3 shows crude migration rates for native Swedes and nine categories of foreign-born and it does so for all three types of migration mentioned above. Information about the average year of immigration for the immigrants and average age for all ten categories are provided in the last two columns in the table. Immigrants arriving a long time ago – as have many Nordic and Western European migrants – have not only spent more time in Sweden, they are also much older than other immigrant categories. The most recent immigrants are the sub-Saharan group, having 1994 as the average year of immigration. This group is on average 16 years younger than the Nordic immigrant group and more than 7 years younger

Table 13.3 Migration rates 2005 to 2008 by country groups, age 16 to 71 in 2005

Country of birth	Changed labour market (LM) region		Changed n'hood within LM region		Moved within n'hood		Total migration		N	Average year of Immigration	Average Age (in 2005)
	Males	Females	Males	Females	Males	Females	Males	Females			
Born in Sweden	5.9	6.3	18.9	19.3	9.1	9.2	33.9	34.8	5,289,773		43
Rest of Nordic countries	4.5	3.9	15.4	14.7	9.7	10.1	29.6	28.7	205,433	1982	52
Rest of Western European countries	5.3	5.1	19.6	17.5	9.1	9.8	34.1	32.5	78,166	1990	47
Eastern European countries	5.5	5.6	23.2	20.9	12.3	11.8	40.9	38.3	215,338	1991	42
Sub-Saharan Africa	8.8	8.0	29.8	23.6	15.3	15.6	53.8	47.2	45,382	1994	35
Western Asia incl Turkey and N Africa	6.7	6.6	28.8	24.5	13.3	13.9	48.8	44.9	199,598	1992	38
Eastern Asia	9.9	7.8	28.6	25.7	10.6	10.6	49.2	44.2	72,028	1990	35
Latin America incl Central America & Mexico	6.7	6.0	31.3	27.6	11.3	11.9	49.4	45.5	55,104	1989	38
North America, Australia, New Zeeland	6.9	6.3	23.6	22.3	9.3	8.7	39.9	37.4	12,727	1992	41
Unknown	6.7	8.1	30.6	22.8	10.4	18.7	47.8	49.6	257	2000	36
Total	5.9	6.2	19.6	19.6	9.4	9.5	35.0	35.3	6,173,806	1990	43

Note: * Migration rate is calculated by dividing the number of people having migrated with the number of people in each population category in 2005 (N). Migration within a neighbourhood is defined as change of coordinates (100m) within a SAMS. Migration within a multifamily house is thus not registered as a move.

Source: Geosweden database. Institute for Housing and Urban Research, Uppsala University.

than the native Swedes. The effect of these compositional differences is clearly visible in the migration values.

Migration by Age and Year of Immigration

Figure 13.2a shows, first of all, that geographical mobility across labour market regions (LM) is age-dependent. Roughly speaking, people aged 25 at the beginning of this three year period are twice as prone as somebody aged above 30 to relocate (for a more in-depth analysis of LM migration in Sweden, see Lundholm 2007). Secondly, LM mobility is also much higher for recently immigrated people. The same tendency holds true also for mobility within LM regions (Figure 13.2b) but not for moving within neighbourhoods (2c). A close reading of the different charts in Figure 13.2 reveals that the 'recent immigration mobility effect' is most noticeable for those in age groups 30 to 40 and seems to be lasting for approximately ten years, in this case for those having immigrated in 1996 or later. This said, the propensity to move is clearly the highest during the first few years after immigration. It has been stated before (Andersson 2000), that new immigrants have a similar internal migratory behavior as young adults, almost irrespective of their age upon arrival: they are pro-urban in their migration directions, they move more often, and they move for pretty much the same reasons as the young, i.e. for better job opportunities and education.

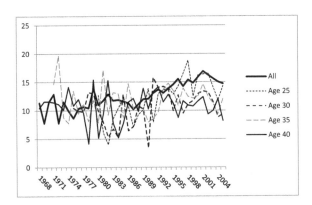

Figure 13.2   Migration frequency for foreign-born residents in Sweden across labour markets Regions (A), Neighbourhoods (B), and within neighbourhoods (C) 2006–2008, by age and year of immigration

Source: Geosweden database. Institute for Housing and Urban Research, Uppsala University. (A) Migration Frequencies (%) across labour market regions (Age in 2005). Average migration distance: 241 km. (B) Migration Frequencies (%) across Neighbourhoods within LM regions (Age in 2005). Average migration distance: 9.6 km. (C) Migration Frequencies (%) within Neighbourhoods (Age in 2005). Average migration distance: 500m.

Migration by Country of Birth, Age and Year of Immigration

Changing labour market region is the least common type of internal migration among the three studied here and it is also the longest in terms of average geographical distance (about 240 km). The propensity to move house across neighbourhoods (9.6 km for intra LM neighbourhood migration) is of course much higher. Focusing on the least common type of relocation, the basic picture derived from Figure 13.2a holds true also for specific ethnic categories; the focus here is on five of the most numerous categories having entered Sweden since the 1970s; people born in Bosnia, Somalia, Iraq, Iran and Chile. Precisely these categories will be in focus also in the multivariate analyses below.

Table 13.4 shows that it is by far more common to change labour market region when you are in your mid-20s compared to later on in life, and it is even more common if you are a relatively recent immigrant to the country. There are however some interesting deviations from this overall picture. Early arrived 25 year old Bosnian and Iraqi immigrants have a much higher LM migration rate 2005 to 2008 compared to more recently arrived from these two countries. It seems to be the case that those having arrived earlier have a mobility pattern more similar to the young Swedes (including moving to university cities and metropolitan regions), while the latecomers on average might be more occupied with the early phase of the integration process (learning Swedish, taking introductory courses, etc). Perhaps even more surprising is the fact that late coming 40 year old immigrants from Somalia have a higher migration rate than is reported by their younger co-nationals. It is difficult to present a plausible explanation for this finding.

Table 13.4   **Age-specific labour market migration rates (2005–2008) for a selection of immigrant categories aged 25, 30, 35, and 40, by period of immigration. Rates as percentage of 2005 population**

Country of birth	Age 25	Age 30	Age 35	Age 40
Sweden	16.7	7.5	3.9	2.7
Bosnia, immigration before 1996	13.8	5.3	3.5	1.9
Bosnia, immigration 1996–2005	6.7	7.6	4.1	5.6
Somalia, immigration before 1996	14.7	5.8	7.2	5.6
Somalia, immigration 1996–2005	16.4	17.2	15.6	17.9
Iraq, immigration before 1996	13.5	7.0	3.5	3.6
Iraq, immigration 1996–2005	10.1	8.3	6.2	6.1
Iran, immigration before 1996	15.8	9.5	5.6	3.2
Iran, immigration 1996–2005	15.4	12.4	7.3	4.4
Chile, immigration before 1996	9.7	6.8	3.0	2.1
Chile, immigration 1996–2005	17.8	11.0	4.7	7.7

Source: Geosweden database. Institute for Housing and Urban Research, Uppsala University.

Before turning to statistical modeling of the likelihood of moving, this descriptive section will finally provide data on migration frequencies by gender and family types.

Migration by Country of Birth and Family Types

Earlier Swedish research found that females migrate more often than men, and singles move more often than couples, and especially couples with children. Table 13.5 confirms these findings but with a couple of exceptions. Females from Somalia and Chile are less mobile than their male countrymen. Chilean females are the least mobile category while Somali females are much more mobile than any category of males, except for the male Somalis. It is however worth noting that Somali females are much less mobile than their male counterparts if they have children.

Table 13.5   **Labour market migration 2005 to 2008 by household composition, sex and country of birth**

Country of birth	Sex	Married/ partner without children	Married/ partner with children	Single parents	Single households	Total
Sweden	Males	3.4	5.3	7.9	8.5	5.8
	Females	3.6	5.8	7.7	9.7	6.3
Bosnia	Males	4.1	6.0	9.1	10.6	6.1
	Females	4.6	7.3	7.8	10.0	6.4
Somalia	Males	9.3	8.9	10.2	15.5	12.4
	Females	9.0	5.0	9.0	14.3	10.1
Iraq	Males	5.9	5.1	7.9	9.4	6.9
	Females	6.8	5.8	7.8	11.1	7.5
Iran	Males	4.8	5.5	7.8	10.1	7.1
	Females	5.5	6.1	6.7	12.2	7.5
Chile	Males	3.9	5.2	5.6	7.4	5.8
	Females	3.3	4.5	3.7	8.2	4.9

Source: Geosweden database. Institute for Housing and Urban Research, Uppsala University.

Multivariate Analyses

With reference to earlier research and the descriptive overview, it is obviously necessary to take a set of demographic attributes into account when trying to understand group variation in migration rates. Descriptive data have documented the key importance of age, gender, family type, and for immigrants also time spent in Sweden. These attributes will be included as independent variables in a model estimating potential factors explaining ethnic variation in mobility across labour market regions and across neighbourhoods within urban areas. The model will also include a set of socio-economic variables, such as educational level (low=12 years or less of schooling, medium=13–14 years, or high=15 years or more[1]), employment status (employed or not), whether one has received cash social allowances (yes or no), disposable income (individualised, classified into quartiles), and settlement region (residing in metropolitan municipality or not and residing in rental-dominated and in the most immigrant dense neighbourhoods or not).

Finally, a dummy intending to take earlier migration history into account will be included. This dummy will separate those who have remained in the same municipality from 1991 to 2005 from those who have moved across a municipality border. About 25 percent of all 5.48 million individuals in the database have made such a move. Referring to the Fischer et al.'s (2000) notion of insider advantages, it is reasonable to expect earlier migrants to show higher rates of especially inter-labour market migration. It is more uncertain whether the migration history dummy will affect also intra-urban mobility. No interactions between ethnic categories and different determinants of migration are in the model. All independent variables refer to the situation in 2005 (pre-migration attributes) and the dependent variable is dichotomous: have moved or not moved 2005 to 2008.

Table 13.6 provides descriptive statistics per ethnic category for all variables used in the models. All in all, the database comprises 5.48 million people, out of which about 15 per cent are foreign-born.

---

1   The Educational register gives detailed and updated accounts of individuals' specific type of education, including the level of education. However, for immigrants, especially those more recently arrived, the register is definitely less reliable. Immigrants' education is validated and transferred to the Swedish classification system but this sometimes takes considerable time. The higher proportion of 'missing educational information' reported in Table 13.6 is illustrative of these problems.

Table 13.6 Descriptive statistics (percentage per ethnic group) for variables used in the logistic regressions

			Country of Birth			
Variable	Sweden	Bosnia	Somalia	Iraq	Iran	Chile
Dependent variables						
Inter-Labour Market region move 2005–2008	6	6	11	7	7	5
Intra-Labour Market region move 2005–2009	25	29	37	37	37	34
Independent variables						
Sex Males	51	50	51	55	54	51
Age Age 16 to 27 in 2005	21	26	39	30	23	22
Age 28 to 49	42	50	54	55	55	52
Age 50+	37	24	8	15	22	26
Year of Immigration Immigrated 0 to 2 years ago	0	3	13	8	5	3
Immigrated 3 to 9 years ago	0	18	36	56	16	11
Immigrated 10 or more years ago	0	79	52	37	80	86
Household Couple without a child	40	54	32	63	41	28
Couple with child(ren)	18	17	6	6	14	20
Single parent	10	8	25	9	14	19
Single	32	21	37	22	31	33
Changed family status 2005–2008	28	29	37	32	33	34

Table 13.6 *Concluded*

Employment	Employed in November 2005	69	57	26	31	51	61
Social allowance	Received social allowance 2005	3	21	56	53	21	15
Educational Level	Low educ level (less than 12 years)	49	43	52	46	32	52
	Medium educ level (12–14 years)	34	40	21	26	42	35
	High educ level (15+ years)	17	11	5	16	22	10
	Missing educational information	1	5	21	12	4	2
Disposable income	Disposable income quartile 1	22	41	69	65	41	33
	Disposable income quartile 2	25	28	20	22	29	30
	Disposable income quartile 3	26	21	8	9	19	25
	Disposable income quartile 4	27	9	2	3	11	12
Residency	Moved municipality 1991–2005	26	0	10	7	35	30
	Live in Metro area 2005	15	26	53	37	40	30
	Live in high immigrant dense neighbourhood (upper decile)	8	50	78	67	45	48
	Live in a rental dominated neighbourhood	8	37	65	48	32	31
	N (age 16 to 71 in 2005)	5,297,918	47,722	11,963	57,213	48,674	25,081

Note: The N might vary somewhat over the different variables. All data refer to December 31 except for income and social allowances (measured for the entire year) and employment, measured during the first week in November. Educ is educational.

Source: Geosweden database. Institute for Housing and Urban Research, Uppsala University.

Labour Market Migration

Table 13.7 shows the model run for labour market mobility 2005 to 2008. As expected, almost all variables are statistically significant, the only exceptions being Bosnian immigrants in Model 1 and 2, Iraqi immigrants in Model 2, and disposable income, 3rd quartile in Model 3. According to Model 1, which only includes information concerning country of birth, Chileans are less likely and immigrants from Iraq, Iran and especially Somalia are more likely than the Swedish-born to have moved across a labour market border 2005 to 2008. Models 2 to 4 are more informative and include also information on each individual's demographic characteristics (Model 2), socio-economic characteristics (Model 3) and information concerning migration history and type of 2005 residency (Model 4). It is of course not surprising that these models are more powerful (note that Log Likelihood decreases and R square increases for each model). Adding demographic information means a lot. The young have a much higher propensity to relocate, as have the recently immigrated and singles. Males are somewhat less likely to move than females.

These demographic features remain also having controlled for socio-economic attributes in Model 3 and migration history and residency in Model 4. It is interesting to note that employed individuals are much less likely to migrate from one labour market to another compared to those not employed in 2005, which is also consistent with the finding that people on social allowances are more prone to migrate. Disposable income contributes only marginally to explain variation in migration propensities. Some of the potential effect from income is probably picked up by the educational variables; those having a high level of education are more likely to have moved. People residing outside of metropolitan municipalities are twice as likely to have moved as those residing within such environments. This is in line with the general assumption that labour market migration primarily goes in the direction of the main urban centers. It is less likely that someone moves if the address in 2005 was in a high immigrant dense area but more likely if the address in 2005 was in a neighbourhood having more than 75 percent rental accommodation.

Finally, earlier migration history has a very big impact. While only one in four in the dataset have conducted a move across a municipality border 1991 to 2005, those who have done so are 2.5 times more likely to be a labour market migrant 2005 to 2008. The four independent variables introduced in the last stage (Model 4) radically increase the odds ratio for moving for all immigrant groups and especially so for the Somali and Bosnian immigrants. When all controls are in the model, Somalia-born immigrants are twice as likely compared to native Swedes to have changed labour market location. The odds ratio for the Bosnian immigrants is 1.6; for the Iraqi about 1.4 and the Iranians 1.25. The only exception is noted for the Chilean immigrants; they exhibit the same likelihood as a native Swede to relocate.

Table 13.7   Logistic regression of the likelihood of inter-labour market migration in Sweden 2005 to 2008, predicted by residents' characteristics

		Model 1	Model 2	Model 3	Model 4
		Odds ratio (exp(β))	Odds ratio (exp(β))	Odds ratio (exp(β))	Odds ratio (exp(β))
Constant		0.065***	0.039***	0.040***	0.034***
Country of Birth	Bosnia	1.036*	1.033	0.945**	1.634***
	Somalia	1.966***	1.286***	1.162***	2.053***
	Iraq	1.198***	1.030	0.860***	1.390***
	Iran	1.218***	1.161***	0.935***	1.260***
	Chile	0.872***	0.847***	0.800***	1.017
	Sweden (ref)				
Demographic characteristics	Females (ref)				
	Males		0.884***	0.944***	0.965***
	Age 16 to 27		4.277***	3.624***	4.004***
	Age 28 to 49 (ref)				
	Age 50 plus		0.521***	0.521***	0.589***
	In Sweden <3years (ref)		1.353***	1.304***	1.289***
	In Sweden 3–9 years				
	In Sweden 10+ years		0.935***	0.959*	0.788***
	Couple without child(ren)		0.902***	0.909***	0.866***
	Couple with child (ref)				
	Single parent		1.113***	1.118***	1.157***
	Single		1.792***	1.712***	1.355***
Socioeconomic characteristics	Employed			0.575***	0.566***
	Not employed (ref)				
	Low level of education (ref)				
	Medium level of educ			1.609***	1.418***
	High level of educ			2.148***	1.770***
	1st disposable income quartile (ref)				
	2nd disp. income quartile			1.122***	1.151***
	3rd disp. income quartile			1.001	1.073***
	4th disp. income quartile			0.980***	1.073***
	Received Social allowance			1.215***	1.177***
	Not Soc allowance (ref)				

Table 13.7 *Concluded*

		Model 1	Model 2	Model 3	Model 4
		Odds ratio (exp(β))	Odds ratio (exp(β))	Odds ratio (exp(β))	Odds ratio (exp(β))
	Did not migrate 91-05 (ref)				
characteristics	In Metro region				0.492***
	Not in Metro region (ref)				
	In immigrant-dense neighbourhood				0.801***
	Not in immigr.-dense neighbourhood (ref)				
	Live in rental dominated neighbourhood				1.286***
	Not in rental dominated neighbourhood (ref)				
Log Likelihood		2519930	2249592	2216026	2151414
R2 (Nagelkerke)		0.000	0.131	0.147	0.177

Note: *=p<0.05; **=p<0.01; ***=p<0.001. All residents aged 16 to 71 in 2005, having a known address in Sweden 2005 and 2008. N= 5,471,446, thereof 335,031 having moved (6.1 per cent).

Source: Geosweden database. Institute for Housing and Urban Research, Uppsala University.

Intra-urban Neighbourhood Migration

Tables 13.8 displays results from the neighbourhood migration model run. The models are exactly the same as applied for the labour migration mobility analysis. Naturally, a much higher proportion has moved (21.7 per cent compared to 6.1 per cent for labour market migration) and all but one independent variable estimate are statistically significant (the sole exception is Somali immigrants in Model 2). Results are partly similar to the ones reported above (a much higher probability of moving if a person is young, single or single parent, and recently immigrated), but there are also some striking differences: employed people are now more likely to have moved and being a metro resident substantially lowers the probability of moving across labour market regions but clearly increases the likelihood of moving house from one neighbourhood to another within these regions. One reasonable interpretation is that many labour market moves originate outside of the metro regions and instead have such regions as destinations. The higher neighbourhood mobility in metro areas could be explained by these regions' higher shares of rental housing, which allows for more mobility due to reduced transaction costs, but could also relate to the fact that it is a tough challenge to find appropriate housing in regions having housing shortage. Households therefore have to move more often in search of the right housing costs, tenure, and location vis-à-vis their work place. The model does incorporate the residents' tenure context and tenure difference between metropolitan municipalities and the rest of Sweden should at least partly be picked up by the dummy variable 'live in rental dominated neighbourhood'.[2] That variable and also the dummy 'live in neighbourhood having high immigrant density' (upper decile in terms of concentration, i.e. more than 30 percent foreign-born) show that people residing in such areas are more likely to have moved neighbourhood than those living elsewhere in 2005. The higher probability of moving if one resides in a high immigrant concentration area is also in contrast to the labour migration result. It does however fit earlier findings that the selective migration going on is driven by a continuous out-migration of people from the areas and a simultaneous influx of more recently arrived immigrants (Andersson and Bråmå 2004). The model requires that all movers in the database are in Sweden in 2005 so the latter category is not well represented in this analysis.

It is worth mentioning that people on social allowance, when everything else is controlled for, exhibit higher probabilities of moving compared to those not having received such support. One might expect poor people to be more stuck in a particular neighbourhood – not least because landlords typically apply rules concerning a certain amount of work income for accepting a new tenant – but results here indicate something else. It is of course often so that people on social allowances also have weak positions in the housing market. Less secure tenure might imply that they have to move more often. Others might have to move in order to reduce housing costs. If such housing related aspects are important for explaining the outcome, the effects should be most pronounced in the intra-urban residential mobility model. This is indeed also the case.

Like for labour market migration, but with the noteworthy exception of the Somalis, immigrant categories show higher probabilities compared to native Swedes of having moved neighbourhood. Values are especially interesting for the Somalis who were found to be twice as likely as native-born Swedes to move between labour market regions but have an odds ratio of 0.70 in the final neighbourhood model. This is not so surprising given the fact that Somali immigrants exhibit a very high level of segregation and are clustered into a few neighbourhoods in the cities where their presence is noticeable. Contrary to many other immigrant groups they so far show no tendency of leaving these concentrations. To generalise: if they move they move within neighbourhoods or to another city.

---

2 Unfortunately, the database does not contain information about housing tenure for individuals in the year 2005. In order to partly control for tenure, information concerning the tenure compositions of neighbourhoods in 2006 have been transferred onto the 2005 SAMS files. It is very unlikely that the tenure mix of a particular SAMS would have changed dramatically during one year but it could of course occur in a few cases.

Table 13.8   Logistic regression of the likelihood of intra-labour market migration in Sweden 2005 to 2008, predicted by residents' characteristics

		Model 1	Model 2	Model 3	Model 4
		Odds ratio (exp(β))	Odds ratio (exp(β))	Odds ratio (exp(β))	Odds ratio (exp(β))
Constant		0.256***	0.219***	0.161***	0.140***
Country of Birth	Bosnia	1.265***	1.345***	1.251***	1.224***
	Somalia	1.577***	0.959*	0.882***	0.700***
	Iraq	1.842***	1.669***	1.464***	1.295***
	Iran	1.876***	1.877***	1.672***	1.469***
	Chile	1.661***	1.634***	1.565***	1.407***
	Sweden (ref)				
Demographic characteristics	Females (ref)				
	Males		0.916***	0.927***	0.939***
	Age 16 to 27		2.868***	3.018***	3.115***
	Age 28 to 49 (ref)				
	Age 50 plus		0.377***	0.399***	0.435***
	In Sweden less than 3 years		1.546***	1.601***	1.644***
	In Sweden 3–9 years (ref)				
	In Sweden 10+ years		0.888***	0.931***	0.876***
	Couple without child(ren)		0.826***	0.837***	0.825***
	Couple with child(ren) (ref)				
	Single parent		1.528***	1.574***	1.522***
	Single		1.946***	1.847***	1.604***
Socioeconomic characteristics	Employed			1.038***	1.051***
	Not employed (ref)				
	Low level of education (ref)				
	Medium level of education			1.314***	1.241***
	High level of education			1.520***	1.335***
	1st disposable income quartile (ref)				
	2nd disp. income quartile			1.070***	1.075***
	3rd disp. income quartile			1.096***	1.112***
	4th disp. income quartile			1.136***	1.143***

Table 13.8   *Concluded*

	Recieved Social allowance			1.503***	1.399***
	Not Soc allowance (ref)				
Geographic location and housing characteristics	Moved between municip. 91 to 05				1.496***
	Did not change municip. (ref)				
	In Metro region				1.201***
	Not in Metro region (ref)				
	In immigrant-dense neighbourhood				1.103***
	Not in immigrant-dense neighbourhood (ref)				
	Live in rental dominated neighbourhood				1.479***
	Not in rental dom. Neighbourhood (ref)				
Log Likelihood		5227431	4639686	4613459	4571066
R2 (Nagelkerke)		0.003	0.172	0.179	0.190

Note: *=p<0.05; **=p<0.01; ***=p<0.001. All labour market stayers aged 16 to 71 in 2005, having a known address in Sweden 2005 and 2008. N=5,144,016, thereof 1,062,485 having moved neighbourhood (21.7 per cent).

Source: Geosweden database. Institute for Housing and Urban Research, Uppsala University.

It is clear that it is not the concentration of immigrants into rental housing and immigrant-dense areas that explain some categories overall higher inter-neighbourhood mobility compared to native Swedes. The likelihood of moving is reduced for all immigrant categories in Model 4 compared to Model 3, i.e. after controlling for neighbourhood characteristics and location.

One might ask whether the similarities and differences found when running the same model for labour market migration and intra-urban migration persist also when analysing intra-neighbourhood mobility? The answer is yes, at least for the ethnic dimension. In short, the results are as follows: about ten percent of all individuals in the database carry out a move within a neighbourhood 2005 to 2008. When all control variables are in the model, all five minority categories exhibit a significantly higher risk of moving; odds ratios are, for Bosnia: 1.5; Somalia: 1.7; Iraq: 1.5; Iran: 1.2; Chile: 1.4. Results differ from the inter-neighbourhood model output in the sense that Somalia-born people show much higher propensity to move (in fact similar to the results from the labour market migration analysis), that both the youngest and the oldest age group are less prone to move compared to individuals aged 28 to 49, and that people with higher income and more education are less likely to move. Living in a rental dominated neighbourhood, in an immigrant-dense neighbourhood, and having social allowances, all contribute to increase the propensity of having moved within a neighbourhood.

Discussion and Conclusions

It is now possible to state that it is indeed the case that when controlling for a range of mobility-related individual attributes, immigrants in Sweden are more prone to migrate. Whether we focus on inter- or intra labour market migration makes some difference but most of the more numerous categories of immigrants tend to move more often than native-born Swedes. This holds true also for intra-neighbourhood moves. The demographic composition of especially newly immigrated people will result in high crude migration rates both due to the fact that they are new in the country and because they are young and often single. These facts alone do not however explain ethnic variations in geographical mobility. This result is in contrast to some of the UK evidence (see Finney and Simpson 2008), where difference in crude migration rates tend to disappear when compositional differences between groups are controlled for.

It is difficult to know whether this discrepancy between the UK and Sweden is due to real differences or differences in data, definitions of ethnic groups and/ or in the construction of the respective logistic regression models applied for analysing data. One obvious and probably the single most important difference is the inclusion of second generation immigrants in the UK case while the Swedish population categories are based on country of birth information (i.e. only includes first generation immigrants). As we have seen, differences in migration rates for natives and immigrants drop when time in the new country increases and there is no reason to believe that differences should persist in the second generation – at least not after having controlled for demographic and socio-economic attributes. Furthermore, the Swedish logistic regression models are applied for larger groups of refugee immigrants (and their families), and it cannot be ruled out that a labour migrant, who is more often recruited into particular job in a specific area, exhibits different migratory behavior. The current Swedish immigration regime allows for labour immigration also from outside the EES treaty area but numbers are still small and labour migrants constitute less than ten percent of residence permissions given to non-Nordic residents during a particular year (Andersson, Magnusson Turner and Holmqvist 2010). The selection of immigrant categories made for the Swedish study is therefore highly relevant for Sweden but not necessarily appropriate for similar analyses in other countries.

The control variables used by Finney and Simpson (2008) are partly similar but certainly not the same and the migration period differs both in length and in real time (2001 in the UK case, 2005 to 2008 in this paper). This study covers migration over a three year period, which of course tends to reduce the (average annual) volume of migrants (less return migrants would be included). This Swedish study also controls for earlier migration, and this could have significant importance. It is clearly the fact that controlling for this means less for intra-urban mobility than it does for inter-urban moves. Immigrants' higher propensity to move across labour market borders is dramatically affected by the inclusion of the Model 4 control variables. If these are taken out of the equation (see Model 3 in Table 13.7), Bosnian, Iraqi, Iranian and Chilean immigrants have lower migration probabilities than have the native Swedes.

Returning to the results of the most comprehensive model outcomes: What is the most plausible explanation for the findings reported? It would of course be easier if demographic and socio-economic differences between categories of natives and immigrants were the only reason for why migration frequencies differ. This is however not the case. When controlling for a range of such migration determinants it turns out they all play a significant role, but they are not enough for explaining outcomes. Could there be a 'cultural' explanation? Yes, it could but before engaging further with this question, other possible explanations should be tested.

Migration researchers are well aware of the fact that migration correlates strongly with demographic events like getting married, having children, divorcing, etc (see for instance Rossi 1955; Clark, Deurloo, and Dieleman et al. 1994, 2003; Clark and Dieleman 1996; Fischer et al. 2000; Feijten and Mulder 2002; Abramsson, Borgegård, and Fransson 2002; Li and Li 2006; and Mulder 2006). Such information is available in the dataset but has not been included as controls in the regression models reported in Tables 13.7 and 13.8. The primary reason is that causality would get blurred as we do not know the correct sequence of events; migration is often triggered by demographic events but could also lead to such events.[3] However, if demographic events are indeed different across ethnic categories, they could potentially explain why migration propensities differ so much. The models have therefore experimentally been run also with demographic events included (a dummy indicating all types of family type changes). About 25 per cent of the population under study experienced change in family type 2005 to 2008. It turns out that odds ratios for the ethnic categories are not much affected (the distribution of demographic events are relatively evenly distributed across ethnic categories) but that such events indeed overall radically increases the risk of migrating across labour market regions (odds ratio 2.5) and from one neighbourhood to another within labour market regions (odds ratio 3.3). Surprisingly, demographic events turn out to be statistically non-significant for explaining migration within neighbourhoods.

---

3 The Swedish registers can be used for studying the sequence of demographic events. However, the comprehensive database Geosweden does not contain specific dates for such events (only yearly information) and the sub-dataset set up for this analysis has data only for 2005 and 2008.

Future research needs to explore two further avenues towards a deeper understanding of ethnic differences in geographical mobility. One is to address the challenge of finding appropriate control groups. It is important to yet again stress the fact that natives always comprise a relatively big proportion of less mobile people, be it because of occupation, insider advantages such as place attachment and denser local social networks, ownership of fixed resources, or regional sentiments and identities. This population segment is by and large left behind when people move from one country to another. In this sense, immigrants are a selected group of (more mobile) people. Using the vocabulary suggested by Fischer et al. (2000), there are reasons to believe that the accumulation of insider-advantages is something that differentiates natives from most categories of immigrants. It might be the case that controlling for migration history in the way done here is not enough in order to fully grasp the importance of differences in local attachment. The accumulation of insider-advantages can be blocked or reduced by the fact that immigrants on average hold less secure positions in the housing market – be it because of discrimination, lack of networks or financial resources – which of course can result in higher levels of migration. Anyhow, developing ways of operationalising the concept of insider-advantages, or location-specific capital, may prove a constructive way of getting a deeper understanding of the findings reported here.

Partly related to this is the need to try to better link and bridge existing divides between research focusing on international and internal migration respectively. As discussed by King, Skeldon and Vullnetari (2008), there are many ways by which these often very different bodies of literature could be developed and linked closer together and there are also many possible trajectories linking individuals' international and internal migration to each other. Although the Swedish registers provide very good opportunities for longitudinal research, they do not contain much useful information on immigrants' trajectories *before* entering the country. Building theory on a more complete life course perspective would necessitate the inclusion of such information, typically by the complementary use of survey methods.

The second aspect concerns differences in labour market attachment, which is not satisfactorily accounted for in this study. The models control for certain aspects, like employment, and whether individuals receive social assistance. There are, however, also other potentially important factors, such as job security and occupation, which might affect geographical mobility. Many immigrants who do work and earn reasonable sums of money are certainly over represented in less secure jobs and therefore need to be geographically flexible to find a job and to stay employed. The welfare state might also be a factor. Although it is the municipalities that more or less run the welfare state, which theoretically might produce local variation in support and service levels, differences in living standard for the unemployed and under-employed are small across the country. Lacking a job does not only increase incentives for moving in search of one, there is no economic penalty of doing so. A substantial part (almost 75 percent) of all

Somali adults lack employment and at the same time they are twice as likely as native Swedes to move from one labour market to another. This needs to be further explored.

If there is some sort of cultural factor involved, differences in family systems, marriages and divorces are candidates for where to look. Despite the fact that longitudinal individual register data provide a good basis for revealing differences across categories of people, such quantitative studies need to be complemented with in-depth qualitative studies of a more ethnographic character.

Chapter 14

Internal Migration and Residential Patterns Across Spain After Unprecedented International Migration

Albert Sabater, Jordi Bayona and Andreu Domingo

Introduction

The strong demand for labour-intensive and low-skilled jobs in low-paid occupational sectors made Spain one of the top destinations for international migrants after the turn of the twenty first century (Baldwing-Edwards and Arango 1999). As a result, during the 2000s Spain became the European country with the largest net absolute migration in the EU, lagging only behind the USA worldwide (OECD 2007). Within this context the number of new migrants rose from around 0.7 million in the year 1999 to 5.7 million at the beginning of 2010, thus changing the proportion of non-nationals from 1.6 to 12.2 per cent of the population. This international migration turnaround in Southern Europe is well documented since the mid-1990s (Aja and Arango 2006, King, Fielding and Black 1997). Although the composition of the foreign population traditionally reflected tourist activities and residential migration from other EU countries, the unprecedented economic boom also changed the makeup of Spain's immigrants. In addition to established inflows from Northern Africa (mostly from Morocco) and Sub-Saharan Africa, recent international migration to Spain has vastly diversified but is still dominated by two geographical origins: Latin America and Eastern Europe. In the context of a preference system that distinctly favours the inflows from Latin America for historical reasons (Izquierdo, López de Lera and Martínez Buján 2003) and Eastern Europe for contextual reasons (Viruela 2008), various bilateral labour contracts have been issued and represent the government's strategy to assist the diversification of inflows away from persons of African origin only, and especially Moroccans, following the riots in the town of El Ejido (Domingo 2006). The evolution of immigration policy in Spain in recent years reveals this trend not only for migrants whose arrival is under the auspices of the Ministry of Labour and Immigration in Spain, but also for undocumented migrants whose entry is illegal or who overstay a visa (Sabater and Domingo 2012). In the meantime, heated policy debates about the social integration of immigrant groups have been raised notably with regard to the clustering of immigrant groups as a new feature of the Spanish geography. The immigrant composition of neighbourhoods and, in particular, the potential impact of segregation on integration and its causes and consequences for social policy have become an issue high on national, regional and local agendas in Spain (Cachón 2003, Aja and Arango 2006). The provision of generous funds to benefit autonomous communities and local authorities to finance integration programmes under the Strategic Plan for Citizenship and Integration (Plan Estratégico de Ciudadanía e Integración 2007–2010) reveals the importance of these debates. Additionally, a major and current policy dilemma is how to deal with the unfavourable economic outlook, possible resentment against immigrant populations, and the danger that existing patterns of inequality and discrimination may amplify over time.

Meanwhile, the existing body of literature on sociospatial integration in Spain in recent years has mostly followed the classic spatial assimilation model, thus focusing on patterns of immigrant concentration and segregation in the main urban settlement areas as a way to reflect discrimination, disadvantage and isolation from opportunities and resources. Hence, various studies have been undertaken in relation to immigrant residential patterns mostly for Madrid (Lora-Tamayo 2001, Martínez del Olmo and Leal Maldonado 2008) and Barcelona (Martori and Hoberg 2004, Bayona 2007, Musterd and Fullaondo 2008). Other studies have pointed out the reasons for residential segregation, particularly the importance of poor and scarce housing in the main Southern European cities (Martínez Veiga 1999, Arbaci 2008) and the urban location of employment opportunities from the perspective of the enclave-economy hypothesis (e.g. for Andalucia, see Arjona 2007; for Madrid, see Riesco 2008; for specific nationalities, see Beltrán, Oso and Ribas 2006). In general, the emergence of new patterns of segregation as a result of international migration has been perceived as an expression of social exclusion, with implications for the development of social policy (Malherios 2002). Whilst it is widely acknowledged that the basic hypothesis of the assimilation model remains valid (Massey and Denton 1988, Simpson 2004), the specialised literature has also highlighted the importance of taking into consideration segmented and diverse possible outcomes (Portes and Zhou 1993, Alba and Nee 1997), economic integration but social encapsulation (Peach 1997, 2005) and even dispersal immediately after arrival (Zelinsky and Lee 1998). The latter idea represents a third model, labelled as 'heterolocalism', which has been developed to supplement and partially replace the older two (assimilationist and pluralist models). According to Zelinsky and Lee (1998: 285): 'Heterolocally inclined individuals and families currently enjoy a much greater range of location options in terms of residence and also economic and social activity than anything known in the past. They become heterolocal by virtue of choosing spatial dispersion, or at most a modest degree of clustering, immediately or shortly after arrival instead of huddling together in spatial enclaves. It would, of course, be naïve to assert that their locational decision-making is completely unencumbered. The tightness of the local housing market, the availability of appropriate economic niches, and the diversity of the local ethnic context all impose some degree of constraint.'

In this current work, we argue that whilst a replication of clustering processes by immigrant groups in Spain is likely to take place, heterolocal tendencies are also plausible for some groups, especially those whose experience of discrimination, disadvantage and self-segregation is lowest. Therefore, a co-existence of clustering resonant of past times with immigrant groups who might choose to settle heterolocally is expected to be found. Hence, the goal of this chapter is to analyse new forms of sociospatial behaviour in Spain beyond the simple examination of clustering arrangements. For this purpose two specific and modest analyses are undertaken: a study of residential segregation and various analyses of internal migration of the largest immigrant groups in the provinces of Madrid and Barcelona. In doing so, this study provides an explanatory account of spatial clustering and dispersal, addressing two simple questions which are not to date fully investigated by the literature: (1) Has residential segregation increased for the ten largest immigrant groups in the provinces of Madrid and Barcelona between 2005 and 2010? (2) Is internal migration re-inforcing concentrations and segregation of the ten largest immigrant groups or dispersing them towards de-segregation in these provinces?

Following this introduction, the chapter outlines the data and methods used. It then provides a snapshot of the significance of recent international migration. It goes on to analyse residential segregation separately for the largest 10 immigrant groups in the provinces of Madrid and Barcelona. The examination of patterns of residence is subsequently examined with analysis of internal migration within the provinces of Madrid and Barcelona. Finally, a section with conclusions and discussion is provided.

Data and Methods

Population data and migration statistics released on a yearly basis by the National Statistics Institute (aka INE) were used for analyses of immigrant settlement and internal migration. Population counts and residential variation statistics are derived from the *Padrón Municipal de Habitantes* or Municipal Registers, which constitute the administrative registers where municipality neighbours and in- and out-migrations are processed. Racial or ethnic categories are not used in administrative or population datasets in Spain and, therefore, the analyses of residential segregation and internal migration were focused on the ten largest immigrant groups, on the basis of nationality, resident in the province of Madrid (Romania, Ecuador, Morocco, Colombia, Peru, Dominican Republic, Bolivia, Argentina, China and Bulgaria) and Barcelona (Morocco, Ecuador, Argentina, Bolivia, Peru, Colombia, China, Romania, Pakistan and Dominican Republic) for the years 2005 and 2010 (see Table 14.1). It is important to note that the great majority of irregular migrants are expected to be included in the analyses as the registration of irregular migrants is seen as a *condition sine qua non* for obtaining legal status in the country (Sabater and Domingo 2012).

Table 14.1   Selected immigrant groups. Spain and provinces of Madrid and Barcelona, 2010

Madrid			Barcelona			Spain		
Top 10	Total	%	Top 10	Total	%		Total	%
Romania	210,822	3.26	Morocco	139,736	2.54	Romania	831,235	1.77
Ecuador	125,469	1.94	Ecuador	69,687	1.26	Morocco	754,080	1.60
Morocco	86,386	1.34	Argentina	22,940	0.42	Ecuador	399,586	0.85
Colombia	69,155	1.07	Bolivia	47,259	0.86	Colombia	292,641	0.62
Peru	62,120	0.96	Peru	31,409	0.57	Bolivia	213,169	0.45
Dom. R.	35,154	0.54	Colombia	33,474	0.61	Bulgaria	169,552	0.36
Bolivia	50,644	0.78	China	37,561	0.68	China	158,244	0.34
Argentina	17,909	0.28	Romania	34,916	0.63	Peru	140,182	0.30
China	42,894	0.66	Pakistan	31,905	0.58	Argentina	132,249	0.28
Bulgaria	32,685	0.51	Dom. R.	17,447	0.32	Dom. R.	91,212	0.19
						Pakistan	56,877	0.12
Subtotal	733,238	11.35	Subtotal	466,334	8.46	Subtotal	3,239,027	6.89
% Non-Spanish	67,90		% Non-Spanish	57,89		% Non-Spanish	56,35	
Non-Spanish	1,079,944		Non-Spanish	805,487		Non-Spanish	5,747,734	
% Total	16.72		% Total	14.62		% Total	12.22	
Spanish	5,378,740		Spanish	4,705,660		Spanish	41,273,297	
Total	6,458,684		TOTAL	5,511,147		Total	47,021,031	

Source: Own elaboration with data from the Population Municipal Register (INE).

The smallest geography at which population data are published is the *Secciones Censales* or Census Output Areas (COAs), with an average population of 1,500 people per unit, whereas migration statistics are only released at municipal level. Whilst the municipal geography has remained unchanged between 2005 and 2010 in the provinces of Madrid and Barcelona with 179 and 311 municipalities respectively, the smallest geographies (COAs) have been affected by boundary changes between 2005 and 2010, so harmonisation of these geographical areas was implemented by updating the 'old' spatial sets (2005 and 2008) to the contemporary zones (2010) using appropriate *ad hoc* Geographical Conversion Tables (GCTs). The GCTs allowed the conversion from the 2005 COAs (3,989 units) and 2008 COAs (4,165 units) to 2010 COAs (4,224 units) in the province of Madrid and the 2005 COAs (3,838) and 2008 COAs (3,971 units) to 2010 COAs (3,563 units). As a result we avoid the complexity of dealing with different units and minimise the possible artefact produced by boundary changes over time. Due to the availability of geographic detail in the data used, the smallest geographies were only used to compute residential segregation across COAs in the provinces of Madrid and Barcelona, and the analysis of internal migration only concentrated on the flows between municipalities in each province. Municipality characteristics as of 2010 produced by INE were added to the migration files where indicated to allow specific analyses of intraprovincial migration from each capital province (Madrid and Barcelona), large municipalities (>100,000 residents) and small-medium sized municipalities (10,000 to 100,000 residents). Since immigrant groups are relatively very small in rural municipalities (<10,000) and this results in less stability, analyses of internal migration including these areas are not shown.

For the calculation of residential segregation, two common measures were used, the Dissimilarity Index (D) and the Isolation Index (xPx^*). D was used to determine the level of evenness with which the Spanish group and each immigrant group separately are distributed across COAs in the provinces of Madrid and Barcelona using the traditional index of dissimilarity (Duncan and Duncan 1955). P^* was used to measure the extent to which immigrant groups are exposed only to one another (xPx^*), thus providing a measure of the average local concentration of each immigrant group across COAs (Lieberson 1963). Since P^* is highly sensitive to group size the outcome was scaled to take into account the group's relative size with the total population in the province. The interpretation of both indices is straightforward as a percentage. They indicate total evenness (D) or the lowest degree of potential contact between members of the same group (P^*) when index values are close to zero and *vice versa* (see Massey and Denton (1988) for a full description of the indices). For the analyses of internal migration two specific measures, migration efficiency ratios (MERs) and net in-migration rates (NIRs), were used in addition to the basic computation of absolute net migration from in- and out-migration. The MER is defined as the net migration of an area (in-migration minus out-migration) divided by the total number of moves within that area (Galle and Williams 1972). The MER is normally expressed as a percentage and assumes values between -100 and +100, with low values indicating little redistribution and *vice versa*. Net migration rates are usually computed by dividing the difference between in-migration and out-migration in an area during a period relative to the total midyear population in that area. Positive net migration indicates net in-migration (NI), while negative net migration indicates net out-migration (NO). The result is then multiplied by 100 to arrive at the crude measure of net in-migration rates (NIRs) or net out-migration rates (NORs). The combination of the index of isolation and the NIR was used to investigate whether groups are moving towards their own concentrations, thus replicating similar studies in Europe (e.g. in Northern Ireland; see Catney 2008).

International Migration

As noted in the introduction, international migration in Spain deservedly grabbed many headlines over the past years. The significance of international inflows was clearly in line with the labour demands of an expanding economy that attracted significant amounts of foreign investment and was able to create more than half of all the new jobs in the European Union (EU) over the five years ending 2005 (Tremlett 2006). However, this process has been clearly reversed as a consequence of the economic recession, with labour migration of low-skilled being the most affected (Beets and Willekens 2009).

Figure 14.1 clearly shows the above mentioned process for the provinces of Madrid and Barcelona, which can be used as a thermometer for Spain as a whole. A peak of international migration was reached in 2007, when the net absolute migration in the province of Madrid and Barcelona went beyond 200,000 (124,600 in Madrid and 81,600 in Barcelona). Since then, new arrivals have decreased dramatically, resulting in a total net gain of 38,800 (32,700 in Madrid and 6,100 in Barcelona) in 2009. In addition, many immigrant groups who came to Spain as foreign workers and with temporary contracts may opt to return, especially those without social capital and social security. Nonetheless, empirical evidence from past recessions in Europe (e.g. the 'oil crisis' in 1973–74) tells us that only a minority, between 10 and 15 per cent, embarked on that journey in the long-term (Dobson, Latham and Salt 2009). So far in Spain, a similar picture is found at the onset of the current recession, as demonstrated by the low response of the Voluntary Return Programme for non-EU migrant unemployed workers (Rogers 2009). According to SEME (2010) the total number of resolved applications has gone from 1,800 in 2008 to 4,000 in 2009, an indisputable increase between the two years, although well below expectations (ILO 2009) considering the extraordinary levels of unemployment, which reached 29.7 per cent among all non-nationals during the fourth quarter of 2009 (INE 2010).

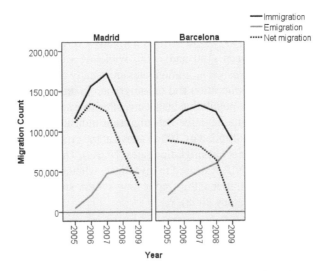

Figure 14.1   International migration of non-nationals. Provinces of Madrid and Barcelona, 2005–2009

Source: Own elaboration with data from the Residence Variation Statistics (INE).

It is clear that after an episode of unprecedented international migration the assessment of settlement patterns and population movement of recently arrived immigrant groups is pivotal for social policy to assist and monitor changes over time and space. The recentness of international migration in Spain allows us to assess whether clustering arrangements follow an initial expansion phase and/or depict heterolocal tendencies in the provinces of Madrid and Barcelona.

Residential Segregation

The purpose of this section is not so much to establish an association between the outcomes of residential segregation and the social integration of different immigrant groups, but simply to identify levels and change of residential segregation over time. For this purpose, Table 14.2 provides the relative information for the provinces of Madrid and Barcelona. In the province of Madrid, the first dimension (evenness) shown by the Dissimilarity Index (*D*) already indicates how the geographical spread varies between immigrant groups. Whilst some groups such as the Chinese and, to a lesser extent, the Bulgarians display moderate-to-high *D* values in 2005 (between 55.6 and 60.6) the majority of other groups show moderate-to-low *D* values in 2005, with the lowest values among the Colombians (34.6) and the Argentineans (38.8). Most importantly, the analysis of *D* values between 2005 and 2010 make visible a general reduction of clustering between the two years. In some cases (Ecuador, Argentina and Colombia) a negative population growth is likely to explain de-segregation, partly due to nationalisations, although the substantial addition of members of other immigrant groups (for example, Romania, Bolivia and China) follows the same decrease of geographical spread versus the Spanish group. The second dimension (isolation) shown by the Isolation Index (*P**) illustrates the paradox of segregation. Whilst *D* values tend to decrease, *P** values show increases in immigrant group isolation, with some exceptions too. This effect becomes clear with immigrant groups whose population growth has been positive between 2005 and 2010 (e.g. Romania and China) and *vice versa* (e.g. Ecuador and Colombia). The latter is likely to be caused by out-migration to other Spanish provinces and/or return migration. Generally, the *P** values highlight how immigrant groups tend to live in areas in which they form a larger share compared to that of the province; for example, residents from Romania constitute 3.3 per cent of the total population in the province but they live in areas in which they average 9.5 per cent in 2010. Although these results highlight an expected tendency of clustering, the values of *P** are clearly not high in absolute terms as immigrant groups with most exposure live in areas where more than 90 per cent of the population are from other groups.

Table 14.2 Residential segregation by selected immigrant groups across Census Output Areas. Provinces of Madrid and Barcelona, 2005–2010

Madrid

Top 10	Year 2005	Year 2010	Change
Dissimilarity			
Romania	49.1	44.6	-4.6
Ecuador	45.5	47.7	2.2
Morocco	51.9	51.7	-0.3
Colombia	34.6	33.7	-0.9
Peru	40.7	38.8	-1.8
Dom. R.(*)	52.3	51.1	-1.2
Bolivia	51.8	48.7	-3.1
Argentina	38.8	36.5	-2.3
China	60.6	53.1	-7.5
Bulgaria	55.6	51.5	-4.0
Isolation			
Romania	6.25	9.53	3.3
Ecuador	6.70	4.62	-2.1
Morocco	3.85	4.50	0.7
Colombia	2.20	1.82	-0.4

Barcelona

Top 10	Year 2005	Year 2010	Change
Dissimilarity			
Morocco	51.8	50.1	-1.7
Ecuador	46.2	46.4	0.2
Argentina	40.3	35.5	-4.8
Bolivia	58.5	48.4	-10.2
Peru	53.0	45.5	-7.4
Colombia	39.8	34.8	-5.0
China	62.9	52.0	-10.9
Romania	55.3	44.3	-11.0
Pakistan(*)	79.0	75.4	-3.6
Dom. R.(*)	57.1	53.6	-3.4
Isolation			
Morocco	7.42	8.66	1.24
Ecuador	4.03	3.50	-0.52
Argentina	1.17	0.76	-0.40
Bolivia	3.18	2.81	-0.37

Table 14.2 *Concluded*

Peru	1.37	1.78	0.4	Peru	1.58	1.44	-0.14
Dom. R.(*)	1.68	1.72	0.0	Colombia	1.18	1.03	-0.15
Bolivia	1.72	2.64	0.9	China	2.97	3.31	0.34
Argentina	0.76	0.52	-0.2	Romania	1.27	1.76	0.50
China	2.16	3.36	1.2	Pakistan(*)	7.69	8.00	0.32
Bulgaria	1.59	1.87	0.3	Dom. R.(*)	1.28	1.28	-0.01
Population				*Population*			
Romania	96,437	210,822	118.6%	Morocco	109,304	139,736	27.8%
Ecuador	173,593	125,469	-27.7%	Ecuador	79,509	69,687	-12.4%
Morocco	69,532	86,386	24.2%	Argentina	26,835	22,940	-14.5%
Colombia	72,636	69,155	-4.8%	Bolivia	20,433	47,259	131.3%
Peru	39,274	62,120	58.2%	Peru	25,360	31,409	23.9%
Dom. R.(*)	30,931	35,154	13.7%	Colombia	30,168	33,474	11.0%
Bolivia	26,589	50,644	90.5%	China	24,084	37,561	56.0%
Argentina	21,367	17,909	-16.2%	Romania	15,904	34,916	119.5%
China	23,924	42,894	79.3%	Pakistan(*)	25,488	31,905	25.2%
Bulgaria	21,843	32,685	49.6%	Dom. R.(*)	15,522	17,447	12.4%
TOTAL	5,964,143	6,458,684	6,271.638	TOTAL	5,226,354	5,511,147	5,416,447

Note: (*) Since 2005 data were not available we have used 2008 data for the calculations.

Source: Own elaboration with data from the Population Municipal Register (INE).

In the province of Barcelona index values of residential segregation of both evenness and isolation are higher than in the province of Madrid. These differences are likely to reflect the impact of where the provincial boundary is drawn. Whilst Barcelona's province includes a significant number of rural areas which are predominantly Spanish (only 78 out of the 311 municipalities in the province have more than 10 per cent of immigrant population), Madrid's province is tightly bound around a dense urban area (136 out of 179 municipalities have more than 10 per cent of immigrant population). As a result one might expect that the inclusion of rural areas is likely to increase the indices of segregation, thus exemplifying that using two different ways of delineating an urban region can also have an impact on the segregation indices. The results of the first dimension (evenness) in the province of Barcelona show, on the one hand, a high level of spatial encapsulation for two recently arrived immigrant groups whose growth has been very significant: the Chinese and the Pakistani (with D values between 62.9 and 79.0). On the other hand, other immigrant groups such as the Moroccans and the Latin American range within the moderate band of segregation (between 39.8 for the Colombians to 58.5 for the Bolivians). The analysis of D values over time highlights once again a general decrease of the level of clustering for every group (excluding the Ecuadorians) for the 2005(08)-2010 period. As in the province of Madrid, the P^* values tend to increase over time for those groups whose population has experienced population growth. However, this is not always the case, for example, the Bolivian, the Peruvian and the Colombian have seen a reduction of their isolation despite their population growth since 2005.

The results shed some light on two important factors underlying the residential accommodation of immigrant groups. First, the analysis of patterns of settlement reveals that the majority of immigrant groups would range within a moderate level of segregation, although there are some exceptions such as the Pakistani group in the province of Barcelona. The latter may constitute the perfect example of what the enclave economy theory suggests: the spatial concentration of the immigrant groups is the result of newcomers that work with co-ethnics, which allows the formation of prosperous business ventures and facilitates the economic progress of the group (Portes and Manning 1986). This is also likely to apply to other immigrant groups such as the Chinese whose main occupations in the service sector (small restaurants, bars and bazaars) initially led to high levels of segregation (Martori, Hoberg and Suriñach, 2006). Second, the analysis exposes how dispersal is not always a function of longer history in Spain, and this is clearly exemplified with the Moroccans, whose large arrivals took place during the 1990s but the level of de-segregation is slower compared to the other groups whose arrival has been more recent (predominantly from Latin America and Eastern Europe). This brings to view a co-existence of different sociospatial behaviours, with Latin American groups being closest perhaps to heterolocalism and Asian groups displaying more economic integration but spatial encapsulation. In general, the combination of D and P^* values suggest the existence of population movement from the existing clusters as each group gains uniformity even with an increase of population size.

Nonetheless, in order to examine such movement properly the following section provides various analyses of internal migration.

Internal Migration

Internal migration has traditionally been an important component of population change in Spain, particularly in areas such as Madrid and Barcelona, where rural-urban migration in the mid-1970s (Santillana 1981, Cabré, Moreno and Pujadas 1985), intra-provincial migration and continued suburban migration since the 1980s has been more pronounced (García-Coll and Stillwell 1999). Although the well known exodus of Spanish nationals from core cites to surrounding municipalities has been unstoppable since then, the boom of international migration played an important role in counteracting the population decline of these core cities. The influx of immigrants in Spain over the past decade has meant that municipalities such as Madrid and Barcelona have mostly grown by immigration (Domingo and Gil Alonso 2007). Population pressure in gateway municipalities with high immigration has also led to a negative residential migration balance (Bayona, Gil-Alonso and Pujadas 2010), and a growing significance of immigrant groups partaking in the population redistribution (Recaño and Domingo 2006). This is clearly visible in Figures 14.2 and 14.3 for the municipalities of Madrid and Barcelona respectively, where immigrant groups are responsible for a negative net migration (-28,000 in Madrid and -25,300 in Barcelona over the study period). Additionally, a close examination illustrates that suburbanisation may not be the only ongoing process in these areas, with reurbanisation operating simultaneously, as demonstrated by the reversal of the migration turnaround for both Spanish and non-Spanish people. This upward trend of entries is mainly associated with a new functional specialisation of the inner city (Musterd 2006).

The detailed analysis of intraprovincial migration of each immigrant group from different municipality types provides more insight into the dynamics of population dispersal and settlement (see Table 14.3). The results for each capital province (Madrid and Barcelona) indicate that after the inevitable clustering in the original areas of settlement, the majority of immigrant groups experienced a net migration loss, with the exception of the Bolivians in Madrid (one of the most recent arrivals from Latin America), whose migration balance is positive by a small margin. The migration efficiency ratio also captures the dispersal movement from the capitals of each province of the ten largest immigrant groups, with the greatest efficiencies in population redistribution found among the Ecuadorians (-27.6), the Chinese (-23.9) and the Bulgarians (-22.7) in Madrid and the Romanians (-24.4), the Ecuadorians (-24.2) and the Dominicans (-20.7) in Barcelona. The picture for large municipalities (>100,000 residents) and small-medium sized municipalities (10,000 to 100,000 residents) is reversed for the majority of groups with more arrivals than departures and, therefore, a general positive net migration in these areas. In some cases there are also signs of dispersal occurring

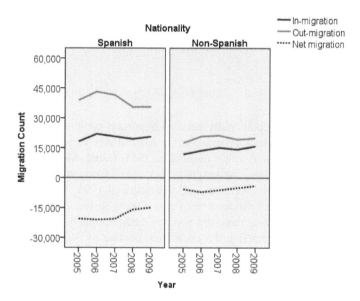

Figure 14.2   Migration between Madrid's municipality and the rest of Madrid's province by nationality, 2005–2009

Source: Own elaboration with data from the Residence Variation Statistics (INE).

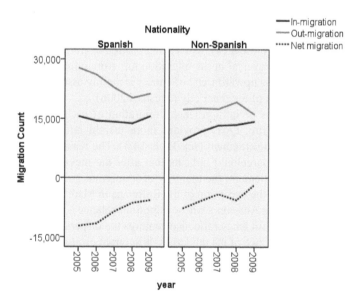

Figure 14.3   Migration between Barcelona's municipality and the rest of Barcelona's province by nationality, 2005–2009

Source: Own elaboration with data from the Residence Variation Statistics (INE).

from large municipalities (the Argentineans in Madrid and Barcelona and the Chinese in Barcelona) and smaller areas (the Bolivians and the Dominicans in Madrid). Although in some cases (e.g. Argentineans) this may be associated with further episodes of suburbanisation from the urban core, the emergence of urban subcentres and the territorial specialization since the late 1990s (Nel·lo 1997) is likely to explain dispersal from other smaller areas. In general, immigrant groups exert an attraction for more in-migration to dense urban areas whether in the form of social networks or family relations principally (Arango 2006). This was previously demonstrated by the segregation indices but also with indicators of movements such as the migration efficiency ratio which suggests that the intensity of arrivals still remains higher in large municipalities compared to smaller areas (10.7 *versus* 4.2 in Madrid, and 5.6 *versus* 5.3 in Barcelona). This is particularly evident with some immigrant groups such as the Ecuadorian in Madrid and Barcelona, whose efficiency in redistributing the population is greatest in large municipalities which, of course, tends to increase their unevenness, as previously shown with the analysis of segregation. The role of occupation specificity is likely to be significant for different immigration groups across municipalities. The bulk of jobs created during the past years, especially construction (mostly among the North African) and domestic services (mostly among the Latin American), have tended to disperse greatly also as a consequence of the high turnover of homes by Spanish nationals. The major access to employment opportunities and, above all, a surplus of housing have been acknowledged as the two pivotal factors influencing moves out from the two gateway municipalities in the provinces of Madrid and Barcelona. Hence, the exponential rise of new housing in Spain over the past decade (García Montalvo 2007, Bielsa and Duarte 2010,) is seen as fundamental for explaining the speed of dispersal of immigrant groups from the most densely populated areas, a process that might have taken many more years in the absence of abundant housing.

Table 14.3 Migration between Madrid's urban areas (>10,000 residents) and the rest of Madrid's province by selected immigrant groups, 2005–2009

Madrid's capital					Barcelona's capital				
Top 10	In-	Out-	Net	MER	Top 10	In-	Out-	Net	MER
Romania	9.342	13.387	-4.045	-17,80	Morocco	4.667	6.377	-1.710	-15,48
Ecuador	9.187	16.206	-7.019	-27,64	Ecuador	7.587	12.444	-4.857	-24,25
Morocco	5.021	6.909	-1.888	-15,83	Argentina	1.778	2.515	-737	-17,17
Colombia	7.289	9.885	-2.596	-15,12	Bolivia	5.771	8.504	-2.733	-19,15
Peru	6.149	8.312	-2.163	-14,96	Peru	3.923	5.185	-1.262	-13,86
Dom. R.	2.780	3.084	-304	-5,18	Colombia	3.975	4.911	-936	-10,53
Bolivia	5.466	5.241	225	2,10	China	6.112	6.217	-105	-0,85
Argentina	1.215	1.792	-577	-19,19	Romania	1.283	2.114	-831	-24,46
China	3.601	5.863	-2.262	-23,90	Pakistan	5.609	7.357	-1.748	-13,48
Bulgaria	1.365	2.167	-802	-22,71	Dom. R.	2.218	3.380	-1.162	-20,76
Total	51.415	72.846	-21.431	-17,25	Total	42.923	59.004	-16.081	-15,78

Madrid's municipalities >100,000					Barcelona's municipalities >100,000				
Top 10	In-	Out-	Net	MER	Top 10	In-	Out-	Net	MER
Romania	14.755	12.785	1.970	7,15	Morocco	11.329	10.671	658	2,99
Ecuador	14.785	8.900	5.885	24,85	Ecuador	14.872	11.832	3.040	11,38
Morocco	10.274	9.059	1.215	6,28	Argentina	1.616	1.643	-27	-0,83
Colombia	9.969	8.717	1.252	6,70	Bolivia	9.730	7.994	1.736	9,79
Peru	7.443	6.160	1.283	9,43	Peru	4.585	4.195	390	4,44

	In-	Out-	Net	MER		In-	Out-	Net	MER
Dom. R.	2.868	2.460	408	7,66	Colombia	4.308	4.113	195	2,32
Bolivia	3.731	3.317	414	5,87	China	10.924	11.417	-493	-2,21
Argentina	1.080	1.081	-1	-0,05	Romania	2.304	2.031	273	6,30
China	5.392	4.050	1.342	14,21	Pakistan	8.739	7.669	1.070	6,52
Bulgaria	2.212	1.923	289	6,99	Dom. R.	3.766	2.879	887	13,35
Total	72.509	58.452	14.057	10,73	Total	72.173	64.444	7.729	5,66

Madrid's municipalities 10,001 to 100,000					Barcelona's municipalities 10,001 to 100,000				
Top 10	In-	Out-	Net	MER	Top 10	In-	Out-	Net	MER
Romania	18.427	17.638	789	2,19	Morocco	24.091	22.568	1.523	3,26
Ecuador	7.085	6.074	1.011	7,68	Ecuador	13.003	11.116	1.887	7,82
Morocco	7.967	7.421	546	3,55	Argentina	4.302	3.749	553	6,87
Colombia	8.316	7.176	1.140	7,36	Bolivia	8.039	7.050	989	6,55
Peru	5.332	4.624	708	7,11	Peru	4.371	3.725	646	7,98
Dom. R.	1.872	2.011	-139	-3,58	Colombia	6.748	6.028	720	5,64
Bolivia	3.543	4.134	-591	-7,70	China	5.922	5.605	317	2,75
Argentina	1.853	1.520	333	9,87	Romania	4.268	3.913	355	4,34
China	2.771	2.131	640	13,06	Pakistan	2.476	1.995	481	10,76
Bulgaria	2.031	1.692	339	9,11	Dom. R.	2.025	1.816	209	5,44
Total	59.197	54.421	4.776	4,20	Total	75.245	67.565	7.680	5,38

Note: ER (Efficiency Ratio).

Source: Own elaboration with data from the Residence Variation Statistics (INE).

Finally, we use Figures 14.4 and 14.5 to answer more specifically whether and to what extent immigrant groups are moving towards their own concentrations in the provinces of Madrid and Barcelona respectively. For this purpose we have computed the net in-migration rate during 2005–2010 for each immigrant group (10) from each capital province to the other municipalities (178 and 310 in the province of Madrid and Barcelona respectively) and using the overall distribution of values of the Index of Isolation into three groups (from low to high). Figure 14.4 shows that the inward movement of immigrant groups from the municipality of Madrid into other municipalities within the province is generally to areas with low concentrations of each group, thus giving more evidence of continued spreading of the ten largest immigrant groups. Hence, those municipalities with low concentrations of immigrant populations gained more through the net balance of in-migration (6.7 per cent) than those municipalities with high concentrations of immigrant populations (4.7 per cent). Dispersal is evident for the majority of immigrant groups (Romanians, Ecuadorians, Moroccans, Colombians, Dominicans, Bolivians and Bulgarians). For some immigrant groups such as the Peruvian and

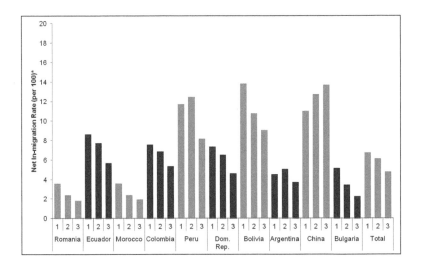

Figure 14.4   Net in-migration 2005–2009 (as percentage of 2010 population) between Madrid's municipality and the rest of Madrid's province by selected immigrant groups and concentration

Note: Net in-migration to municipalities outside Madrid's municipality. The overall distribution of values of the Index of Isolation has been divided into tertiles (low-medium-high). All municipalities (also rural) with at least a population of 100 for each immigrant group are included.

Source: Own elaboration with data from the Residence Variation Statistics and the Population Municipal Register (INE).

the Argentinean the pattern of dispersal is slower, as suggested by the greatest gains in the municipalities with medium concentrations of each group. In fact, only the Chinese exhibit a movement into areas with a relatively high concentration of the same group, an outcome that as noted above would go in line with their high population redistribution in all sorts of municipalities in the province of Madrid. Figure 14.5 also demonstrates a steady movement of de-concentration towards municipalities with low average concentrations of each group in the province of Barcelona. Overall the inward movement of immigrant groups also suggests that municipalities with low concentrations of immigrant populations received more through the net balance of in-migration (7.9 per cent) than municipalities with high concentrations of immigrant populations (2.9 per cent). This can be seen as more evidence that despite the initial demographic consequences of immigration that generally lead to a modest clustering of immigrant groups, immediate spatial dispersion (i.e. heterolocalism) has occurred to a certain extent in the provinces of Madrid and Barcelona, and is constrained by the tightness of the housing market and the availability of economic niches for immigrant groups in each province.

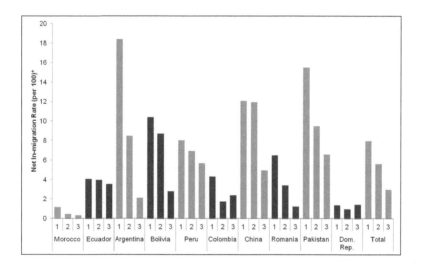

Figure 14.5   Net in-migration 2005–2009 (as percentage of 2010 population) between Barcelona's municipality and the rest of Barcelona's province by selected immigrant groups and concentration

Note: Net in-migration to municipalities outside Barcelona's municipality. The overall distribution of values of the Index of Isolation has been divided into tertiles (low-medium-high). All municipalities (also rural) with at least a population of 100 for each immigrant group are included.

Source: Own elaboration with data from the Residence Variation Statistics and the Population Municipal Register (INE).

Discussion and Conclusion

The current investigation has made some progress in addressing the two initial research questions. The evidence suggests that (1) residential segregation has *not* increased for the great majority of immigrant groups (with the exception of Ecuadorians) in the provinces of Madrid and Barcelona between 2005 and 2010, and (2) internal migration is *not* re-enforcing concentrations of the ten largest immigrant groups but dispersing them towards de-segregation. Using segregation indices with measures of internal migration has proved to be useful and demonstrates the validity of these two complementary angles to examine patterns and processes of settlement. In general, the empirical evidence would suggest that heterolocalism may be a valid model for the majority of the ten largest immigrant groups in the provinces of Madrid and Barcelona, although the co-existence of clustering resonant of past times is also found (e.g. Pakistani and Chinese).

Despite the positive messages from the current investigation one has to be cautious for several reasons. First, we think that although the model of heterolocalism seems to apply to a great extent to the largest groups in the provinces of Madrid and Barcelona, there are other immigrant groups with African origin, especially Sub-Saharan nationalities, that are known to have high levels of segregation and little dispersion from the original areas of settlement (Domingo and Sabater 2011), thus reflecting the importance of continuous disadvantage and discrimination as a generator of residential segregation. The empirical evidence also suggests that even though immigrants tend to arrive at traditional gateway areas from where dispersal generally occurs, new immigrants also migrate directly to new destinations and, therefore, the exit story from traditional areas of settlement might no longer capture the full story of dispersal of immigrant groups (Bayona, Gil-Alonso and Pujadas 2010). From this perspective, we are prone to believe that immigrant residential patterns are more likely to reflect a bimodal pattern, whereby some immigrants tend to cluster whilst others exhibit dispersal after arrival. Second, we believe that the unprecedented influx of international migration and settlement of immigrant groups has also taken place during an exceptional period of economic growth in Spain which contributed a great deal to what some scholars call as an 'institutionally generated migration' (Brama 2006: 29). The vast supply of housing in Spain met an ideal scenario: the high turnover of homes by Spanish nationals, the demand from immigration and continued low mortgage rates. Such factors, combined with the geographically specific labour demands, are likely to be key explanations for the rapid speed of dispersal of immigrant groups both in Madrid and Barcelona. However, the financial crisis is already acting differently for immigrant groups who increasingly face more difficulties than Spanish nationals in meeting their rent and mortgage payments, thus restricting their freedom of movement considerably. Within this context, we already know that the drop in internal mobility in Spain is of the same magnitude as the decrease in international migration during the same period (Domingo and Sabater 2009, Domingo and Recaño 2010). Since immigrant groups are overrepresented in sectors which have been particularly hard hit, especially in labour-intensive industries such as construction and the service sector, internal mobility generated by labour demand has also suffered dramatically. If the economic recession has clearly decreased internal mobility, it is still unclear whether there will be some possible spatial effects with consequences on the existing residential concentration and segregation of immigrant groups. Certainly more research will be needed to determine the aftermath of the crisis and its consequences for the sociospatial behavior of different immigrant groups. Of course, we still do not know whether the model of heterolocalism in Spain might even be mitigated. One may speculate that this gloomy period may impose more selectivity on internal migration with perhaps more immigrant groups staying put in distressed neighborhoods in the original areas of settlement.

Acknowledgements

Albert Sabater's research is funded under the Juan de la Cierva Fellowship Programme of the Ministry of Science and Innovation of Spain. Jordi Bayona's research is funded under the Juan de la Cierva Fellowship Programme of the Ministry of Science and Innovation of Spain. Andreu Domingo's research is supported by the R+D+R grant 2008–2010 of the Ministry of Science and Innovation entitled 'Differential sociodemograhphic behaviour and social integration of immigrant populations and their descendants in Spain'. The official population and flow figures are managed by the Spanish National Statistics Institute (aka INE). We would like to acknowledge the help of the Editors (Nissa Finney and Gemma Catney) and Douglas Massey who provided insightful comments and suggestions for improvement.

Chapter 15

Minority Internal Migration in Europe: Research Progress, Challenges and Prospects

Nissa Finney and Gemma Catney

Introduction

The collection of original analyses of minority internal migration in countries across Europe presented in this volume represents the state of current knowledge in this field. The chapters examine the experience of thirteen countries, which vary considerably in terms of their economic situations, policy and political contexts, immigration histories and minority experiences. Minority internal migration is examined in Belgium, London, Portugal (Lisbon), The Netherlands (Amsterdam), Israel, Scotland, Germany, Turkey, Greece (Athens), Britain, Italy, Sweden and Spain. This concluding chapter aims to complement Chapter 1 by revisiting the common themes outlined, summarising the findings of the book, and considering some of the important differences between chapters, thus identifying priorities, prospects and challenges for minority internal migration research.

Progress in Minority Internal Migration Research

The concerns of the chapters in this collection can be characterised by five research questions, all empirically addressed through quantitative secondary data analysis: How residentially mobile are minorities compared to the majority population? What are the determinants/characteristics of the migration of minorities compared to the majority? How do the geographies of minority internal migration compare to patterns for the majority? Is internal migration racialised (for example, in terms of processes such as so-called 'White flight' or 'minority self-segregation')? How can residential patterns be interpreted in terms of social integration? These questions are cross-cut by the three themes outlined in Chapter 1. Foremost is the central and framing theme of ethnic or immigrant integration (or incorporation, or assimilation), which is accompanied by consideration of inter-generational change and the importance of place.

With these questions and themes in mind this section reviews the contributions made by the book. The investigations of the levels of mobility of minorities/ immigrants generally confirm what we can expect from the extant literature: minorities are more residentially mobile than the majority population (Andersson,

Recaño-Valverde and de Miguel-Luken, Sirkeci, Cohen and Can, Malheiros). This is theorised as being related to their migration history (Andersson, Vidal and Windzio); and, in the initial period after immigration, their search for suitable housing and work (Malheiros).

Contrary to this, however, not all chapters found minorities to be more residentially mobile than others. Two chapters in particular provide examples of this. In Israel, Cohen, Hefetz and Czamanski found Arabs to be less mobile than Jewish citizens, which they interpret as a result of restrictive planning and settlement practices. In her examination of the student population in Britain, Finney identifies minority students (Blacks and Asians) as being less mobile than their White counterparts, challenging the norms of student mobility. These findings highlight the importance of national ethno-political contexts, and the variability in internal migration for population sub-groups.

The scale or distance of migration was found to have some effect on the levels of mobility of minorities and this is not consistent across countries: in Germany, minorities were more mobile over short distances but not longer distances (Vidal and Windzio) whereas in Sweden immigrants were more internally mobile than the Swedish-born at several geographical scales (Andersson).

Inter-generational patterns in levels of mobility also vary between chapters. In Germany the second generation were found to be more mobile than the first, linked to educational moves and inter-generational social mobility (Vidal and Windzio). In Sweden, however, differences in mobility decrease as time living in the country increases and generational convergence is expected (Andersson). In Britain, immigrant students of all ethnic groups are more mobile than their UK-born counterparts (Finney).

Minority internal migrants were found by a number of chapters to have the same characteristics as internal migrants of the majority population (Andersson, Vidal and Windzio, Manley and Catney, Sirkeci, Cohen and Can) and some have argued that it is socio-demographic and economic rather than 'ethnic factors' that are important for understanding residential mobility (Malheiros, Manley and Catney, Andersson). Some characteristics of the more mobile appear persistently important across countries. In particular, renters were very residentially mobile (confounded by high levels of renting amongst immigrants) (de Valk and Willaert, Malheiros, Recaño-Valverde and de Miguel-Luken) and higher education was associated with higher mobility (Cohen, Hefetz and Can, Recaño-Valverde and de Miguel-Luken). De Valk and Willaert and Finney illustrate the importance of taking lifecourse factors into account to understand migration of migrants and natives, and they identify the need to examine whether lifecourse transitions are associated with migration in the same way across ethnic/immigrant groups.

The chapters present mixed results regarding the association between gender and residential mobility. For example, Recaño-Valverde and de Miguel-Luken find females to be less mobile than males in Spain and Portugal and particularly in Italy; Finney finds no gender differences in student migration for any ethnic group in Britain; Cohen, Hefetz and Czamanski find Arab women to be more mobile than Arab men. The Israeli findings on gender are particularly interesting and are interpreted as a result of ongoing patrilocalism together with an effect of increased female education. It is apparent that the gendered nature of migration takes different forms in different places and for different population sub-groups. Thus, this is one realm of migration theory where particular attention should be paid to local specificities.

A number of the chapters are concerned with the geographies of minority internal migration. In their chapter on Spain, Sabater, Bayona and Domingo focus on the geographical distribution of recent migrants in the provinces of Barcelona and Madrid. They conclude that recent immigrants are already de-concentrating and dispersing from their neighbourhoods of original settlement and that processes of heterolocalism can be identified for the majority of immigrant groups. In a similar vein, Malheiros identifies immigrant residential de-concentration in relation to the urban fragmentation thesis in Lisbon. This dispersal can alternatively be interpreted as suburbanisation, a process of spatial integration which has been seen to increase with time of residence in the country and over immigrant generations, especially for those who enter into relationships with natives (de Valk and Willaert, Musterd and van Gent).

In some cases, suburbanisation was evident for the majority population but not for minorities. In Scotland, minorities were more likely than Whites to migrate into and around the large urban centres of Glasgow and Edinburgh, most likely because of the housing, job and educational opportunities on offer in these cities (Manley and Catney). In Athens, immigrants' relocation showed preference for the inner suburbs of the city, whilst Greeks were relocating primarily to more peripheral suburbs (Kandylis and Maloutas).

De Valk and Willaert draw to our attention the importance of considering neighbourhood characteristics and perceptions in understanding geographies of internal migration. For example, neighbourhood dissatisfaction is a good predictor of out migration for all groups (de Valk and Willaert). Ethnic composition of neighbourhood is also important in understanding mobility. In Brussels, having more co-ethnics in the neighbourhood was associated with lower levels of mobility (de Valk and Willaert), and clustering was found to be persistent for some minority groups in Britain, Spain and Greece (Stillwell and McNulty, Sabater, Bayona and Domingo, Kandylis and Maloutas). Kandylis and Maloutas remind us that clustering does not necessarily constitute self-segregation and that ethnically mixed residential areas are not necessarily spaces of lower inequality for members of all ethnic groups.

In any neighbourhood, internal migration is just one component of population change, along with natural change and international migration. Stillwell and McNulty remind us of the need to consider the dynamics of all elements of population change. They found that for Asians and Blacks in London, areas of co-ethnic concentration experienced net internal out migration and replacement immigration; this was the case to a lesser extent for Whites and, for the Chinese group, immigration was dominant. Immigrant destination may, however, be diversifying from traditional gateway areas (Stillwell and McNulty, Sabater, Bayona and Domingo). In London some ethnic groups were found to immigrate to areas of co-ethnic residence (Pakistani, Other South Asian, Bangladeshi, Whites from the USA) but this was less the case for other groups. However, within these areas of co-ethnic concentration, the destinations of minority internal migrants were neighbourhoods of (relatively) low co-ethnic concentration (Stillwell and McNulty).

Questioning the meaning of minority internal migration patterns from the perspective of ethnic integration leads us to consider their relation to social integration. Several chapters have considered how mobility intersects with other social and economic realms which may be seen to represent dimensions of long-term integration, namely housing, employment and education. Musterd and van Gent, de Valk and Willaert, Malheiros, Kandylis and Maloutas, Vidal and Windzio and Andersson consider the significance of the housing market structure in ethnically differentiated migration patterns and the potential for discrimination against immigrants operating in housing markets. In Amsterdam, although non-Western immigrants are less likely than the native Dutch to obtain single-family housing in low density areas outside the region, there is evidence of spatial 'assimilation' and increased housing market access between first and second generations (Musterd and van Gent). In the German case, it is suggested that high minority mobility is linked to the need for housing re-adjustment because minorities are less able to satisfy housing needs in the first instance and, possibly, because of simultaneous experiences of structural discrimination in the housing market (Vidal and Windzio). Of course, housing markets change over time: the unprecedented large-scale immigration to Spain since the 1990s took place during a period of exceptional economic growth. The vast supply of housing in Spain was a result of the high turnover of homes by Spanish nationals, the demand from immigration, and continued low mortgage rates, none of which may continue in economic recession (Sabater, Bayona and Domingo).

The association between engagement with the labour market and migration is considered by Andersson and Malheiros. Andersson points to labour market attachment as a key factor in determining migration, and particularly the higher migration of minorities and Malheiros identifies work as the main reason for residential move. Sirkeci, Cohen and Can, using somewhat unique Turkish Census data on motivations for internal migration, show that work-related reasons drove a higher proportion of moves for the Turkish-born than for the foreign-born.

Finney discusses migration in relation to education, highlighting that residential mobility for young adults is intertwined with residential norms during Higher Education. The 'norm' of students as mobile does not apply across ethnic groups. How ethnic differences in the migration of students are indicative of inequalities or integration is uncertain and debateable.

Clearly the associations between residential mobility and social and economic integration are complex. Kandylis and Maloutas, in their study of Athens, find that better socio-economic conditions seem to be more clearly linked with relocation in the case of the less mobile (and relatively deprived) groups of Pakistani and Filipino rather than for the more mobile groups, as may be expected. They conclude that residential mobility may indicate social integration for some, but for others it may be a mechanism by which to lessen 'socio-spatial restrictions'. This serves to highlight the point that, without exception in the contributions to this book, there is diversity of experience both between countries and between immigrant/ minority groups within countries. What is common to all of the chapters is that whilst distinctions can be drawn between minority and majority populations, there is considerable variation in the residential mobility experiences of minorities in each national context. Differing motivations for migration, combined with how these may interplay with particular national political and economic contexts, make it difficult to generalise about experiences. This is highlighted in particular by Vidal and Windzio and Stillwell and McNulty who group minorities on the basis of their immigration histories.

The chapter on Israel provides the clearest example of the need to take into account specific migration histories and national political contexts in understanding ethnic differences in residential mobility (Cohen, Hefetz and Czamanski). The authors examined the mobility of Arabs, a native homeland minority, and found it to be low in comparison with Israeli Jews, though there were variations between religious groups. The residential experience of Arabs in Israel cannot be understood without consideration of the position of this minority group within Israeli society, and the geo-political organisation of the country.

A further aspect that influences the experiences of minority internal migration is the long-term history of immigration, particularly whether the primary concern is with post-immigration settlement (countries of recent immigration) or differential mobility of minorities in the longer term (countries of established immigration). Post-immigration settlement is the concern of the chapters on Spain, Italy, Portugal (Sabater, Bayona and Domingo; Recaño-Valverde and de Miguel-Luken; Malheiros), Athens (Kandylis and Maloutas) and Turkey (Sirkeci, Cohen and Can) – countries or cities where immigration over the last decade has been historically high and in some cases unprecedented. The Netherlands, Belgium, Germany, Sweden and Britain (including Scotland) can be considered to be established immigration countries. Of course, 'established' and 'recent' immigration experiences are not mutually exclusive; the countries with established immigration have also had net gain from immigration in recent years. Nevertheless, the long-term history of immigration does shape the national contexts of dominant political debates, past and present policies of integration, and equality and public attitudes towards immigration, minorities and cohesion. Thus, it can be helpful to distinguish the chapters on the basis of whether their experiences are characterised by 'recent' or 'established' immigration. Through comparison, this enables us to comment on the extent to which patterns of post-immigration settlement and integration from previous decades are being repeated. This book's chapters suggest that to some extent this is the case (for example in terms of dispersal from immigrant settlement areas), but there are also important specificities to the experiences of contemporary immigrants (for example, the pioneering of new settlement neighbourhoods).

Methods of Minority Internal Migration Research

A diverse range of secondary data has been used in this volume to examine minority internal migration. Each chapter has used existing (as opposed to purposely commissioned) quantitative datasets. These can be divided into five types as shown in Table 15.1: national censuses (aggregate and microdata), international census microdata, national sample surveys, population registers and other administrative data.

There is a great variety and volume of data in European countries which allows for analyses of minority internal migration. The comparability, at a broad level, in the type of data collected in different countries is also notable. Thus, the analyses presented in this collection can address similar themes with analogous approaches. However, none of the datasets used were designed specifically for the study of migration. As a result, there are limitations to the study of migration in terms of how migration is defined, the co-variates available, the time dimensions of the data, and the geographical scales of analysis that are possible. This poses challenges and constraints for the individual country analyses, as the authors have discussed, and makes international comparison particularly difficult. In this regard, the International Public Use Microdata, derived from national censuses, is a very valuable resource for comparative demographic studies (made use of by Recaño-Valverde and de Miguel-Luken for Italy, Portugal and Spain).

Table 15.1   Data sources for investigating minority internal migration

Data	Chapter (1st Author)	Country
Census (aggregate and microdata)	Vidal, Sirkeci, Stillwell, Cohen, Kandylis, Finney, Manley, de Valk, Recaño-Valverde	Germany, Turkey, Britain, Israel, Greece, Scotland, Belgium, Spain, Italy, Portugal
International Census microdata	Recaño-Valverde	Spain, Italy, Portugal
Longitudinal national survey	Vidal	Germany
National/regional population register	Musterd, Sabater, Andersson, de Valk	Netherlands, Spain, Sweden, Belgium
Other administrative data	Finney	Britain

Table 15.2 summarises how internal migration has been defined for the analyses in each chapter. All of the chapters, whether using aggregate or individual level datasets, identify internal migration as a transition; that is, the place of residence is different at one point in time compared with another. This is in contrast to migration being measured as an event, where each move would be recorded (see Bell et al. 2002). Four of the chapters (Cohen, Hefetz and Czamanski, Manley and Catney, Recaño-Valverde and de Miguel-Luken, Finney) define migration at the individual level, identifying whether or not an individual changed their place of residence between given time points. A more common approach in these chapters, however, has been to identify moves between defined sub-national geographical areas which vary between countries in scale and in the political or administrative function of their boundaries. The years under consideration and the length of the transition periods (time points between which migration is defined) also vary, though most chapters consider moves between intervals of one to five years.

Table 15.2   Measures of internal migration across Europe

Level of migration analysis	Chapter (1st Author)	Country	Sub-national geography (if applicable)	Time(s) over which migration is defined (transition period)
Individual	Cohen	Israel	n/a	Pre 1990/1990–1995
	Manley	Scotland	Short distance/Long distance	1990–1991/2000–2001
	Recaño-Valverde	Spain, Portugal, Italy	n/a	2000–2001
	Finney	Britain	n/a	2000–2001/2006–2009
Aggregate	De Valk	Belgium	Municipality	2001–2006
	Stillwell	Britain	District and Ward	2000–2001
	Malheiros	Portugal	Region	First residence – 2010
	Musterd	Netherlands	Neighbourhood	1999–2006
	Vidal	Germany	County	2005–2010/1984–2010
	Sirkeci	Turkey	District/village/province	1995–2000
	Kandylis	Greece	Municipality	1996–2001
	Andersson	Sweden	Labour Market region/ Neighbourhood	2005–2008
	Sabater	Spain	Municipality	2005–2010

Not only is migration defined differently in the various national and regional datasets, but the definition of ethnicity or immigrant origin also varies. This reflects differences in immigration histories and political contexts for immigration. Table 15.3 summarises the ways in which ethnicity and immigrant status are identified in the chapters and reveals the diversity of measuring diversity across Europe. In continental Europe there is tendency for minorities to be identified using birthplace and/or citizenship. Britain and Israel are notable for their conceptualisation and measurement of minorities on the basis of ethnicity and ethnic identity which is self-defined in the datasets used in chapters on those countries. Israel, as already noted, is a unique case in that Arabs constitute a native or homeland minority. In Britain, the measurement of ethnic group sits alongside the measurement of birthplace in the Census. An ethnic group question was first asked in 1991 with the primary motivation of identifying inequalities and discrimination (see Finney and Simpson (2009) for an overview of the evolution of the British Census question on ethnic group). The concept of ethnicity is less specific than that of immigrant based on birth in another country. Analyses of ethnic group differences in internal migration thus invite interpretations that not only incorporate immigrant experience but also ethnic differences in, for example, culture, religion, lifestyle, social status, discrimination and demographics.

Table 15.3   The diversity of measuring diversity

Ethnic/immigrant measurement	Chapter (1st Author)	Country
Birthplace (and/or grand/ parental birthplace)	Vidal, Sirkeci, Stillwell, Kandylis, Andersson, Recano-Valverde	Germany, Turkey, Britain, Greece, Netherlands, Spain, Sweden, Italy, Portugal
Nationality/Citizenship	Vidal, Sirkeci, Kandylis, Sabater, Andersson, de Valk	Germany, Turkey, Greece, Spain, Sweden, Italy, Portugal, Belgium
Country of origin	Stillwell, Kandylis, Finney	Britain, Greece
Moved from other country	Stillwell, Finney	Britain
Self-identified ethnic/ ethno-religious group	Stillwell, Finney, Manley, Cohen	Britain, Scotland, Israel
Native/homeland minority	Cohen	Israel

Theoretical Challenges and Prospects for Minority Internal Migration Research

The research in this collection represents an emerging field that has been propelled largely by political concerns. The development of minority internal migration research has been enabled by large scale secondary data on migration and minorities becoming available in the last few decades. This review of the state of knowledge reveals that we know quite a lot about *who* minority internal migrants are, and variations in migration propensities; *what* characteristics are associated with migration across minority and majority groups; and something about the *where* of minority internal migration for individual countries. However, we know relatively little about *why* there are differences in mobility between immigrant groups, or how minority status shapes migration experiences; or about the *when* (temporality), *how* (processes, for example, decision making) or *so what?* (experiences and social impacts) of minority internal migration. With this in mind, this section assesses potential future directions for minority internal migration research and theoretical arenas to which it can contribute.

It can be argued without difficulty that the theoretical focus of minority internal migration research has to date been narrowly framed by debates on ethnic segregation. The centrality of integration theory to the collection is not surprising given the longstanding, if subsidiary, interest in internal migration in the larger body of work on ethnic segregation, from the work of the Chicago school (Park, McKenzie and Burgess 1925) onwards; and the political context in the early years of the twenty first century across much of Europe characterised by concerns about how to manage growing ethnic diversity nationally and locally.

However, the topic has the potential to contribute to a theoretically broader and more diverse set of debates; and in doing so to reinvigorate 'segregation' and 'integration' theories with more nuanced understandings of population change, and ones that pay attention to the specificities of experiences in Europe as opposed to North America. We propose six areas of contribution and briefly discuss each in turn.

Migration Studies

Minority internal migration research can engage more with core debates in migration studies. That the potential for this has not yet been realised is perhaps a reflection of the tendency for migration (and population change) literatures in a broad sense to be separate from residential segregation literatures. Minority internal migration studies in this volume and elsewhere (such as Simpson 2004, Catney 2008 and Sabater 2010) have begun to bridge this gap in relation to ethnic segregation. However, three avenues of migration scholarship in particular seem ripe for contributions from minority internal migration studies: the selectivity of migrants; the connections between immigration and internal migration; and transnationalism.

Concern about migrant selectivity is at the core of minority internal migration research which has demonstrated that there is something about immigrant experience or minority ethnic group identification that affects experiences of internal migration. Relatively little is understood, however, about the role of ethnicity in migration, or the mechanisms and processes associated with ethnic identity that impact on migration. In her chapter on migration of students in Britain, Finney identifies theories from education literature that can inform understandings of ethnic differences in migration. It is likely that both a greater focus on specific (ethnic) sub-populations and social realms (education, housing, employment, for example), together with an interdisciplinary perspective, is necessary to improve theoretical understandings of the links between mobility and ethnicity.

It is important also to pay attention to the multi-dimensionality of ethnicity and how ethnic identity intersects with other social markers, particularly gender, class and religion. Some studies have attempted to address issues of intersectionality in understanding internal migration; these include Munoz's (2011) study of mobility of South-Asian ethno-religious groups in Scotland and Catney and Simpson's (2010) examination of the association between class and ethnicity in migration from urban centres, which concluded that there is a social gradient to migration from traditional immigrant gateway areas regardless of ethnicity.

It has long been recognised that, in terms of the people and processes involved, the distinction between internal and international migration is not as clear as this binary categorisation implies. King and Skeldon (2010) re-ignited this debate and called for theoretical development that bridges the gap between internal and international migration, and that is concerned less with the demarcation of national boundaries. Minority internal migration research is ideally situated to meet this challenge, being as it is a study of the continued mobility of immigrants. It is readily applicable to the question of how the international migrant experience is transferred to sub-national mobility.

Furthermore, addressing questions about the information, communication and social networks of minority internal migrants; the experiences of post-immigration mobility; and the migration decision making processes, can contribute to debates about transnationalism (which are concerned with inter-national connectivity and the diminishing significance of state boundaries). Whilst internal migration in this book is specifically concerned with movement contained within nation states, both the flows of people and the causes and consequences of these movements operate within the broader geo-political context of Europe and beyond.

Ethnic Integration

As discussed above, ethnic or immigrant integration (or 'assimilation') theory has been the dominant framework for research on minority internal migration. However, we suggest that minority internal migration research can contribute to integration theory in additional ways, and these are possible avenues for the future development of the field. In particular, longitudinal analysis of the coincidence of internal migration – and the geographies of that migration – with other life changes would illuminate the question of what constitutes integration in terms of migration patterns. This involves a re-examination of the meaning of ethnic concentrations, of what constitutes 'good segregation' and 'bad segregation' (Peach 1996); and a re-thinking of 'segmented assimilation' in spatial terms (Portes and Zhou 1993). It also requires an inter-generational perspective, comparing experiences of immigrants with those of their children. A clearer understanding of the meaning, causes and consequences of ethnic population change and the geographies of population diversity will develop debates about the impact of immigration that are pertinent politically across Europe.

Lifecourse Pathways

Understanding the differing migration experiences of minority groups can be seen through the lens of diverse lifecourse pathways and de-standardisation of the lifecourse. This perspective, that views migration in the context of other life events (or transitions) across the span of individuals' lives, is becoming increasingly important in migration research (Wingens et al. 2011). In addition, examining how migration experiences and decisions through the lifecourse are influenced by minority or immigrant status will contribute to our understanding of the meaning of ethnicity in relation to migration. A lifecourse perspective also encourages engagement with the temporal dimensions of migration, viewing migration as a process or transition, rather than as an isolated event. Taking a lifecourse perspective allows the longer-term causes and consequences of migration to be identified, feeding into debates about how experiences in early life affect later life chances and outcomes.

Neighbourhood Studies

All studies of internal migration are, by definition, geographical, in that they consider a spatial event. Some are explicitly concerned with geography when they consider the places (or types of place) people move to and from. We know that migration changes the character of neighbourhoods along ethnic and other social lines. A body of work on urban change is concerned with these issues (for example, Bailey 2012) and minority internal migration research can usefully contribute. In particular, it can assess the impact of selective migration in ethnic terms on neighbourhoods change; and the impact of ethnically differentiated neighbourhood change on residents' experiences, (that is, the 'effect' of neighbourhood ethnic group population change; see Galster 2007 and Laurence and Heath 2008). Improved understanding of the links between ethnicity, neighbourhood dynamics and individual and neighbourhood outcomes can enable constructive engagement with debates about whether regeneration and development policies should be targeted towards people or places.

Demographic Change

Minority internal migration research can contribute to theories of demographic change at individual and societal levels. At the individual level, it can enhance understanding of how demographic decisions and the inter-relations between fertility, mortality and migration differ between minority and immigrant groups. As discussed in the section on lifecourse pathways, these differences may be expected, and could have implications for planners and providers of health and housing services. At a societal level, minority internal migration studies can reveal the causes and consequences of changes to the ethnic/immigrant population structure of sub-national areas. Sub-national projections of population with an ethnic group dimension, which are important for planning purposes (and increasing so as developed societies become more ethnically mixed), will depend on accurate estimates of internal migration by ethnic group (Rees 2008).

Minority internal migration research can engage in key debates in Demography and Population Studies about the balance of population and resources, and particularly about the challenges of 'overpopulation' and population ageing. Specifically, it can contribute to the growing interest in regional variation in population dynamics and the need to understand not just the number of people and where they are but who is where and how this is changing. By engaging with demographic debates, minority internal migration research can contribute to work that attempts to bring migration into theories of demographic change and transition.

Ethnic Inequalities

How do ethnic differences in migration patterns represent differing barriers and opportunities for movement? Do they reflect inequalities in, for example, housing, employment and education? Do they represent discrimination? And what are the consequences of inequalities in mobility? If the notion of mobility capital is employed, are minority groups particularly advantaged or disadvantaged in certain life outcomes by their relatively high or low mobility? These questions have yet to be addressed by minority internal migration research but have the potential to illuminate ethnic inequality from a migration perspective, inform equality policies, and re-frame understandings of migration in terms of equality.

There is clearly considerable scope for the development of minority internal migration research and for its contribution to the theoretical development of a number of politically relevant fields. In all of these realms the historical, political and policy contexts are important and theories will need to take into account the unique political and economic structures of Europe and its constituent countries, as well as the context of varying national and international policies on immigration and mobility.

Methodological Challenges and Prospects for Minority Internal Migration Research

Many of the questions raised and potential theoretical contributions outlined above cannot be addressed with the (type of) data and methods used in the chapters of this book. To a large extent, the extant work has made best use of existing quantitative data, but it is restricted by the limited migration information available. Two solutions are apparent: making use of new, forthcoming data sources; and creating new data. We focus here on prospects for the UK but similar opportunities exist in other countries in Europe. In terms of new quantitative data for secondary analysis in the UK, for example, the UK Household Longitudinal Survey (UKHLS) (known as 'Understanding Society') offers great promise because of its ethnic minority boost sample, backwards incorporation of the British Household Panel Study, and module on migration history with questions on internal and international migration. The national censuses of a number of countries conducted in the early 2010s (2011 in the UK) will provide an update to the studies presented in this collection and opportunity through census longitudinal studies to incorporate a time dimension. In the UK, for example, ethnic group information will be available from the 1991, 2001 and 2011 Censuses, allowing investigation of trends over three decades. Census data have the advantage of sufficient numbers to enable analysis for small geographical areas.

In addition to making use of new data, there may be existing data which can be accessed and used for studies of minority internal migration. For example, in England and Wales the National Pupil Database, a census of all school pupils,[1] records pupils' home address, ethnicity and other individual information which is available to researchers in an anonymised form. Although the data are primarily intended for research on education, they have potential for studies of migration, particularly those interested in children and families for small areas with a longitudinal dimension (Simpson, Marquis and Jivraj 2010).

If we are to fully address the questions of the 'why', 'how', 'when' and 'so what' of minority internal migration and fulfil the research agenda, the creation of new data seems essential. Much of the theoretical contributions outlined above, including understanding migrants' social networks, migration decision making and migration experiences, will require primary data collection, both qualitative and quantitative. A mixed methods approach, such as 'facet' methodology (Mason and Dale 2011) may be particularly appropriate for this topic, where there is a desire to understand large scale patterns, processes and trends as well as individual decisions and experiences. The collection of primary data is clearly more costly (in terms of money, time and human resources) than secondary data analysis, and the degree to which this is practical will depend

---

1 The National Pupil Database excludes pupils of independent (non-state funded) schools.

upon how strongly a case can be made for the need and impact of minority internal migration research.

One area with considerable potential for methodological development is the international comparison of minority internal migration, or, rather, an integrated international investigation. This will pose challenges of how to compare and account for varying national (and sub-national) economic, political and cultural contexts; how to use data from differing sources in a combinatory or complementary manner; how to create comparable measures and definitions of minorities and migration; and how to understand the implications of differing definitions for the research findings. A project that may assist in the challenges of such a study is the international database of internal migration data (Bell et al. 2002).

There is also scope for minority internal migration research to be innovative in terms of the methods of analysis that are used. The chapters in this collection have used standard quantitative approaches but future work might consider the benefits of spatial analysis techniques (Lloyd 2011), advanced longitudinal approaches and statistical methods for distinguishing between the effects of individuals and places (including multilevel analysis, see Goldstein 2011). Advanced qualitative methods, including visual and participatory approaches (Mason and Dale 2011) also have much to offer and a mixed methods framework including quantitative and qualitative techniques seems appropriate for studying patterns of population change alongside minority migration experiences and motivations.

Conclusion

Minority internal migration research in Europe is emerging as a field with a great deal to contribute to wider debates about integration, population change and the impact of immigration, as the chapters in this volume illustrate. The body of work has illuminated much about ethnically differentiated levels and geographies of internal migration and characteristics of migrants, and has considered the importance and meaning of residential mobility for social integration. The chapters in this book have illustrated that there are some commonalities in the residential experiences of immigrants and minorities in different European countries, but also some differences which, for interpretation, require us to consider the specific political, economic and historical experiences of each national context.

This chapter has identified six theoretical arenas to which minority internal migration research can contribute: migration studies (particularly selectivity of migration; the relation between international and internal migration; and transnationalism), ethnic integration, lifecourse pathways, neighbourhood studies, demographic change and ethnic inequalities. Contributing to these debates will require minority internal migration scholars to make full use of quantitative data that become available, and consider the gathering of new data, both qualitative and quantitative. A mixed methods framework is particularly appropriate for advancing our understanding of large scale patterns, processes and trends, as well as individual migration decisions and experiences. Internationally comparative studies are crucial for the theoretical development and integrated study of migration, and to understand the effects of continental economic and political change on population movement.

We hope that this book has demonstrated the richness of minority internal migration research in Europe, provided a review of the state of knowledge, and identified key elements of an agenda for future research that will encourage scholars to build our understanding of the patterns, causes and consequences of the ethnic dimensions of internal migration.

References

Aalbers, M.B. 2005. Place-based social exclusion: redlining in the Netherlands. *Area*, 37(1), 100–109.

Abramsson, M., Borgegård, L-E. and Fransson, U. 2002. Housing careers: immigrants in local Swedish housing markets. *Housing Studies*, 17(3), 445–464.

Abu-Tabih, L. 2009. On collective national rights, civil equality and women's rights: Arabs women in Israel and the negation of their rights to choose their place of residence. *Theory and Criticism*, 34, 43–70. [In Hebrew].

Ahmed, A. and Hammarstedt, M. 2008. Discrimination in the rental housing market: a field experiment on the internet. *Journal of Urban Economics*, 64, 362–372.

Aja, E. and Arango, J. (eds). 2006. *Veinte años de Inmigración en España*. Barcelona: CIDOB.

Akgunduz, A. 1998. Migration to and from Turkey, 1783–1960: types, numbers and ethno-religious dimensions. *Journal of Ethnic and Migration Studies*, 24(1), 97–120.

Alba, R. 2005. Bright vs. blurred boundaries: second generation assimilation and exclusion in France, Germany and the United States. *Ethnic and Racial Studies*, 1, 20–49.

Alba, R. and Nee, V. 1997. Rethinking assimilation theory for a new era of immigration. *International Migration Review*, 31(4), 826–874.

Alba, R.D. and Nee, V. 2003. *Remaking the American Mainstream: Assimilation and the New Immigration*. Cambridge, MA: Harvard University Press.

Allen, C. 2008. *Housing Market Renewal and Social Class*. London: Routledge.

Allen, J., Barlow, J., Leal J., Maloutas T. and Padovani, L. 2004. *Housing and Welfare in Southern Europe*. Oxford: Blackwell.

Allison, P.D. 2005. *Fixed Effects Regression Methods for Longitudinal Data Using SAS*. Cary, NC: SAS Institute Inc.

Amersfoort, H.V. 1990. La répartion spaciale des minorities ethniques dans un état providence: les leçons des Pays Bas 1970–1990, *Espaces, Populations, Sociétés*, 2, 302–323.

Andersson, R. 1998. Socio-spatial dynamics: ethnic divisions of mobility and housing in Post-Palme Sweden. *Urban Studies*, 35(3), 397–428.

Andersson, R. 2000. Segregerande urbanisering? Geografisk rörlighet i Sveriges storstadsregioner, in *Hemort Sverige*. Norrköping: Board of Integration, 149–182.

Andersson, R. 2008. Neighbourhood effects and the welfare state. Towards a European research agenda? *Schmollers Jahrbuch*, 128, 1–14.

Andersson, R. 2012. Understanding ethnic minorities' settlement and geographical mobility patterns in Sweden using longitudinal data, in *Minority Internal Migration in Europe*, edited by Nissa Finney and Gemma Catney. Aldershot: Ashgate.

Andersson, R. and Bråmå, Å. 2004. Selective migration in Swedish distressed neighbourhoods: can area-based urban policies counteract segregation processes? *Housing Studies,* 19(4), 517–539.

Andersson, R., Magnusson Turner, L. and Holmqvist, E. 2010. *Contextualising Ethnic Residential Segregation in Sweden: Welfare, Housing and Immigration-Related Policies.* [Online]. Available at: http://blogs.helsinki.fi/nodesproject/files/2010/12/Sweden.pdf.

Antonopoulou, S. 1991. *Post-war Transformation in the Greek Economy and the Residential Phenomenon.* Athens: Papazisis. [In Greek].

Arango, J. 2000. Becoming a country of Immigration at the end of the Twentieth Century: the Case of Spain, in *Eldorado or Fortress? Migration in Southern Europe*, edited by R. King, G. Lazaridis and C. Tsardanidis. Macmillan: London, 253–276.

Arapoglou, V. 2006. Immigration, segregation and urban development in Athens. The relevance of the LA debate for Southern European metropolises. *The Greek Review of Social Research*, 121C, 11–38.

Arapoglou, V., Kavoulakos, K., Kandylis, G. and Maloutas, T. 2009. The new social geography of Athens. Immigration, diversity and conflict. *Syghrona Themata*, 107, 57–67. [In Greek].

Arapoglou, V.P. and Sayas, J. 2009. New facets of urban segregation in Southern Europe. Gender, migration and social class change in Athens. *European Urban and Regional Studies*, 16(4), 345–62.

Arbaci, S. 2008. (Re)Viewing ethnic residencial segregation in Southern European cities: housing and urban regimes as mechanisms of marginalisation. *Housing Studies,* 23(4), 589–613.

Arbaci, S. and Malheiros, J. 2010. De-Segregation, peripheralisation and the social exclusion of immigrants. Southern European cities in the 1990s. *Journal of Ethnic and Migration Studies*, 36(2), 227–55.

Arbaci, S. and Rae, I. (forthcoming). Mixed tenure neighbourhoods in London: policy myth or effective device to alleviate deprivation. *International Journal of Urban and Regional Research.*

Arjona, A. 2007. Ubicación espacial de los negocios étnicos en Almería. Formación de enclaves económicos étnicos? *Estudios geográficos,* 68(263), 391–415.

Audenino, P. and Tirabassi, M. 2008. *Migrazioni Italiane. Storia e Storie dall'Ancien re'gime a Oggi.* Milan: Bruno Mondadori.

Bagguley, P. and Hussain, Y. 2007. *The Role of Higher Education in Providing Opportunities for South Asian Women.* Bristol: Policy Press/Joseph Rowntree Foundation.

Bailey, A. 2009. Population geography: lifecourse matters. *Progress in Human Geography,* 33(3), 407–418.

Bailey, A. and Boyle, P. 2004. Untying and retying family migration in the New Europe. *Journal of Ethnic and Migration Studies,* 30(2), 229–241.

Bailey, N. 2012. How spatial segregation changes over time: sorting out the sorting processes. *Environment and Planning A,* in press.

Baldwin-Edwards, M. and Arango, J. 1999. *Immigrants and the Informal Economy in Southern Europe.* London: Frank Cass.

Ball, S.J., Reay, D. and David, M. 2002. 'Ethnic Choosing': minority ethnic students, social class and higher education choice. *Race, Ethnicity and Education,* 5(4), 333–357.

Barata Salgueiro, T. 1997. Lisboa, metrópole policêntrica e fragmentada. *Finisterra – Revista Portuguesa de Geografia,* 32(63), 179–190.

Barata Salgueiro, T. 2000. *Fragmentação e exclusão nas metrópoles. Sociedade e Território,* 30, 16–26.

Barringer, H., Gardner, R. and Levin, M.J. 1993. *Asians and Pacific Islanders in the United States.* New York: The Russell Sage Foundation.

Bartel, A.P. 1989. Where do the new U.S. immigrants live? *Journal of Labor Economics,* 7(4), 371–91.

Bartel, A.P. and Koch, M.J. 1991. Internal Migration of US Immigrants, in *Immigration, Trade, and the Labor Market,* edited by J.M. Abowd and R.B. Freeman. Chicago: The University of Chicago Press, 121–134.

Baumel, Y. 2002. The military government and its cancellation, 1948–1968. *HaMizrach HaChadash* [The New East], 43, 133–156. [In Hebrew].

Bayona, J. 2007. La segregación residencial de la población extranjera en Barcelona: una segregación fragmentada? *Scripta Nova,* 11(235). Available at: http://www.ub.es/geocrit/sn/sn-235.htm.

Bayona, J., Gil-Alonso, F. and Pujadas, I. 2010. *Dinámica Residencial de la Población Extranjera en las Grandes Ciudades Españolas: Suburbanización Entre Expansión Económica y Crisis (1999–2009).* Madrid: Grupo de Estudios sobre Desarrollo Urbano, Documento de Trabajo, 11.

Beckers, P. 2011. *Essays on Immigrant Socio-economic Integration for the Case of the Netherlands: Implications for Economic Outcomes in the Labour Market and in Self-employment.* Maastricht: Dissertation, University of Maastricht.

Beenstock, M. 1997. The internal migration of immigrants: Israel 1969–1972. *The Quarterly Review of Economics and Finance,* 37(S1), 263–284.

Beets, G. and Willekens, F. 2009. The global economic crisis and international migration: an uncertain outlook. *Vienna Yearbook of Population Research,* 2009(7), 19–37.

Bell, M., Blake, M., Boyle, P., Duke-Williams, O., Rees, P., Stillwell, J. and Hugo, G. 2002. Cross-national comparison of internal migration: issues and measures. *Journal of the Royal Statistical Society A,* 165(3), 435–464.

Bell, M. and Muhidin, S. 2009. *Cross-National Comparisons of Internal Migration.* Human Development Research Paper 2009/30, United Nations Development Programme Human Development Reports Research Paper, New York: Human Development Reports Office.

Bell, M. and Rees, P. 2006. Comparing migration in Britain and Australia: harmonisation through use of age-time plans. *Environment and Planning A*, 38(5), 959–988.

Beltrán, J., Oso, L. and Ribas, N. 2006. *Empresariado étnico en España*. Madrid: Ministerio de Trabajo y Asuntos Sociales, Observatorio Permanente de la Inmigración.

Bengtsson, T. 1991. Cohort size and propensity to migrate. Internal migration in Sweden between 1961 and 1988, by age and sex. *Cahiers Quebecois de Demographie*, 20(1), 51–68.

Bergström, L, Van Ham, M. and Manley, D. 2010. *Neighbourhood Reproduction through Neighbourhood Choice*. Paper to the ENHR Annual Conference Istanbul, 4–7 July 2010.

Bevilacqua, P., De Clementi, A. and Franzina, E. (eds) 2002. *Storia dell'Emigrazione Italiana*. Partenze/Roma: Donzelli.

Bielsa, J. and Duarte, R. 2010. Size and linkages of the Spanish construction industry: key sector or deformation of the economy? *Cambridge Journal of Economics,* 35(2), 317–334.

Bilsborrow, R.E. 1993. Issues in the measurement of female migration in developing countries, in *Internal Migration of Women in Developing Countries: Proceedings of the United Nations Expert Group Meeting on the Feminization of Internal Migration*. New York: United Nations, 116–130.

Boal, F.W. 1999. From undivided cities to undivided cities: assimilation to ethnic cleansing. *Housing Studies*, 14(5), 585–600.

Bolt, G., Özüekren, A.S. and Phillips, D. 2010. Linking integration and residential segregation. *Journal of Ethnic and Migration Studies*, 36(2), 169–86.

Bolt, G. and van Kempen, R. 2010. Ethnic segregation and residential mobility: Relocations of minority ethnic groups in the Netherlands. *Journal of Ethnic and Migration Studies*, 36(2), 333–54.

Bolt, G., van Kempen, R. and van Ham, M. 2008. Minority ethnic groups in the Dutch housing market. Spatial segregation, relocation dynamics and housing policy. *Urban Studies*, 45(7), 1359–1384.

Böltken, F., Gatzweiler, H-P., Meyer, K. 2002. Räumliche integration von Ausländern und Zuwanderern, in *Internationale Wanderungen und räumliche Integration, Vol.8*, edited by Bundesamt für Bauwesen und Raumordnung, 397–415.

Bonvalet, C., Carpenter, J. and White, P. 1995. The residential mobility of ethnic minorities: a longitudinal analysis. *Urban Studies*, 32(1), 87–103.

Borjas, G.J. 1989. Economic theory and international migration. *International Migration Review*, 23(3), 457–485.

Bowes, A., Dar, N. and Sim, D. 1997. Tenure preference and housing strategy : an exploration of Pakistani experiences. *Housing Studies*, 12(1), 63–84.

Boyle, P. 1995. Public housing as a barrier to long-distance migration. *International Journal of Population Geography*, 1(2), 147–164.

Boyle, P., Cooke, T.J., Halfacree, K. and Smith, D. 2001. A cross-national comparison of the impact of family migration on women's employment status. *Demography*, 38(2), 201–213.

Boyle, P., Cooke, T., Halfacree, K. and Smith, D. 2003. The effect of long-distance family migration and motherhood on partnered women's labour-market activity rates in Great Britain and the USA. *Environment and Planning A*, 35, 2097–2114.

Boyle, P., Halfacree, K., Robinson, V. 1998. *Exploring Contemporary Migration*. Harlow: Prentice Hall.

Bråmå, Å. 2006. *Studies in the Dynamics of Residential Segregation. Geografiska Regionstudier No 67*. Uppsala: Dissertation, Dept. of Social and Economic Geography, Uppsala University.

Bråmå, Å. 2006. 'White flight'? The production and reproduction of immigrant concentration areas in Swedish cities, 1990–2000. *Urban Studies*, 43(7), 1127–1146.

Bråmå, A. 2008. Dynamics of ethnic residential segregation in Göteborg, Sweden, 1995–2000. *Population, Space and Place*, 14(2), 101–117.

Bråmå, Å, and Andersson, R. 2010. Who leaves rental housing? Examining possible explanations for ethnic housing segmentation in Uppsala, Sweden. *Journal of Housing and the Built Environment* 25(3), 331–352.

Brooks, R. 2003. Young people's Higher Education choices: the role of family and friends. *British Journal of Sociology of Education*, 24(3), 283–297.

Burgess, E. 1928. Residential segregation in American cities, *Annals of the American Association of Political and Social Science*, 14, 105–115.

Bryant, I. (ed.) 1995. *Vision, Invention, Intervention: Celebrating Adult Education: Proceedings of the Standing Conference on University Teaching and Research in the Education of Adults (Great Britain)*. Southampton: Dept. of Adult Continuing Education, University of Southampton.

Cabré, A., Moreno, J. and Pujadas, I. 1985. Cambio migratorio y reconversión territorial en España. *Revista Española de Investigaciones Sociológicas*, 32, 43–65.

Cachón, L. 2003. La inmigración en España: los desafíos de la construcción de una nueva sociedad. *Migraciones*, 14, 219–304.

Caglayan, S. 2007. *Bulgaristan'dan Türkiye'ye Göçler*. Unpublished PhD thesis, Izmir, Turkey Ege University.

Cangiano, A. 2008. Foreign migrants in southern European countries: evaluation of recent data, in *International Migration in Europe: Data, Models and Estimates*, edited by J. Raymer and F. Willekens. Chichester: Wiley, 89–114.

Cantle, T. 2001. *Community Cohesion: A Report of the Independent Review Team*. London: Home Office.

Carella, M. and Pace, R. 2001. Some Migration Dynamics Specific to Southern Europe: South-North and East-West Axis. *International Migration*, 39(4), 63–99.

Carneiro, R. (Editor). 2006. *A Mobilidade Ocupacional do Trabalhador Migrante em Portugal*. Lisbon: Direcção-Geral de Estudos, Estatística e Planeamento, Ministério do Trabalho e Solidariedade Social.

Castles, S. and Miller, M.J. 2009. *The Age of Migration*. 4th Edition. Basingstoke: Palgrave.

Catney, G. 2008. *Internal migration, community background and residential segregation in Northern Ireland*. Unpublished PhD thesis, Belfast: Queen's University.

Catney, G., Finney, N. and Twigg, L. 2011. Diversity and the complexities of ethnic integration in the UK: Guest editors' introduction. *Journal of Intercultural Studies*, 32(2), 107–114.

Catney, G. and Simpson, L. 2010. Settlement area migration in England and Wales: assessing evidence for a social gradient. *Transactions of the Institute of British Geographers*, 35(4), 571–584.

Cavounidis, J. 2002. Migration in Southern Europe and the case of Greece. *International Migration*, 40(1), 45–70.

Cavounidis, J. 2004. Migration to Greece from the Balkans. *South Eastern Europe Journal of Economics*, 2, 35–59.

Central Bureau of Statistics. 1996. *National Census of Housing and Population, Public Use File*. Jerusalem: Central Bureau of Statistics.

Central Bureau of Statistics. 2008. *Press Release*. [Online]. Available at: http://www.cbs.gov.il/reader/newhodaot/hodaa_template.html?hodaa=200811198

Central Bureau of Statistics. 2010. *The Population of Israel (Demographic Characteristics): 1999–2009*. Statisti-lite series. Jerusalem: Central Bureau of Statistics.

Chant, S. 1992. (ed.) *Gender and Migration in Developing Countries*. London: Belhaven Press.

Champion, A.G. 1989. *Counterurbanization: The Changing Pace and Nature of Population Deconcentration*. New York: Routledge.

Champion, T. 1996. Internal migration and ethnicity in Britain, in *Social Geography and Ethnicity in Britain: Geographical Concentration and Internal Migration (Ethnicity in the 1991 Census), Volume 3*, edited by P. Ratcliffe. London: HMSO, 135–173.

Champion, T. 2001. Urbanization, suburbanization, counterurbanization and reurbanization, in *Handbook of Urban Studies*, edited by R. Paddison. London: Sage, 143–161.

Champion, T. 2005. Population movement within the UK, in *Focus on People and Migration*, edited by R. Chappell. Basingstoke: Palgrave, 92–114.

Charles, C.Z. 2003. The dynamics of racial residential segregation. *Annual Review of Sociology*, 29(1), 167–207.

Clapham, D. 2005. *The Meaning of Housing*. Bristol: Policy Press.

Clapham, D, and Kintrea K. 1984. Allocation systems and housing choice. *Urban Studies,* 21(3), 261–269.

Clark, W.A.V. 1992. Residential preferences and residential choices in a multiethnic context. *Demography,* 29(3), 451–466.

Clark, W.A.V. 1998. Mass migration and local outcomes: is international migration to the United States creating a new urban underclass? *Urban Studies*, 35(3), 371–382.

Clark, W.A.V., Deurloo, M.C. and Dieleman F.M. 1986. Residential mobility in Dutch housing markets. *Environment and Planning A,* 18(6), 763–788.

Clark, W.A.V., Deurloo, M.C. and Dieleman, F.M. 1994. Tenure changes in the context of micro-level family and macro-level economic shifts, *Urban Studies,* 31(1), 137–154.

Clark, W.A.V., Deurloo, M.C. and Dieleman, F.M. 2003. Housing careers in the United States, 1968–93: modelling the sequencing of housing states. *Urban Studies,* 40(1), 143–160.

Clark, W.A.V., Deurloo, M.C. and Dieleman, F.M. 2006. Residential mobility and neighbourhood outcomes. *Housing Studies*, 21(3), 323–342.

Clark, W.A.V. and Dieleman, F.M. 1996. *Households and Housing. Choice and Outcomes in the Housing Market.* New Brunswick: Centre for Urban Policy Research, State University of New Jersey.

Clark, W. and Drever, A. 2000. Residential mobility in a constrained housing market: implications for ethnic populations in Germany. *Environment and Planning A*, 32, 833–846.

Cohen, J.H., Rodriguez, L. and Fox, M. 2008. Gender and migration in the Central Valleys of Oaxaca. *International Migration*, 46(1), 79–101.

Cohen, J.H. and Sirkeci, I. 2011. *Cultures of Migration, the Global Nature of Contemporary Mobility*. Austin, TX, US: University of Texas Press.

Cohen-Eliya, M. 2003. Discrimination against Arabs in Israel in public accommodations. *NYU Journal of International Law and Politics*, 36(4), 717–748.

Coleman, D. 1994. Trends in fertility and intermarriage among immigrant populations in Western Europe as measures of integration. *Journal of Biosocial Science*, 26(1), 107–136.

Coleman, D. 2004. Partner choice and the growth of ethnic minority populations, *Bevolking en Gezin*, 33, 7–34.

Communities and Local Government. 2008. *Managing the Impacts of Migration: A Cross-Government Approach.* London: Department of Communities and Local Government.

Connor, H. Tyers, C., Modood, T. and Hillage, J. 2004. *Why the Difference? A closer look at higher Education Minority Students and Graduates, Research Report No 552,* Brighton: Institute for Employment Studies.

Conway, D. 1980. Step-Wise migration: toward a clarification of the mechanism. *International Migration Review*, 14(1), 3–14.

Corgeau, D. 1973a. Migrants et migrations. *Population*, 28(1), 95–129. Published in English in 1979 in *Population, Selected Papers*, 3, 1–35.

Courgeau, D. 1973b. Migrations et découpages du territoire. *Population*, 28(3), 511–538.

Corkill, D. 2001. Economic migrants and the labour market in Spain and Portugal. *Ethnic and Racial Studies*, 24(5), 828–844.

Crul, M. and Schneider, J. 2010. Comparative integration context theory: participation and belonging in new diverse European cities. *Ethnic and Racial Studies*, 33(7), 1249–1268.

DaVanzo, J. 1981. Repeat migration, information costs, and location-specific capital. *Population and Environment*, 4(1), 45–73.

De Lannoy, W., Lammens, M., Lesthaeghe, R. and Willaert, D. 1999. Brussel in de jaren negentig en na 2000. Een demografische doorlichting, in *Het statuut van Brussel - Bruxelles et son statut,* edited by E. Witte, A. Alen, H. Dumont and R. Ergec. Brussels: Deboeck & Larcier, 101–154.

De Miguel, V., Solana, M. and Pascual, À. 2007. *Las Redes de Apoyo: el Tejido Social Básico Para la Acomodación de los Extranjeros*. Madrid: Fundación BBVA.

Dedeoglu, S. 2011. Survival of the excluded: Azerbaijani immigrant women's survival strategies and industrial work in Istanbul. *Migration Letters*, 8(1), 26–33.

Dennett, A. and Stillwel, J. 2010a. Internal Migration Patterns by Age and Sex at the Start of the 21st Century, in *Technologies for Migration and Commuting Analysis: Spatial Interaction Data Applications* edited by J. Stillwell, O. Duke-Williams and A. Dennett . Hershey: IGI Global, 153–174.

Dennett, A. and Stillwell, J. 2010b. Internal migration in Britain, 2000–01, examined through an area classification framework. *Population, Space and Place*, 16(6), 517–538.

Denton, N.A. and Massey, D.S. 1988. Residential segregation of Blacks, Hispanics, and Asians by socioeconomic status and generation. *Social Science Quarterly*, 69(4), 797–817.

DESA-DAW (Division of Economic and Social Affairs, Division for the Advancement of Women) 2009. *2009 World Survey on the Role of Women in Development Women's Control over Economic Resources and Access to Financial Resources, including Microfinance*. New York: United Nations.

Deshingkar, P. 2006. Internal migration, poverty and development in Asia: including the excluded. *IDS Bulletin*, 37: 88–100.

Deurloo, R. and Musterd, S. 2001. Residential profiles of Surinamese and Moroccans in Amsterdam. *Urban Studies,* 38(3), 467–485.

De Valk, H. 2007. Living arrangements of migrant and Dutch young adults: The family influence disentangled. *Population Studies,* 61(2), 201–217.

De Valk, H.A.G., Huisman, C. and Noam, K.R. 2011. *Migration Patterns and Immigrant Characteristics in North-Western Europe.* NIDI: The Hague (Report for CELADE/UN regional office).

De Valk, H., Liefbroer, A.C., Esveldt, I., and Henkens, K. 2004. Family formation and cultural integration among migrants in the Netherlands. *Genus*, 55, 9–36.

Diaz-Más, P. 1992. *Sephardim: The Jews from Spain*. Chicago: University of Chicago Press.

Dinçer, B., Özaslan, M. ve Kavasoglu, T. 2003. *Illerin ve Bölgelerin Sosyo-Ekonomik Gelismislik. Siralamasi*. Ankara: Devlet Planlama Teskilati.

DO. 2008. *En Rapport från DO:s Särskilda Arbete under åren 2006–2008 Kring Diskriminering på Bostadsmarknaden. DO:s rapportserie 2008:3*. Stockholm: The Equality Ombudsman.

Dobson, J., Latham, A. and Salt, J. 2009. On the move? Labour migration in times of recession. What can we learn from the past, Policy Network Paper. Available at http://www.policy-network.net/publications/publications.aspx?id=2688.

Doff, W. and Kleinhans R. 2011. Residential outcomes of forced relocation: lifting a corner of the veil on neighbourhood selection. *Urban Studies*, 48(4), 661–680.

Domingo, A. 2002. Reflexiones demográficas sobre la inmigración internacional en los países del sur de la unión europea, in *La Inmigración en España. Contextos y Alternativas. III Congreso de la Inmigración en España. Vol. II: Ponencias*, edited by F.J. García and C. Muriel. Granada: Laboratorio de Estudios Interculturales, Universidad de Granada, 197–212.

Domingo, A. 2006. Tras la retórica de la hispanidad: la migración latinoamericana en España entre la complementariedad y la exclusión, in *Panorama actual de las migraciones en América Latina*, edited by A. Canales. Guadalajara (México): Asociación Latinoamericana de Población, Universidad de Guadalajara, 21–44.

Domingo, A. and Gil-Alonso, F. 2007. Immigration and changing labour force structure in the Southern European Union. *Population (English edition)*, 62(4), 709–727.

Domingo, A. and Recaño, J. 2010. La inflexión del ciclo migratorio internacional en España: impacto y consecuencias demográficas, in *La Inmigración en Tiempos de Crisis. Anuario de la Inmigración en España*, edited by E. Aja, J. Arango and J. Oliver. Barcelona: Fundació CIDOB, 182–207.

Domingo, A. and Sabater, A. 2009. *Impacte de la crisi econòmica en la immigració internacional a Catalunya l'any 2008*. Barcelona: Fundació Jaume Bofill.

Domingo, A. and Sabater, A. 2011. *Segregació i enclavaments subsaharians a la cruïlla dels discursos institucionals a Catalunya*. Barcelona: Direcció General per a la Immigració, Generalitat de Catalunya.

Dorling, D. and Thomas, B. 2004. *People and Places: A 2001 Census Atlas of the UK*. Bristol: Policy Press.

Drever, A. 2004. Separate spaces, separate outcomes? Neighbourhood impacts on minorities in Germany. *Urban Studies*, 41, 1423–1439.

Drever, A. and Clark, W. 2002. Gaining access to housing in Germany: the foreign minority experience. *Urban Studies*, 39, 2439–2454.

Duke-Williams, O. 2010. Interaction data: confidentiality and disclosure, in *Technologies for Migration and Commuting Analysis: Spatial Interaction Data Applications*, edited by J. Stillwell et al. Hershey: IGI Global, 51–68.

Duncan, O.B. and Duncan, B. 1955. A methodological analysis of segregation indexes, *American Sociological Review,* 20, 210–217.

Dustman, C., Fabbri, F. and Preston, I. 2005. *The Impact of Immigration on the UK Labour Market, CReAM Discussion Paper Series CDP No 01/05*. London: Centre for Research and Analysis of Migration, UCL.

Ekberg, J. 2000. Invandring och befolkningsutveckling. *HemortSverige,* Norrköping: Board of Integration, 296–302.

EKKE-ESYE. 2005. *Census data panorama, 1991–2001. Database and Mapping Application in the Institute of Urban and Rural Sociology*. Athens: National Centre for Social Research.

Ellis, M. and Wright, R. 1998. The balkanization metaphor in the analysis of US immigration, *Annals of the Association of American Geographers*, 88(4), 686–698.

Ellis, M., and Wright, R. 2005. Assimilation and differences between the settlement patterns of individual immigrants and immigrant households. *Proceedings of the National Academy of Sciences of the United States of America*, 102(43), 15325–15330.

Ellis, M. and Goodwin-White, J. 2006. 1.5 Generation Internal Migration in the U.S.: Dispersion from States of Immigration? *International Migration Review*, 40(4), 899–926.

Equality Challenge Unit. 2009. *Statistical Report 2009* [Online] Available at: http://www.ecu.ac.uk/publications/equality-in-he-stats-09.

Faggian, A., McCann, P. and Sheppard, S. 2006. An analysis of ethnic differences in UK graduate migration behaviour. *The Annals of Regional Science*, 40(2), 461–471.

Farwick, A. 2009. Internal Migration. Challenges and Perspectives for the Research Infrastructure. *RatSWD Working Papers Series*, No.97 (July, 2009).

Feijten, P. and Mulder, C.H. 2002. The timing of household events and housing events in the Netherlands: a longitudinal perspective, *Housing Studies,* 17(5), 773–792.

Feijten, P. and Van Ham, M. 2009. Neighbourhood change... Reason to leave? *Urban Studies,* 46(10), 2103–2122.

Feng, G., Boyle, P., van Ham, M. and Raab, G. 2010. Neighbourhood ethnic mix and the formation of mixed-ethnic unions in Britain, in *Ethnicity and Integration Understanding Population Trends and Processes Volume 3*, edited by J. Stillwell and M. van Ham. Dordrecht: Springer, 83–104.

Fernández, P.P. Arlinda Garcia Coll and Ángeles Asensio Hita. 2006. *La movilidad laboral y geográfica de la población extrangera en España*. Madrid: Ministério de Trabajo y Asuntos Sociales.

Findlay, A., Mason, C., Houston, D., Harrison, R., and McCollum, D. 2008. Getting off the escalator? A study of Scots out-migration from a global city. *Environment and Planning A*, 40(9), 2169–85.

Findlay, A., Short, D. and Stockdale, A. 2000. The labour market impacts of migration in and to rural Scotland. *Applied geography*, 20(4), 333–348.

Finney, N. and Simpson, L. 2008. Internal migration and ethnic groups: evidence for Britain from the 2001 Census. *Population, Space and Place*, 14(1), 63–83.

Finney, N., and Simpson, L. 2009a. *'Sleepwalking to segregation'? Challenging Myths about Race and Migration.* Bristol: The Policy Press.

Finney, N., and Simpson, L. 2009b. Population dynamics: the roles of natural change and migration in producing the ethnic mosaic, *Journal of Ethnic and Migration Studies*, 35(9), 1479–1496.

Finney, N. 2010. Ethnic group population change and integration: a demographic perspective on ethnic geographies, In *Ethnicity and Integration Understanding Population Trends and Processes Volume 3,* edited by J. Stillwell and M. van Ham. Dordrecht: Springer, 27–45.

Finney, N. 2011a. Understanding ethnic differences in migration of young adults within Britain from a lifecourse perspective. *Transactions of the Institute of British Geographers,* 36(3), 455–470.

Finney, N. 2011b. Educational constraints of immobility? Examining ethnic differences in student migration in Britain using Census microdata. *Documents d'Anàlisi Geogràfica,* 57(3).

Fischer, P.A., Holm, E., Malmberg, G. and Straubhaar, T. 2000. *Why do People Stay? Insider Advantages and Immobility. HWWA Discussion paper 112.* Hamburg: Hamburg Institute of International Economics.

Fonseca, M.L. 1990. *População e Território: do país à Área Metropolitana de Lisboa.* Lisbon: Centro de Estudos Geográficos.

Frey, W.H. 1995. Immigration and internal migration 'flight' from US metropolitan areas: toward a new demographic balkanization, *Urban Studies,* 32(4–5), 733–757.

Frey, W.H. 1996. Immigration, domestic migration, and demographic balkanization in America: new evidence for the 1990s, *Population and Development Review*, 2(4), 741–763.

Frey, W.H. 2003. *Metropolitan Magnets for International and Domestic Migrants.* Brookings Census 2000. Washington DC: Brookings Institution Centre on Urban and Metropolitan Policy.

Frey, W.H. and Liaw, K-L. 1998. The impact of recent immigration on population redistribution within the United States, in *The Immigration Debate Studies on the Economic, Demographic and Fiscal Effects of Immigration,* edited by J.P. Smith and B. Edmonston. Washington DC: National Academy Press, 388–448.

Frey, W.H. and K.L. Liaw 2005. *Interstate Migration of Hispanics, Asians and Blacks: Cultural Constraints and Middle Class Flight. Research Report 05-575.* Population Studies Center: University of Michigan Institute for Social Research, Ann Arbour. [Online]. Available at: http://citeseerx.ist.psu.edu/viewdoc/download?doi=10.1.1.128.8096&andrep=rep1&andtype=pdf.

Friedrichs, J. 1998. Ethnic segregation in Cologne, Germany, 1984–1994. *Urban Studies*, 35, 1745–1764.

Fussell, E., Gauthier, A.H. and Evans, A. 2007. Heterogeneity in the transition to adulthood: the cases of Australia, Canada, and the United States. *European Journal of Population,* 23, 239–414.

Galle, O.R. and Williams, M.W. 1972. Metropolitan migration efficiency. *Demography,* 9, 655–664.

Galster, G. 1988. Residential segregation in American cities: a contrary view. *Population Research and Policy Review,* 7(2), 93–112.

Galster, G. 2007. Should policymakers strive for neighborhood social mix? An analysis of the Western European evidence base. *Housing Studies,* 4, 523–546.

Gans, P. 1987. Intraurban migration of foreigners in Kiel since 1972. In *Foreign Minorities in Continental European Cities,* edited by G. Glebe and J. Loughlin. Wiesbaden: Steiner-Verlag, 116–138.

Gans, P. 1990. Changes in the structure of foreign population of West Germany since 1980. *Migration,* 7, 25–49.

Gans, H.J. 1997. Toward a reconciliation of 'assimilation' and 'pluralism': the interplay of acculturation and ethnic retention. *International Migration Review,* 31(4), 875–892.

García-Coll, A. and Stillwell, J. 1999. Inter-provincial migration in Spain: temporal trends and age-specific patterns. *International Journal of Population Geography,* 5(2), 97–115.

García Montalvo, J. 2007. Algunas consideraciones sobre el problema de la vivienda en España. *Papeles de Economía Española,* 113, 138–153.

Gastner, M.T. and Newman, M.E. 2004. Diffusion-based method for producing density-equalizing maps. *Proceedings of the National Academy of Sciences of the United States of America,* 101(20), 7499–7504.

Gauthier, A.H. 2007. Becoming a young adult: an international perspective on the transition to adulthood. *European Journal of Population,* 23, 217–223.

Gedik, A. 1997. Internal migration in Turkey, 1965–1985: Test of conflicting findings in the literature. *Review of Urban & Regional Development Studies,* 9(2), 170–179.

Ghanem, A. 2001. *The Arab-Arab Minority in Israel, 1948–2000* (SUNY Series in Israeli Studies). Albany: State University of New Press.

Glebe, G. 1997. Housing and segregation of Turks in Germany, in *Turks in European Cities: Housing and Urban Segregation,* edited by A.S. Özüekren, and R. van Kempen Utrecht: European Research Centre on Migration and Ethnic Relations, 122–157.

Goldschneider, C. and Goldschneider, F.K. 1988. Ethnicity, religiosity and leaving home: the structural and cultural bases of traditional family values. *Sociological Forum,* 3(4), 525–547.

Goldschneider, F.K. and Goldschneider, C. 1997. The historical trajectory of the Black family: ethnic differences in leaving home over the Twentieth Century. *The History of the Family,* 2(3), 295–307.

Goldstein, S. 1958. *Patterns of Mobility, 1910–1950.* Philadelphia: Pennsylvania Press.

Goldstein, H. 2011. *Multilevel Statistical Models*. 4th Edition. Chichester: Wiley.

Gonen, A. 1998. Settlement of Immigrants: Geographic Patterns. In *Profile of an Immigration Wave: The Absorption Process of Immigrants from the Former Soviet Union, 1990–1995*, edited by M. Sicron and Leshem, E. Jerusalem: Magness Press, 232–269. [In Hebrew].

Greenwood, M. 1985 Human migration: theory, models and empirical studies. *Journal of Regional Science*, 25(4), 521–544.

Grüner, S. 2010. The others don't want… small scale segregation: hegemonic public discourses and racial boundaries in German neighbourhoods. *Journal of Ethnic and Migration Studies*, 36(2), 275–292.

Gubhaju, B. and De Jong, G.F. 2009. Individual versus household migration decision rules: gender and marital status differences in intentions to migrate in South Africa. *International Migration*, 47(1), 31–61.

Guest, P. 1999. *The dynamics of internal migration in Vietnam*. UN Development Program Discussion Paper. Hanoi: UNDP Country Office.

Gurak, D.T. and Kritz, M.M. 2000. The interstate migration of U.S. immigrants: individual and contextual determinants. *Social Forces*, 78, 1017–1039.

Hall, M. 2009. Interstate migration, spatial assimilation, and the incorporation of US immigrants. *Population, Space and Place*, 15, 57–77.

Hammar, T. 1993. The 'Sweden-wide strategy' of refugee dispersal. In *Geography and Refugees*, edited by R. Black and V. Robinson. London: Belhaven, 105–117.

Hamnett C. 1991. The relationship between residential migration and housing tenure in London, 1971–1981: a longitudinal analysis. *Environment and Planning A*, 23, 1147–1162.

Hamnett, C. and Butler, T. 2010. The changing ethnic structure of housing tenures in London, 1991–2001. *Urban Studies*, 47(1), 55–74.

Hannerberg, D., Hägerstrand, T. and Odeving, B. (eds). 1957. *Migration in Sweden, Lund Studies in Geography, Ser. B., Human Geography*, No. 13, Lund: Royal University of Lund.

Hanushek, E. and Quiqley, J. 1978. Housing disequilibrium and residential mobility, in *Population Mobility and Residential Change*, edited by W.A.V. Clark and E.G. Moore. Evanston, Ill: Northwestern University.

Hatziprokopiou, P. 2003. Albanian immigrants in Thessaloniki, Greece. Processes of economic and social incorporation. *Journal of Ethnic and Migration Studies*, 29(6), 1033–1057.

Heath, A.F., Rothon C. and Kilpi E. 2008. The second generation in Western Europe: education, unemployment, and occupational attainment. *Annual Review of Sociology*, 34(1), 211–235.

HEFCE. 2009. *Strategic Plan 2006–2011*. [Online]. Available at: http://www.hefce.ac.uk/pubs/.

HEFCE. 2010. *Student Ethnicity. Profile and Progression of Entrants to Full-time, First Degree Study, Issues paper 2010/13*. [Online]. Available at: http://www.hefce.ac.uk/pubs/

Hempstead, K. 2007. Mobility of the foreign-born population in the United States, 1995–2000: the role of gateway states. *International Migration Review,* 41(2), 466–479.

Herod, A. 2011. *Scale.* London: Routledge.

Holdsworth, C. 2006. 'Don't you think you're missing out, living at home?' student experiences and residential transitions. *The Sociological Review,* 54(3), 485–519.

Holdsworth, C. 2009. 'Going away to uni': mobility, modernity, and independence of English higher education students. *Environment and Planning A,* 41,1849–1864.

Home Office. 2006. *A Points Based System: Making Migration Work for Britain.* CM 6741, Norwich: HMSO.

Houseman, G. 1981. Access of minorities to the suburbs. *The Urban Social Change Review,* 14(1), 11–20.

Houston, D. 2010. Changing housing segregation and ethnic disadvantage in Dundee. *Scottish Geographical Journal,* 126(4), 285–298.

Hugo, G.J. 1993. Migrant women in developing countries, in *Internal Migration of Women in Developing Countries: Proceedings of the United Nations Expert Group Meeting on the Feminization of Internal Migration.* New York: United Nations, 47–73.

Iceland, N. and Nelson, K.A. 2008. Hispanic segregation in metropolitan America: exploring the multiple forms of spatial assimilation. *American Sociological Review,* 73(5), 741–765.

ILO 1969. *International Standard Classification of Occupations, Revised Edition 1968.* Geneva: International Labour Office.

ILO. 2009. *Voluntary Return Programme for non-EU migrant workers in Spain.* Available at http://www.ilo.org/dyn/migpractice/migmain.showPractice?p_lang=en&p_practice_id=29.

INE. 2010. *Encuesta de Población Activa.* Madrid: Instituto Nacional de Estadística.

Instituto de Habitação e Reabilitação Urbana (IHRU). 2008. *Contributos para o Plano Estratégico de Habitação, 2008–2013, Sumário Executivo para Debate Público – Abril de 2008.* Lisbon: Instituto de Habitação e Reabilitação Urbana.

Iosifides, T. and King, R. 1996. Recent immigration to Southern Europe: the socio-economic and labour market contexts. *Journal of Area Studies,* 4(9), 70–94.

Ireland, P. 2008. Comparing responses to ethnic segregation in urban Europe. *Urban Studies,* 45(7), 1333–1358.

Izquierdo, A., López de Lera, D. and Martínez Buján, R. 2003. The favorites of the twenty-first century: Latin American immigration in Spain. *International Journal of Migration Studies,* 149, 98–125.

Jakobsson, A. 1969. *Omflyttningar I Sverige 1950–1960. Komparativa studier av migrationsfält, flyttningsavstånd och mobilitet.* Monografiserie 5 i anslutning till Folk- och Bostadsräkningen 1965. Stockholm: Statistics Sweden.

Jamal, A. 2009. The contradictions of state-minority relations in Israel: the search for clarifications. *Constellations*, 16(3), 493–508.

Jargowsky, P. 2009. Immigrants and neighbourhoods of concentrated poverty: Assimilation or stagnation? *Journal of Ethnic and Migration Studies*, 35(7), 1129–51.

Johnston, R., Poulsen, M.F., Forrest, J. 2005. On the measurement and meaning of segregation: a response to Simpson. *Urban Studies*, 42(7), 1221–1227.

Johnston, R., Poulsen, M. and Forrest, J. 2010. Moving on from indices, refocusing on mix: on measuring and understanding ethnic patterns of residential segregation, *Journal of Ethnic and Migration Studies,* 36(4), 697–706.

Jolly, S. and Reeves, H. 2005. *Gender and Migration.* Brighton: Institute of Development Studies.

Jones, H. and Davenport, M. 1972. The Pakistani community in Dundee. *Scottish Geographical Magazine*, 88(2), 75–85.

Kalra, V.S. and Kapoor, N. 2009. Interrogating segregation, integration and the community cohesion agenda. *Journal of Ethnic and Migration Studies*, 35(9), 1397–1415.

Kandylis, G. and Maloutas, T. 2012. Here for good: immigrants' residential mobility and social integration in Athens during the late 1990s, in *Minority Internal Migration in Europe,* edited by Nissa Finney and Gemma Catney. Aldershot: Ashgate.

Kandylis, G. Maloutas, T. and Sayas, J. 2012. Inequality and diversity: Socio-ethnic hierarchy and spatial organization in Athens. *European Urban and Regional Studies*, 19(3), 267–286.

Karpat, K.H. 1985. The Ottoman Emigration to America, 1860–1914. *International Journal of Middle East Studies*, 17(2), 175–209.

Karpat, K.H. 1995. The Turks of Bulgaria: The struggle for national-religious survival of a Muslim minority. *Nationalities Papers*, 23(4), 725–749.

Karsten, L. 2011. Children's social capital in the segregated context of Amsterdam: an historical-geographical approach. *Urban Studies*, 48(8), 1651–1666.

Kearney, R.N. and Miller, B.D. 1984. Sex-differences in patterns of internal migration in Sri Lanka. Syracuse University, *Working Paper* 44(July).

Kelly, E. 2000. Asylum seekers in Scotland: Challenging racism at the heart of government. *Scottish Affairs*, 33, Autumn.

Kemper, F-J. 2008. Residential mobility in East and West Germany: mobility rates, mobility reasons, reurbanization. *Zeitschrift für Bevölkerungswissenshaft*, 33(3–4), 293–314.

King, R. 2000. Southern Europe in the changing global map of migration, in *Eldorado or Fortress? Migration in Southern Europe*, edited by R. King, G. Lazaridis and C. Tsardanidis. London: Macmillan, 1–26.

King, R. with Black, R., Collyer, M., Fielding, A. and Skeldon, R. 2010. *The Atlas of Human Migration Global Patterns of People on the Move.* London: Earthscan.

King, R., Fielding, A. and Black, R. 1997. The international migration turnaround in Southern Europe, in *Southern Europe and the New Immigrants*, edited by R. King and R. Black. Portland: Sussex Academic Press, 1–25.

King, K. and Skeldon, R. 2010. 'Mind the Gap' Integrating approaches to internal and international migration. *Journal of Ethnic and Migration Studies*, 36(10), 1619–1646.

King, R., Skeldon, R. and Vullnetari, J. 2008. *Internal and International Migration: Bridging the Theoretical Divide*. Working Paper No 52, Sussex: University of Sussex, Sussex Centre for Migration Research.

King, R. and Zontini, E. 2000. The role of gender in the South European immigration model. Papers. *Revista de Sociologia*, 60, 35–52.

Kirchner, J. 2007. The declining social rental sector in Germany. *European Journal of Housing Policy*, 7, 85–101.

Kirisci, K. 2000. Disaggregating Turkish citizenship and immigration practices. *Middle Eastern Studies,* 36(3), 1–22.

Kirisci, K. 2007. Turkey: a Country of Transition from Emigration to Immigration. *Mediterranean Politics*, 12(1), 91–97.

Koehler, J., Laczko, F., Aghazarm, C. and Schad, J. 2010. *Migration and the Economic Crisis in the European Union: Implications for Policy*. Brussels: International Organization for Migration (IOM).

Krahn, H. and Derwing, T.M. 2005. The retention of newcomers in second and third-tier Canadian cities. *International Migration Review*, 39(4), 872–894.

Kritz, M.M., Gurak, D.T. 2001. The impact of immigration on the internal migration of natives and immigrants. *Demography*, 38(1), 133–145.

Kritz, M.M. and Nogle, J.M. 1994. Nativity concentration and internal migration among the foreign-born. *Demography*, 31(3), 509–524.

Kulu, H. and F. Billari 2004. Multilevel analysis of internal migration in a transitional country: the case of Estonia. *Regional Studies*, 38(6), 679–696.

Kulu, H. Boyle, P.J. and Andersson, G. 2009. High suburban fertility: evidence from four Northern European countries. *Demographic Research*, 21(31), 915–944.

Labrianidis, L. and Lyberaki, A. 2001. *Albanian Immigrants in Thessaloniki*. Thessaloniki: Paratiritis. [In Greek].

Laurence, J. and Heath, A. 2008. *Predicting Community Cohesion: Multilevel Modeling of the 2005 Citizenship Survey*. London: Department for Communities and Local Government.

Lavrentiadou, M. 2009. Immigrants geographic distribution and segregation in Greece, in *Issues of Immigrants Social Integration*, edited by A. Kontis. Athens: Papapazisis, 315–11. [In Greek].

Lawton, R. 1959. Irish immigration to England and Wales in the mid-nineteenth century. *Irish Geography*, 4(1), 35–54.

Lazaridis, G. 1996. Immigration to Greece. A critical evaluation of Greek policy. *Journal of Ethnic and Migration Studies*, 22(2), 335–48.

Lazaridis, G. and Psimmenos, I. 2001. Migrant flows from Albania to Greece: economic, social and spatial exclusion, in *The Mediterranean Passage. Migration and New Cultural Encounters in Southern Europe,* edited by R. King. Liverpool: Liverpool University Press.

Leon, D.A. and Strachan, D.P. 1993. Socioeconomic characteristics of internal migrants in England and Wales, 1939–71. *Environmental and Planning A,* 25(10), 1441–1451.

Lesthaeghe, R. 2000. Transnational islamic communities in a multilingual secular society, in *Communities and Generations: Turkish and Moroccan Populations in Belgium,* edited by R. Lesthaeghe. Brussels: VUB University Press, 1–58.

Li, S.M. and Li, L. 2006. Life course and housing tenure change in urban China: a study of Guang-zhou. *Housing Studies,* 21(5), 653–670.

Liaw, K.L. and Frey, W.H. 1998. Destination choices of the 1985–90 young adult immigrants to the United States: importance of race, educational attainment, and labour market forces. *International Journal of Population Geography,* 4(1), 49–61.

Lieberson, S. 1963. *Ethnic Patterns in American Cities.* New York: The Free Press of Glencoe.

Lievens, J. 2000. The third wave of immigration from Turkey and Morocco. Determinants and characteristics, in *Communities and Generations: Turkish and Moroccan Populations in Belgium,* edited by R. Lesthaeghe. Brussels: VUB University Press, 95–128.

Lipshitz, G. 1991. Ethnic differences in migration patterns – disparities among Arabs and Jews in the peripheral regions of Israel. *The Professional Geographer,* 43(4), 445–456.

Lloyd, C.D. 2011. *Local Models for Spatial Analysis.* 2nd Edition. Boca Raton: CRC Press.

Long, J.F. and Boertlein, C.G. 1990. *Comparing migration measures having different intervals.* Current Population Reports, Series P-23, 1–11, Washington DC: US Bureau of the Census.

Lora-Tamayo, G. 2001. *Extranjeros en Madrid capital y en la Comunidad. Informe 2000.* Madrid, Cáritas Dicesana de Migraciones-ASTI.

Lundholm, E. 2007. Are movers still the same? Characteristics of interregional migrants in Sweden 1970–2001. *Tijdschriftvooreconomische en Sociale Geografie,* 98(3), 336–348.

Lustick, I. 1980. *Arabs in the Jewish State: Israel's Control of a National Minority.* Austin: University of Texas Press.

Luyten, S. and Van Hecke, E. 2007. *De Belgische Stadsgewesten 2001.* Statistics Belgium Working Paper 14, Brussels: Statistics Belgium.

MacDonald, J. and MacDonald, L.D. 1964. Chain migration, ethnic neighborhood formation and social networks, *The Milbank Memorial Fund Quarterly,* 42(1), 82–97.

Malheiros, J. 2002. Ethni-cities: residential patterns in Northern European and Mediterranean Metropolises – implications for policy design. *International Journal of Population Geography*, 8(2), 89–106.

Malheiros, J. and Baganha, M.I. 2001. Imigração ilegal em Portugal: padrões emergentes em inícios do séc. XXI. *Janus 2001 – Anuário de Relações Exteriores*, 190–191.

Malheiros, J. and Fonseca, L. (2011), *Acesso à Habitação e Problemas Residenciais dos Imigrantes em Portugal*. Lisbon: ACIDI.

Malheiros, J. and Vala, F. 2004. Immigration and city change: the Lisbon Metropolis at the turn of the twentieth century. *Journal of Ethnic and Migration Studies*, 30(6), 1067–1085.

Maloutas, T. 1990. *Athens, Housing, Family*. Athens: Eksadas. [In Greek].

Maloutas, T. 2000. *Social and Economic Atlas of Greece: The Cities*. Athens-Volos: University of Thessally-NCSR. [In Greek].

Maloutas, T. 2004. Segregation and residential mobility. Socially entrapped social mobility and its impact on segregation in Athens. *European Urban and Regional Studies*, 11(2), 171–87.

Maloutas, T. 2007. Segregation, social polarization and immigration in Athens during the 1990s. Theoretical expectations and contextual difference. *International Journal of Urban and Regional Research*, 31(4), 733–758.

Manley, D. and Catney, G. 2012. 'One Scotland'? Internal Ethnic Minority Migration in a devolved state, in *Minority Internal Migration in Europe*, edited by Nissa Finney and Gemma Catney. Aldershot: Ashgate.

Marcuse, P. and Van Kempen, R. (eds). 2000. *Globalizing Cities: A New Spatial Order?* Oxford: Blackwell.

Martínez del Olmo, A. and Leal Maldonado, J. 2008. La segregación residencial, un indicador espacial confuso en la representación de la problemática de los inmigrantes económicos; El caso de la Comunidad de Madrid. *Architecture, City and Environment*, 8, 53–64.

Martínez Veiga, U. 1999. *Pobreza, Segregación y Exclusión Social*. Barcelona: Icaria, Institut Catala de Antropología.

Martori, J.C. and Hoberg, K. 2004. Indicadores cuantitativos de la segregación residencial. El caso de la población inmigrante en la ciudad de Barcelona. *Scripta Nova*, 8(169). Available at http://www.ub.edu/geocrit/sn/sn-169.htm.

Martori, J.C., Hoberg, K. and Suriñach, J. 2006. Población inmigrante y espacio urbano. Indicadores de segregación y pautas de localización. *Revista Eure*, 32(97), 49–62.

Marvakis, A. 2004. Social integration or social apartheid?, in *Greece of Immigration. Social Participation, Rights and Citizenship*, edited by M. Pavlou and D. Christopoulos. Athens: Kritiki, 88–120. [In Greek].

Mason, J. and Dale, A. 2011. *Understanding Social Research. Thinking Creatively About Method*. London: Sage.

Massey, D. 1985. Ethnic residential segregation: a theoretical and empirical synthesis. *Sociology and Social Research*, 69, 315–350.

Massey, D. 1986. The settlement process among Mexican migrants to the United States. *American Sociological Review,* 51(5), 670–8.

Massey, D. 1994. *Space, Place and Gender.* Minneapolis: University of Minnesota Press.

Massey, D. 2008. *New Faces in New Places: The Changing Geography of American Immigration.* New York: Russell Sage Foundation.

Massey, D.S., Arango, J., Hugo, G., Kouaouci, A., Pellegrino, A. and Taylor, J.E. 1993. Theories of international migration: a review and appraisal. *Population and Development Review,* 19(3), 431–466.

Massey D.S. and Denton, N.A. 1987. Trends in the residential segregation of Blacks, Hispanics, and Asians: 1970–1980. *American Sociological Review,* 52(6), 802–825.

Massey, D. and Denton, N. 1988. The dimensions of residential segregation. *Social Forces,* 67, 281–315.

Masters, B.A. 2004. *Christians and Jews in the Ottoman Arab World: The Roots of Sectarianism.* Cambridge: Cambridge University Press.

McCarthy, J. 1996. *Death and Exile: The Ethnic Cleansing of Ottoman Muslims, 1821–1922.* Princeton, NJ: Darwin Press.

McGarrigle, J. 2010. *Understanding Processes of Ethnic Concentration and Dispersal. South Asian Residential Preferences in Glasgow.* Amsterdam: IMISCOE/Amsterdam University Press Dissertation Series.

Meir, A. 1997. *As Nomadism Ends: The Israeli Bedouin of the Negev.* New York: Westview Press.

Mendez, P. 2009. Immigrant residential geographies and the 'spatial assimilation' debate in Canada, 1997–2007. *International Migration and Integration,* 10, 89–108.

Mingione, E. 1995. New aspects of marginality in Europe. In *Europe at the Margins – New Mosaics of Inequality,* edited by D. Sadler and C. Hadjimichalis. Chichester: Wyley, 15–32.

Minnesota Population Center. 2009. *Integrated Public Use Microdata Series— International: Version 5.0.* Minneapolis: University of Minnesota.

Mitchell, B.A., Wister, A.V. and Gee, E.M. 2004. The ethnic and family nexus of homeleaving and returning among Canadian young adults. *The Canadian Journal of Sociology,* 29(4), 543–575.

Mocetti, S. and Porello, C. 2010. How does immigration affect native internal mobility? New evidence from Italy. *Regional Science and Urban Economics,* 40, 427–439.

Módenes, J.A. (ed.) 2002. *Trabajo y Residencia Como Factores de las Migraciones Internas: un Estudio Comparativo Europeo.* Bellaterra: Centre d'Estudis Demogràfics.

Molina, I. 1997. *Stadens Rasifiering. Etnisk Boendesegregation i Folkhemmet. Geografiska Regionstudier No 32.* Uppsala: Kulturgeografiska institutionen, Uppsala universitet.

Moore, E.G. and Rosenberg, M.W. 1995. Modelling migration flows of immigrant groups in Canada. *Environment and Planning A*, 27(5), 699–714.

Morelli, A. 2004. *Histoire des Étrangers et de l'immigration en Belgique de la Préhistoire à Nos Jours.* Brussels: Couleurs livre.

Morrison, P.S. and Clark, W.A.V. 2011. Internal migration and employment: macro flows and micro motives. *Environment and Planning A*, 43(8), 1948–1964.

Mulder, C.H. 1993. Log-rate models for synchronized events in the life course: the case of marriage and migration. In *Quantitative Geographical Methods, Applied in Demography and Urban Planning Research*, edited by W.F. Sleegers and A.L.J. Goethals. Amsterdam: Netherlands Universities Institute for Coordination of Research in Social Sciences [SISWO], 69–84.

Mulder, C.H. 1993. *Migration Dynamics: A Lifecourse Approach.* Amsterdam: Thesis Publishers.

Mulder, C.H. 2006. Population and housing, *Demographic Research*, 15, 401–412.

Mulder, C.H. 2007. The family context and residential choice: a challenge for new research. *Population, Space and Place*, 13, 265–278.

Muñoz, S.-A. 2010. Geographies of faith: the differing residential patterns of the Indian-Hindu, Indian-Sikh and Indian-Muslim populations of Dundee and Glasgow. *Population, Space and Place*, 16(4), 269–285.

Muñoz, S.-A. 2011. Ethno-faith-burbs: religious affiliation and residential patterns of the Indian ethnic populations of Dundee and Glasgow. *Journal of Intercultural Studies*, 32(2), 115–131.

Muñoz-Pérez, F. and Izquierdo, A. 1989. L'Espagne, pays d'immigration. *Population*, 44(2), 157–289.

Musterd, S. 2003. Segregation and integration. A contested relationship. *Journal of Ethnic and Migration Studies*, 29(4), 623–41.

Musterd, S. 2005. Social and ethnic segregation in Europe: levels, causes and effects. *Journal of Urban Affairs*, 27(3), 331–348.

Musterd, S. 2006. Segregation, urban space and the resurgent city. *Urban Studies*, 43(8), 1325–1340.

Musterd, S., Andersson, R., Galster, G. and Kauppinen, T. 2008. Are immigrants' earnings influenced by the characteristics of their neighbours? *Environment and Planning A*, 40(4), 785–805.

Musterd, S. and De Vos, S. 2007. Residential dynamics in ethnic concentrations. *Housing Studies*, 22(3), 333–353.

Musterd, S. and Fullaondo, A. 2008. Ethnic segregation and the housing market in two cities in Northern and Sourthern Europe: The cases of Amsterdam and Barcelona. *Arquitecture, City and Environment*, 8, 93–114.

Musterd, S. and Ostendorf, W. (eds) 1998. *Urban Segregation and the Welfare State: Inequality and Exclusion in Western Cities.* London: Routledge.

Musterd, S. and Ostendorf, W. 2009. Spatial segregation and integration in the Netherlands. *Journal of Ethnic and Migration Studies*, 35(9), 1515–1532.

Nagel, J. 1994. Constructing ethnicity: creating and recreating ethnic identity and culture. *Social Problems*, 41(1), 152–176.

Naredo, J.M. and Montiel Márquez, A. 2011. *El Modelo Inmobiliario Español y su Culminación en el Caso Valenciano*. Barcelona: Icaria editorial.

Nazroo, J. and Karlsen, S. 2003. Patterns of identity among ethnic minority people: diversity and commonality. *Ethnic and Racial Studies*, 26(5), 902–930.

Neckerman, K.M., Carter, P. and Lee, J. 1999. Segmented assimilation and minority cultures of mobility. *Ethnic and Racial Studies*, 22(6), 945–965.

Nel·lo, O. 1997. *Las Grandes Ciudades Españolas: Dinámicas Urbanas e Incidencia de las Políticas Estatales*. European Institute for Comparative Urban Research, Rotterdam: Erasmus University Rotterdam.

Newbold, K.B. 1996. Internal migration of the foreign-born in Canada. *International Migration Review*, 30(3), 728–47.

Newman, D. 2000. Internal migration in Israel: from periphery to center, from rural to urban. In *Still Moving: Recent Jewish Migration in Comparative Perspective*, edited by D.J. Elazar and M. Weinfeld. Brunswick, NJ: Transaction Publishers, 205–228.

Nilsson, K. 2003. Moving into the city and moving out again: Swedish evidence from the cohort born in 1968. *Urban Studies*, 40(7), 1243–1258.

Nivalainen, S. 2004. Determinants of family migration: short moves vs. long moves. *Journal of Population Economics*, 17, 157–175.

Nogle, J.M. 1994. Internal migration for recent immigrants to Canada. *International Migration Review*, 28(1), 31–48.

OECD. 2007. *International Migration Outlook*. 2007 edition. Paris: SOPEMI, OECD.

OECD. 2011. *OECD Regions at a Glance 2011*, OECD Publishing. [Online]. http://dx.doi.org/10.1787/reg_glance-2011-en.

ONS. 2010. *Migration Statistics 2009, Statistical Bulletin*. [Online]. Available at: http://www.statistics.gov.uk/pdfdir/miga1110.pdf.

Openshaw, S. 1984. *The Modifiable Areal Unit Problem*, Norwich: CATMOG38, Geo Books.

Ozacky-Lazar, S. and Kabna, M. 2008. *Between Vision and Reality: The Vision Papers of the Arabs in Israel, 2006–2007*. Jerusalem: The Citizens' Accord Forum. [In Hebrew].

Özüekren S., Erzog-Karahan, E. 2010. Housing experiences of Turkish (im) migrants in Berlin and Istanbul: internal differentiation and segregation. *Journal of Ethnic and Migration Studies*, 36(2), 355–372.

Özüekren, A.S. and Van Kempen, R. 2002. Housing careers of minority ethnic groups: experiences, explanations and prospects. *Housing Studies*, 17(3), 365–379.

Pacione, M. 2005a. The changing geography of ethnic minority settlement in Glasgow, 1951–2001. *Scottish Geographical Journal*, 121(2), 141–161.

Pacione, M. 2005b. The geography of religious affiliation in Scotland. *The Professional Geographer*, 57(2), 1467–9272.

Pan Ké Shon, J-L. 2010. The ambivalent nature of ethnic segregation in France's disadvantaged neighbourhoods. *Urban Studies*, 47(8), 1603–1623.

Park, R.E., Burgess, E.W. and McKenzie, R.D. 1925. *The City: Suggestions for Investigation of Human Behaviour in the Urban Environment*. London: The University of Chicago Press.

Parla, A. 2007. Irregular Workers or Ethnic Kin? Post-1990s Labour Migration from Bulgaria to Turkey. *International Migration*, 45(3), 157–181.

Peach, C. 1996a. Does Britain have ghettoes? *Transactions, Institute of British Geographers*, 22(1), 216–235.

Peach, C. 1996b. Good segregation, bad segregation. *Planning Perspectives*, 11, 379–398.

Peach, C. 1997. Pluralist and assimilationist models of ethnic settlement in London 1991. *Tijdschrift Voor Economische en Sociale Geografie*, 88(2), 130–134.

Peach, C. 1998. South Asian and Caribbean minority housing choice in Britain. *Urban Studies*, 35(10), 1657–1680.

Peach, C. 2000. The consequences of segregation, in *Ethnicity and Housing: Accommodating Differences*, edited by F.W. Boal. Aldershot: Ashgate, 10–23.

Peach, C. 2005. The ghetto and the ethnic enclave, in *Desegregating the City: Ghettos, Enclaves and Inequalities*, edited by D.P. Varady. Albany, State University of New York Press, 31–48.

Peach, C. 2009. Slippery segregation: discovering or manufacturing ghettos? *Journal of Ethnic and Migration Studies*, 35(9), 1381–1395.

Peach, C. 2010. 'Ghetto-Lite' or missing the G-Spot? A reply to Johnston, Poulsen and Forrest. *Journal of Ethnic and Migration Studies,* 36(9), 1519–1526.

Peixoto, J. 2002. Strong market, weak state: the case of recent foreign immigration in Portugal. *Journal of Ethnic and Migration Studies*, 28(3), 483–497.

Peixoto, J. and Figueiredo, A. 2007. Imigrantes brasileiros e mercado de trabalho em Portugal. In *Imigração Brasileira em Portugal*, org. J. Malheiros. Lisbon: ACIDI: 87–112.

Peloe, A. and Rees, P. 1999. Estimating ethnic change in London, 1981–91, using a variety of census data, *International Journal of Population Geography*, 5, 179–194.

Pentzopoulos, D. 1962. *The Balkan Exchange of Minorities and its Impact upon Greece*. Paris, The Hague: Mouton.

Permentier, M., Bolt, G. and Van Ham, M. 2011. Determinants of neighbourhood satisfaction and perception of neighbourhood reputation. *Urban Studies*, 48(5), 977–996.

Phillips, D. 1998. Black minority ethnic concentration, segregation and dispersal in Britain, *Urban Studies*, 35(10), 1681–1702.

Phillips, D. 2006a. Moving towards integration: the housing of asylum seekers and refugees in Britain, *Housing Studies*, 21(4), 539–54.

Phillips, D. 2006b. Parallel lives? Challenging discourses of British Muslim self-segregation. *Environment and Planning D: Society and Space,* 24(1), 25–40.

Phillips, D. 2007. Ethnic and racial segregation: a critical perspective. *Geography Compass,* 1(5), 1138–1159.

Phillips, D. 2010. Minority ethnic segregation, integration and citizenship: a European perspective. *Journal of Ethnic and Migration Studies*, 36(2), 209–25.

Phillips, T. 2005. *After 7/7: Sleepwalking to Segregation.* Speech given at Manchester Council for Community Relations, 22 September 2005. Available at: http://www.cre.gov.uk/Default.aspx.LocID-0hgnew07s.RefLoc ID-0hg00900c002.Lang-EN.htm

Pinkster, F.M. 2009. *Living in Concentrated Poverty.* Amsterdam: Faculteit der Maatschappij - en Gedragswetenschappen, Universiteit van Amsterdam.

Plane, D.A. and Heins, F. 2003. Age articulation of US inter-metropolitan migration flows. *The Annals of Regional Science*, 37(1), 107–130.

Portes, A. 1997. Immigration theory for a new century: Some problems and opportunities. *International Migration Review,* 31(4), 799–825.

Portes, A. 1998. Social capital: its origins and applications in modern Sociology. *Annual Review of Sociology*, 24, 1–24.

Portes, A. and Bach, R.L. 1985. *Latin Journey: Cuban and Mexican Immigrants in the United States.* Berkeley: University of California Press.

Portes, A. and Manning, R. 1986. The immigrant enclave: theory and empirical examples, in *Competitive ethnic relations*, edited by J. Nagel and S. Olzak. Orlando, Florida: Academic Press, 47–68.

Portes, A. and Rumbaut, R.G. 2005. Introduction: the second generation and the children of immigrants longitudinal study. *Ethnic and Racial Studies,* 28(6), 983–999.

Portes, A. and Sensenbrenner, J. 1993. Embeddedness and immigration: notes on the social determinants of economic action. *The American Journal of Sociology,* 98(6), 1320–1350.

Portes, A. and Zhou, M. 1993. The new second generation: segmented assimilation and its variants. *Annals of the American Academy of Political and Social Science*, 530, 74–96.

Psimmenos, I. 2000. The making of periphractic spaces: the case of Albanian undocumented female immigrants in the sex industry of Athens, in *Gender and Migration in Southern Europe: women on the move*, edited by F. Anthias and G. Lazaridis. Oxford: Berg, 81–102.

Rabe B. and Taylor, M. 2010. Residential mobility, neighbourhood quality and life-course events. *Journal of the Royal Statistical Society Series A (Statistics in Society),* 3(173), 531–555.

Rabinowitz, D. 2001. The Arab Citizens of Israel, the concept of trapped minority and the discourse of transnationalism in anthropology. *Ethnic and Racial Studies*, 24(1), 64–85.

Ram, B. and Shin, Y.E. 1999. Internal migration of immigrants, in *Immigrant Canada: Demographic, Economic, and Social Challenges,* edited by L. Driedger and S.S. Halli. Toronto: University of Toronto Press, 148–162.

Ratcliffe, P. (ed.) 1996. *Social Geography and Ethnicity in Britain: Geographical spread, Spatial Concentration and Internal Migration, Volume 3 of Ethnicity on the 1991 Census*, London: HMSO.

Ravenstein, E. 1885. The laws of migration, *Journal of the Statistical Society,* 46, 167–235.

Reay, D., Davies, J., David, M. and Ball, S.J. 2001. Choices of degree or degrees of choice? Class, 'race' and the Higher Education choice process. *Sociology,* 35(4), 855–874.

Rebhun, U. 2006. Nativity Concentration and Internal Migration among the Foreign-Born in Israel, 1990–1995. *Revue européenne des migrations internationales,* 22(1), 107–132.

Recaño, J. 2003. La movilidad geográfica de la población extranjera en España: un fenómeno emergente. *Cuadernos de Geografía,* 72, 135–156.

Recaño, J. and Domingo, A. 2006. Evolución de la distribución territorial y la movilidad geográfica de la población extranjera en España, in *Veinte Años de Inmigración en España,* edited by E. Aja, J. Arango and J. Oliver. Barcelona: Fundació CIDOB, 1–37.

Recaño J. and Roig, M. 2006. *The Internal Mobility of Foreigners in Spain.* Paper presented to the 2006 EAPS Conference, Liverpool, June 2006.

Rees, P. 2008. What happens when international migrants settle? Projections of ethnic groups in *United Kingdom regions in International migration in Europe: Data, Models and Estimates,* edited by J. Raymer and F. Willekens. Chichester: John Wiley and Sons, 329–358.

Rees, P. and Butt, F. 2004. Ethnic change and diversity in England, 1981–2001, *Area,* 36(2), 174–186.

Rees, P. and Kupiszewski, M. 1999. *Internal Migration and Regional Population Dynamics in Europe: a Synthesis.* Strasbourg: Council of Europe Publishing.

Rees, P., Wohland, P., Norman, P. and Boden, P. 2011. A local analysis of ethnic group population trends and projections for the UK. *Journal of Population Research,* 28(2–3): 149–184.

Rees, P.H. and Phillips, D. 1996. Geographical spread, spatial concentration and internal migration. In *Ethnicity in the 1991 Census, Volume Three: Social Geography and Ethnicity in Britain: Geographical Spread, Spatial Concentration and Internal Migration,* edited by P. Ratcliffe. London: HMSO, 23–109.

Reher, D.S. and Silvestre, J. 2009. Internal migration patterns of foreign-born immigrants in a country of recent mass immigration: evidence from new micro data for Spain. *International Migration Review,* 43(4), 815–849.

Rérat, P. 2011. The new demographic growth of cities: the case of reurbanisation in Switzerland. *Urban Studies,* in press.

Rex, J. and Moore, R. 1967. *Race, Community and Conflict: A Study of Sparkbrook.* London: Oxford University Press.

Ribas-Mateos, N. 2004. How can we understand immigration in Southern Europe? *Journal of Ethnic and Migration Studies,* 30(6), 1045–1063.

Riesco, A. 2008. Repensar la sociología de las economías étnicas? El caso del empresario inmigrante en Lavapiés. *Migraciones,* 24, 91–134.

Robinson, V. 1992. The internal migration of Britain's ethnic population, In *Migration Processes and Patterns Volume 1, Research Progress and Prospects*, edited by T. Champion and T. Fielding. London: Belhaven Press, 188–200.

Robinson, V., Andersson, R. and Musterd, S. 2003. *Spreading the 'Burden'? A Review of Policies to Disperse Asylum Seekers and Refugees*. Bristol: Policy Press.

Rogers, A., Anderson, B. and Clark, N. 2009. *Recession, Vulnerable Workers and Immigration*. Background report, Oxford: Compas, Oxford University.

Rogers, A. and Henning, S. 1999. The internal migration patterns of the foreign-born and native-born populations in the United States: 1975–80 and 1985–90. *International Migration Review*, 33(2), 403–29.

Rogers, A. and Willekens, F.J. 1986 (eds) *Migration and Settlement: A Multiregional Comparative Study*. Boston, Reidel.

Ronald, R. 2008. *The Ideology of Home Ownership: Homeowner Societies and the Role of Housing*. Basingstoke: Palgrave Macmillan.

Rose, D. 1984. Rethinking gentrification: beyond the uneven development of Marxist urban theory. *Environment and Planning D: Society and Space*, 2(1), 47–74.

Rossi, P.H. 1955. *Why Families Move*. New York: Macmillan.

Rossi, P.H. 1980. *Why Families Move*. 2nd edition. London: Sage.

Rüger, H., Tarnowski, A. and Erdmann, J. 2011. Migration und Berufsmobilität. Sind Migranten mobiler für den Beruf als Deutsche? *Hamburg Review of Soc. Sc.*, 5(3), 26–51.

Ruppenthal, S. and Lück, D. 2009. Jeder fünfte Erwerbstätige ist aus beruflichen Gründen mobil. Berufsbedingte räumliche Mobilität im Vergleich. *Informationssystem Soziale Indikatoren* (ISI), 42, 1–5.

Sabater, A. 2010. Ethnic residential segregation change in England and Wales, in *Ethnicity and Integration Understanding Population Trends and Processes Volume 3*, edited by J. Stillwell and M. van Ham. Dordrecht: Springer, 47–62.

Sabater, A. and Domingo, A. forthcoming 2012. A new immigration regularisation policy: the Settlement Programme in Spain. *International Migration Review*, 46(1).

Sabater, A. and Simpson, L. 2009. Enhancing the population census: a time series for sub-national areas with age, sex and ethnic group dimensions in England and Wales, 1991–2001. *Journal of Ethnic and Migration Studies*, 35(9), 1461–1478.

Saka, B. 2010. *Interregionale Mobilität in Westdeutschland – Einheimische Deutsche und Bevölkerung mit Migrationshintegrund im Vergleich*. Bremen: Thesis, University of Bremen.

Sampson, R.J. and Sharkey, P. 2008. Neighborhood selection and the social reproduction of concentrated racial inequality. *Demography*, 45(1), 1–29.

Santillana, I. 1981. Los determinantes económicos de las migraciones internas en España, 1960–1973. *Cuadernos de Economía*, 4(25), 381–407.

Sassen, S. 1994. *Cities in a World Economy.* Thousand Oaks, C.A: Pine Forge Press.

Sayas, J.P. 2006. Urban sprawl in peri-urban coastal zones. *The Greek Review of Social Research*, 121C, 71–104.

Schaake, K., Burgers, J. and Mulder, C.H. 2010. Ethnicity at the individual and neigborhood level as an explanation for moving out of the neighborhood. *Population Research and Policy Review,* 29(4), 593–608.

Schmidt, C. 1997. Immigrant performance in Germany: labor earnings of ethnic German migrants and foreign guest-workers. *The Quarterly Review of Economics and Finance*, 37(1), 379–397.

Schneider, N., Limmer, R., Ruckdeschal, K. 2002. *Mobil, flexibel, gebunden.* Frankfurt: Campus Verlag.

Schönwälder, K. 2001. *Einwanderung und Ethnische Pluralität. Politische Entscheidungen und öffentliche Debatten in Grossbritannien und der Bundesrepublik von der 1950er bis zu den 1970er Jahren.* Essen: Klartext.

Schönwälder, K. and Söhn, J. 2009. Immigrant settlement structures in Germany: general patterns and urban levels of concentration of major groups. *Urban Studies,* 46(7), 1439–1460.

Schündeln, M. 2007. *Are Immigrants More Mobile than Natives? Evidence from Germany.* IZA discussion paper No. 3226, Bonn: IZA.

SEME. 2010. *Retorno Voluntario de Atención Social, Secretaría de Estado de Inmigración y Emigración.* Madrid: Ministerio de Trabajo e Inmigración.

Semyonov, M, Glikman, A. and Krysan, M. 2007. Europeans' preferences for ethnic residential homogeneity: cross-national analysis of response to neighborhood ethnic composition. *Social Problems,* 54(4), 434–453.

Singelmann, J. 1993. Levels and Trends of Female Internal Migration in Developing Countries: 1960–1980, in *Internal Migration of Women in Developing Countries: Proceedings of the United Nations Expert Group Meeting on the Feminization of Internal Migration.* New York: United Nations, 77–115.

Simon, A. 2010. Do ethnic groups migrate towards areas of high concentration of their own group within England and Wales? in *Understanding Population Trends and Processes Volume 3: Ethnicity and Integration*, edited by J. Stillwell and M. van Ham. Dordrecht: Springer, 133–152.

Simon, A. 2011. White mobility and minority ethnic concentrations: Exploring internal migration patterns for electoral wards in England and Wales using the 2001 Census. *Journal of Intercultural Studies*, 32(2), 173–188.

Simpson, L. 2004. Statistics of racial segregation: measures, evidence and policy. *Urban Studies*, 41(3), 661–681.

Simpson, L. 2007. Ghettos of the mind: the empirical behaviour of indices of segregation and diversity. *Journal of the Royal Statistical Society series A,* 170(2), 405–424.

Simpson, L. and Akinwale, B. 2007. Quantifying stability and change in ethnic group. *Journal of Official Statistics*, 23(2), 185–208.

Simpson, L., and Finney, N. 2009. Spatial patterns of internal migration: evidence for ethnic groups in Britain. *Population, Space and Place*, 15(1), 37–56.

Simpson, L, Gavalas, V. and Finney, N. 2008. Population dynamics in ethnically diverse towns: the long-term implications of immigration. *Urban Studies*, 45(1), 163–184.

Simpson, L., Marquis N. and Jivraj, S. 2010. International and internal migration measured from the School Census in England. *Population Trends*, 140, 106–124.

Singer, A. Hardwick, S.W. and Brettell, C.B. 2008. *Twenty-First Century Gateways: Immigrant Incorporation in Suburban America*. Washington, D.C.: Brookings Institute Press.

Sinning, M. 2010. Homeownership and economic performance of immigrants in Germany. *Urban Studies*, 47(2), 387–409.

Sirkeci, I. 2006. *The Environment of Insecurity in Turkey and the Emigration of Turkish Kurds to Germany*. New York: Edwin Mellen Press.

Sirkeci, I. 2009. *Improving the Immigration and Asylum Statistics in Turkey [Türkiye'de Uluslararası Göç ve Sığınma İstatistiklerinin Geliştirilmesi]*. Ankara, Turkey: Turkish Statistical Institute.

Sirkeci, I., Cohen, J.H. and Yazgan, P. 2012. The Turkish culture of migration: flows between Turkey and Germany, socio-economic development and conflict. *Migration Letters*, 9(1), 33–46.

Skifter Andersen, H. 2008. Why do residents want to leave deprived neighbourhoods? The importance of residents' subjective evaluations of their neighbourhood and its reputation. *Journal of Housing and the Built Environment*, 23, 79–101.

Skifter Andersen, H. 2010. Spatial assimilation in Denmark? Why do immigrants move to and from multi-ethnic neighbourhoods? *Housing Studies*, 25(3), 281–300.

Skeldon, R. 2006. Interlinkages between internal and international migration and development in the Asian region, *Population, Space and Place*, 12(1), 15–30.

Smith, D.P. 2005. 'Studentification': the gentrification factory? in *Gentrification in a Global Context: The New Urban Colonialism*, edited by R. Atkinson and G. Bridge. Abingdon: Routledge, 72–89.

Smith, D.P. and Holt, L. 2007. Studentification and 'apprentice' gentrifiers within Britain's provincial towns and cities: extending the meaning of gentrification. *Environment and Planning A*, 39, 142–161.

Solé, C. 2004. Immigration policies in Southern Europe. *Journal of Ethnic and Migration Studies*, 30(6), 1209–1221.

South, S.J. and Crowder, K.D. 1998. Leaving the hood: residential mobility between black, white, and integrated neighbourhoods. *American Sociological Review*, 63(1), 17–26.

South, S.J., Crowder, K., and Chavez, E. 2005. Geographic mobility and spatial assimilation among U.S. Latino immigrants. *International Migration Review*, 39(3), 577–607.

South, S., Crowder, K. and Chavez, E. 2005. Migration and spatial assimilation among U.S. Latinos: classical versus segmented trajectories. *Demography*, 42(3), 497–521.

South, S.J., Crowder, K. and Pais, J. 2008. Inter-neighborhood migration and spatial assimilation in a multi-ethnic world: comparing Latinos, Blacks and Anglos. *Social Forces*, 87(1), 415–443.

South, S.J. and Deane, G.D. 1993. Race and residential mobility: individual determinants and structural constraints. *Social Forces*, 72(1), 147–167.

Statistisches Bundesamt Deutschland. 2009. *Bevölkerung und Erwerbstätigkeit. Bevölkerung mit Migrationshintergrund – Ergebnisse des Mikrozensus 2005.* Wiesbaden: Statistisches Bundesamt.

Statistisches Bundesamt Deutschland. 2011. *Datenreport 2011: Der Sozialbericht für Deutschland.* Wiesbaden: Statistisches Bundesamt.

Statistics Sweden. 2004. *Efterkrigstidens Invandring och Utvandring.* Stockholm: Statistiska centralbyrån.

Stillwell, J. 2010a. Internal migration propensities and patterns of London's ethnic groups, in *Technologies for Migration and Commuting Analysis: Spatial Interaction Data Applications,* edited by J. Stillwell et al. Hershey: IGI Global, 175–195.

Stillwell, J. 2010b. Ethnic population concentration and net migration in London. *Environment and Planning A*, 42(6), 1439–1456.

Stillwell, J. and Dennett, A. (2012). A comparison of internal migration by ethnic group in Great Britain using a district classification, *Journal of Population Research*, 29(1), 23–44.

Stillwell, J. and Duke-Williams, O. 2005. *Ethnic population distribution, immigration and internal migration in Britain: What evidence of linkage at the district scale,* Paper to the British Society for Population Studies Annual Conference, University of Kent, Canterbury, 12–14 September 2005.

Stillwell, J., Duke-Williams, O. and Dennett, A. 2010. Interaction data: definitions, concepts and sources, in *Technologies for Migration and Commuting Analysis: Spatial Interaction Data Applications,* edited by J. Stillwell et al. Hershey: IGI Global, 1–30.

Stillwell, J. and Hussain, S. 2008. *Ethnic Group Migration Within Britain During 2000–01: A District Level Analysis.* Working Paper 08/2. Leeds: School of Geography, University of Leeds.

Stillwell, J. and Hussain, S. 2010a. Ethnic internal migration in England and Wales: spatial analysis using a district classification framework, in *Ethnicity and Integration Understanding Population Trends and Processes Volume 3,* edited by J. Stillwell and M. van Ham. Dordrecht: Springer, 105–132.

Stillwell, J. and Hussain, S. 2010b. Exploring the ethnic dimension of internal migration in Great Britain using migration effectiveness and spatial connectivity, *Journal of Ethnic and Migration Studies*, 36(9), 1381–1403.

Stillwell, J., Hussain, S. and Norman, P. 2008. The internal migration propensities and net migration patterns of ethnic groups in Britain. *Migration Letters,* 5(2), 135–150.

Stillwell, J. and McNulty, S. 2011. *Immigrants to England and Wales: Where do they come from and where do they settle? An analysis of 2001 Census immigration by ethnic group.* Working Paper 11/03. Leeds: School of Geography, University of Leeds.

Stillwell, J. and Phillips, D. 2006. Diversity and change: understanding the ethic geographies of Leeds. *Journal of Ethnic and Migration Studies*, 32(7), 1131–52.

Stillwell, J.C.H. and Hussain, S. 2008. *Internal migration of Britain's ethnic groups, UPTAP Research Findings.* Leeds: School of Geography, University of Leeds.

Stjernström, O. (2011) Databasen ASTRID och befolkningsgeografi – exemplen integration och barnfamiljers geografi. *Geografiska Notiser*, 69(2), 79–86.

Stockdale, A., Findlay, A. and Short, D. 2000. The repopulation of rural Scotland: opportunity and threat. *Journal of Rural Studies*, 16(2), 243–257.

Strassburger, G. 2004. Transnational ties of the second generation: marriages of Turks in Germany, in *Transnational Social Spaces. Agents, Networks and Institutions*, edited by T. Faist and E. Ozveren. Aldershot: Ashgate, 211–223.

The Migration Observatory 2010. Thinking Behind the Numbers: Understanding Public Opinion on Immigration in Britain. Oxford: COMPAS.

Tienda, M. and Wilson, F.D. 1992. Migration and the earnings of Hispanic men. *American Sociological Review*, 57(5), 661–678.

Tremlett, G. 2006. Economic statistics. *The Guardian*, 26 July 2006. Available at http://www.guardian.co.uk/world/2006/jul/26/spain.gilestremlett.

TurkStat (Turkish Statistical Institute) 2011. Population, demography, housing and gender statistics. [Online]. Available at http://www.turkstat.gov.tr.

TurkStat (Turkish Statistical Institute) 2010. Address based population registration system results of 2010. *Press Release,* No: 19, January 28. [Online]. Available at: http://www.turkstat.gov.tr/PreHaberBultenleri.do?id=842.

Unal, B. 2011. Sustainable illegality: Gagauz women in Istanbul. *Migration Letters*, 8(1), 17–25.

United Nations 1993. *Internal Migration of Women in Developing Countries: Proceedings of the United Nations Expert Group Meeting on the Feminization of Internal Migration.* New York: United Nations.

Van Criekingen, M. 2009. Moving in/out of Brussels' historical core in the early 2000s: migration and the effects of gentrification. *Urban Studies*, 46(4), 825–848.

Van der Haegen, H. and Pattyn, M. 1980. An operationalization of the concept of city region in West-European perspective: the Belgian city regions. *Tijdschrift voor Economische en Sociale Geografie*, 71(2), 70–77.

Van Gent, W.P.C. 2010. Housing policy as a lever for change? The politics of welfare, assets and tenure. *Housing Studies*, 25(5), 735–753.

Van Ham, M. 2002. *Job Access, Workplace Mobility, and Occupational Achievement*. Delft: Eburon.

Van Ham, M. and Clark, W.A.V. 2009. Neighbourhood mobility in context: household moves and changing neighbourhoods in the Netherlands. *Environment and Planning A,* 41(6), 1442–1459.

Van Ham, M. and Feijten P.M. 2008. Who wants to leave the neighbourhood? The effect of being different from the neighbourhood population on wishes to move. *Environment and Planning A,* 40(5), 1151–1170.

Van Ham, M., Findlay, A. Manley, D., and Feijten, P. 2010 *Social Mobility: Is There an Advantage in Being English in Scotland?* IZA Discussion Paper No. 4797, Bonn: IZA.

Van Ham M., Mulder C.H. and Hooimeijer P. 2001. Spatial flexibility in job mobility: macro-level opportunities and micro-level restrictions. *Environment and Planning A,* 33(5), 921–940.

Van Kempen, R. 2002. The academic formulations: explanations for the partitioned city, in *Of States and Cities, the Partitioning of Urban Space,* edited by P. Marcuse and R. Van Kempen. Oxford: Oxford University Press, 35–56.

Van Kempen, R. and Murie, A. 2009. The new divided city: changing patterns in European cities. *Tijdschrift voor Economische en Social Geografie,* 100(4), 377–398.

Van Kempen, R. and Ozuekren, A.S. 1998. Ethnic segregation in cities: new forms and explanations in a dynamic world. *Urban Studies,* 35(10), 1631–1656.

Van Kempen, R. and Özüekren, A.S. 2002. The housing experiences of minority ethnic groups in western European welfare states, in *'Race', Housing and Social Exclusion,* edited by P. Somerville and A. Steele. London: Jessica Kingsley Publishing, 291–311.

Vanneste, D., Thomas, I. and Goossens, L. 2007. *Woning en Woonomgeving in België.* Sociaal-economische enquête 2001, Monografie nr. 2. Brussels: FOD Economie.

Vasileva, D. 1992. Bulgarian Turkish Emigration and Return. *International Migration Review,* 26(2), 342–352.

Viruela, R. 2008. Población rumana y búlgara en España: Evolución, distribución geográfica y flujos migratorios. *Cuadernos de Geografía,* 84, 169–194.

Wacquant, L. 2004. *Punir les Pauvres. Le Nouveau Gouvernement de l'Insécurité Sociale.* Paris, editions Dupuytren.

Wacquant, L. 2008. *Urban Outcasts: A Comparative Sociology of Advanced Marginality.* Cambridge: Polity Press.

Walker, R., Ellis, M. and Barff, R. 1992. Linked migration systems, immigration and internal labor flows in the United States, *Economic Geography,* 68, 234–248.

Waters, M.C. 1994. Ethnic and racial identities of second generation black immigrants in New York City. *International Migration Review,* 28(4), 795–820.

Willaert, D. and Deboosere, P. 2005. *Buurtatlas van de Bevolking van het Brussels Hoofdstedelijk Gewest bij de Aanvang van de 21e Eeuw.* Dossier nr. 42. Brussel: Brussels Instituut voor Statistiek en Analyse, Ministerie van het Brussels Hoofdstedelijk Gewest.

Wingens, M., Windzio, M., de Valk, H. and Aybek, C. 2011. (eds) *A Life-Course Perspective on Migration and Integration.* London: Springer.

White, M.J., Biddlecom, A.E. and Guo, S. 1993. Immigration, naturalization, and residential assimilation among Asian Americans in 1980. *Social Forces,* 72 (1), 93–117.

Wright, R., Ellis, M. and Parks, V. 2005. Re-placing whiteness in spatial assimilation research. *City and Community,* 4(2), 111–135.

Yiftachel, O. 2006. *Ethnocracy: Land and identity politics in Israel/Palestine.* Philadelphia: The University of Pennsylvania Press.

Zelinsky, W. and Lee, B.A. 1998. Heterolocalism: an alternative model of sociospatial behaviour of immigrant ethnic communities. *International Journal of Population Geography,* 4, 281–298.

Zhou, M. and Logan, J. 1991. In and out of Chinatown: residential mobility and segregation of New York City's Chinese. *Social Forces,* 70(2), 387–407.

Zohry, A. 2005. Interrelationships between internal and international migration in Egypt: a pilot study. [Online] Report from *Development Research Centre on Migration, Globalisation and Poverty, University of Sussex.* Available at: http://www.dfid.gov.uk/R4D/PDF/Outputs/migrationglobpov/aymanreport. pdf.

Zorlu, A. 2008. *Who Leaves the City? The Influence of Ethnic Segregation and Family Ties.* IZA discussion paper No. 3343, Bonn: IZA.

Zorlu, A. 2009. Ethnic differences in spatial mobility: the impact of family ties. *Population, Space and Place,* 15(4), 323–342.

Zorlu, A. and Latten, J. 2009. Ethnic sorting in the Netherlands. *Urban Studies,* 46(9), 1899–1923.

Zorlu, A., and Mulder, C.H. 2008. Initial and subsequent location choices of immigrants to the Netherlands. *Regional Studies,* 42(2), 245–264.

Index

Schündeln, M., 157
Scotland
    attitudes to immigration, 129
    change in minority populations, 134,
        135
    characteristics of minorities, 135, *136*,
        137
    counterurbanisation, 149
    data and methods, 131–4
    destination for migration, *143*, 144–5,
        146, 147
    distance moved in migration, *138*,
        138–9, *140–1*, 141–2, *142*, 144
    educational levels, *136*, 137, 139, 142
    ethnic group, 139, 144
    ethnic minority population, 127–8
    government of, 128–9
    immigration flows to, 128
    immigration history in UK, 127
    limitations of contemporary research,
        127
    limitations of study, 149
    propensity for migration, 137–8, *138*,
        140–1, 141–4, *142–3*
    research questions, 129–30
    socio-economic/demographic
        indicators, 135, *136*, 137
    tenure of housing, 142
    urban centres, preference for, 149
    work and migration, 138–9
security as motivation to move in Turkey,
    190, *191*
segregation, residential
    Athens, 195, 197–8
    Britain, 41
    future research, 322–3
    Germany, low levels of, 156
    and integration, 198
    literature on, 14
    self-segregation, 92–3
    Spain, 294, 299, *300–1*, 302–3, **308**,
        308–11, **309**
    Sweden, 268–70, *269*
    theory on, 92–4
self-segregation, 92–3
Sensenbrenner, J., 91
Sharkey, P., 92
Sheppard, S., 220

Shin, Y.E., 176
Short, D., 149
Simpson, L., 63, 91, 130, 288
Skeldon, R., 290, 322
social capital, 91
social fragmentation in urban areas, 66–7
social integration as theme of book, 5–7.
    see also segregation, residential
social networks, 118, 198, 241
socio-economic status
    Athens, 199, 207–12, *208, 209, 210, 211*
    Scotland, 135, *136*, 137
Söhn, J., 156
South Asians, patterns of mobility in
    Portugal, *82*, **83**, *83*, 84, *85*
Southern Europe
    age, 251, *251, 252*, 253, 255, 257
    concentrations as influence on
        mobility, 241
    countries studied, reasons for, 241
    data sources, 242–3, *244*, 245, *246,
        247–8*, 249–50
    distance moved in migration, 255, *256*,
        257
    economic motivation for immigration,
        239–40
    education, 258–9
    employment status, 259–60
    factors explaining immigration, 239–40
    gender, 250–1, *251, 252*, 253, 255
    geography, 239
    growth of immigration in, 239
    home ownership, 259
    marital status, 258
    place of birth, 257–8
    previous research, 240–1, 242
    progressive ageing, 240
    tenure of housing status, 259
Spain. *see also* Southern Europe
    data and methods, 295–7, *296*
    employment status, 305
    heterolocalism, 294, 310
    housing, 305, 310
    immigration policy, 293–4
    internal migration, 303, **304**, 305,
        306–7, **308**, 308–11, **309**
    international migration to, 293, 297–9,
        298